5

W9-ALZ-897

Medical Imaging Signals and Systems

Medical Imaging Signals and Systems

Jerry L. Prince
Electrical and Computer Engineering
Whiting School of Engineering
Johns Hopkins University

Jonathan M. Links
Environmental Health Sciences
Bloomberg School of Public Health
Johns Hopkins University

PEARSON
Prentice
Hall

Upper Saddle River, New Jersey 07458

Library of Congress Cataloging-in-Publication Data

Prince, Jerry L.
    Medical imaging signals and systems / Jerry L. Prince, Jonathan M. Links.
        p. cm.
    Includes bibliographical references and index.
    ISBN 0-13-065353-5
    1. Diagnostic imaging–Digital techniques. 2. Signal processing–Digital techniques. I.
    Links, Jonathan M. II. Title.
    RC78.7.D53P755 2006
    616.07′54–dc22

                                                       2004060008

Vice President and Editorial Director, ECS: *Marcia J. Horton*
Acquisitions Editor: *Dorothy Marrero*
Editorial Assistant: *Richard Virginia*
Executive Managing Editor: *Vince O'Brien*
Managing Editor: *David A. George*
Production Editor: *Kevin Bradley*
Director of Creative Services: *Paul Belfanti*
Art Director: *Jayne Conte*
Cover Designer: *Bruce Kenselaar*
Art Editor: *Greg Dulles*
Manufacturing Buyer: *Lisa McDowell*
Senior Marketing Manager: *Holly Stark*

PEARSON
Prentice Hall

© 2006 Pearson Education, Inc.
Pearson Prentice Hall
Pearson Education, Inc.
Upper Saddle River, NJ 07458

All rights reserved. No part of this book may be reproduced in any form or by any means, without permission in writing from the publisher.

Pearson Prentice Hall® is a trademark of Pearson Education, Inc.

The author and publisher of this book have used their best efforts in preparing this book. These efforts include the development, research, and testing of the theories and programs to determine their effectiveness. The author and publisher make no warranty of any kind, expressed or implied, with regard to these programs or the documentation contained in this book. The author and publisher shall not be liable in any event for incidental or consequential damages in connection with, or arising out of, the furnishing, performance, or use of these programs.

Printed in the United States of America

13 14 15 16 17 18 19 20 V092 18 17 16 15 14 13

ISBN 0-13-065353-5

Pearson Education Ltd., *London*
Pearson Education Australia Pty. Ltd., *Sydney*
Pearson Education Singapore, Pte. Ltd.
Pearson Education North Asia Ltd., *Hong Kong*
Pearson Education Canada Inc., *Toronto*
Pearson Educación de Mexico, S.A. de C.V.
Pearson Education—Japan, *Tokyo*
Pearson Education Malaysia, Pte. Ltd.
Pearson Education, Inc., *Upper Saddle River, New Jersey*

To our families
Carol, Emily, Ben, Mark, and David
Laura, Annie, and Beth
who help us see what's important and what's not

Contents

3

Image Quality 63

Part II   Radiographic Imaging   103

4

Physics of Radiography   106

5

Projection Radiography 135

6

Computed Tomography 181

11

Ultrasound Imaging Systems   347

Part V   Magnetic Resonance Imaging   379

12

Physics of Magnetic Resonance   381

13

Magnetic Resonance Imaging 409

Preface

This book has developed over the past 14 years, during which time we have taught a course on medical imaging systems at Johns Hopkins University. This course started out as a survey course, and then evolved according to our mutual interests and inclinations into a course that emphasizes the *signals and systems* aspects—or more precisely, the signal processing aspects—of medical imaging.

With signal processing as the fundamental viewpoint, this book covers the most important *imaging modalities* in radiology: projection radiography, x-ray computed tomography, nuclear medicine, ultrasound imaging, and magnetic resonance imaging. The authors expect the reader to be familiar with signals and systems, which are usually covered in the sophomore year of most engineering curricula, and with elementary probability. Freshman courses in physics, chemistry, and calculus are also assumed.

The book is organized into parts emphasizing key overall conceptual divisions, as follows. Part I introduces basic imaging principles, including an introduction to medical imaging systems in Chapter 1, a review of signal processing in Chapter 2 (with emphasis on two-dimensional signals), and a discussion of image quality in Chapter 3. Our presentation of the theory of medical imaging systems is strongly based on continuous signals; however, a development of discrete signals is included to permit discussions on sampling and implementation. Issues of image quality, including resolution, noise, contrast, geometric distortion, and artifacts are described in general context here, but revisited within each modality in subsequent chapters.

Part II describes key modalities in radiographic imaging. It begins in Chapter 4 with a brief presentation of the physics of radiography, including the generation and detection of ionizing radiation and its effect on the human body. Chapter 5 describes projection radiography systems, including chest x-ray and fluoroscopy systems. As in all subsequent chapters, coverage focuses on signals, including only enough physics and biology to motivate the modality and provide a model for the analysis. Chapter 5 also presents the mathematics of *projection imaging*, a very fundamental idea in medical imaging. Chapters 6 covers x-ray computed tomography, expanding on the instrumentation and mathematics of projection imaging and introducing the concept of image reconstruction in medical imaging. Computed tomography produces true *tomograms* (images of cross sections of the body) rather than projections of the body.

Part III presents the physics and modalities of nuclear medicine imaging. Chapter 7 describes the physics of nuclear medicine, focusing primarily on the concept of radioactivity. The major modalities in nuclear medicine imaging are described in Chapter 8, which covers planar scintigraphy, and Chapter 9, which covers emission computed tomography.

Part IV covers ultrasound imaging. It begins in Chapter 10 with a brief presentation of the physics of sound, and continues in Chapter 11 with the various imaging modes offered within this rich modality. Part V covers magnetic resonance imaging. Chapter 11 presents the physics of nuclear magnetic resonance, and Chapter 12 continues with a presentation of various magnetic resonance imaging techniques.

We have used drafts of this book for a one-semester upper-level/graduate course on medical imaging systems. In order to cover the material in one semester, we routinely skip some material in the book and we move at a very brisk pace. We feel that this book could be used in a two-semester course as well, perhaps by covering Parts I–III in the first semester and Parts IV–V in the second semester. A two-semester approach would allow instructors to use supplementary materials for additional depth or to present current research topics.

Medical imaging is very visual—just ask any radiologist. Although the formalism of signals and systems is mathematical, we understand the advantages offered through visualization. Therefore, the book contains many images and diagrams. Some are strictly pedagogical, offered in conjunction with the exposition or an example problem. Others are motivational, revealing interesting features for discussion or study. Special emphasis is made to provide biologically relevant examples including images, so that the important context of medical imaging can be appreciated by students.

Many students, friends, colleagues, and teaching assistants contributed to this book through discussions and critiques. Xiao Han, Li Pan, and Vijay Parthasarathy contributed problems and solutions, and Aaron Carass fixed many LaTeX and CVS problems. A special note of appreciation is due to Xiaodong Tao, who created and solved many of the examples and problems. We would also like to thank Rama Chellappa, Brian Caffo, and Sarah Ying, who provided comments and suggestions on several parts of the book. We also thank William R. Brody, who inspired the creation of the course out of which this book grew and who also wrote one of the chapters in our original course notes. Finally, we convey special thanks to Elliot R. McVeigh and John I. Goutsias, who co-taught our course at Hopkins during the early years and helped draft the original version of this book.

JERRY L. PRINCE
JONATHAN M. LINKS

Basic Imaging Principles

Overview

What does the human body look like *on the inside*? The smart answer: It depends on how you look at it. The most direct way to look inside the human body is to cut it open, for example, through surgery. A refinement of this procedure might be to use an endoscope, essentially a light tube that is "threaded" through the body, which conveys an image to a display device. Both methods offer direct optical viewing but also involve cutting the body, putting something into it, or both. These are *invasive techniques*, which have the potential to cause damage or trauma to the body.

The beauty of medical imaging is that we can see inside the human body in ways that are less invasive than surgery or endoscopy. In some cases—e.g., magnetic resonance imaging (MRI) and ultrasound imaging—the methods are completely *noninvasive* and risk-free so far as we know. In other cases—e.g., projection radiography, x-ray computed tomography (CT), and nuclear medicine—there is some risk associated with the radiation exposure, even though these methods are considered noninvasive as well.

Fundamentally, these medical imaging techniques mean that we do not need to cut the body or put a physical device into it in order to "see inside." Of perhaps even greater importance, these techniques allow us to see things that are not visible to the naked eye in the first place. For example, functional magnetic resonance imaging (fMRI) allows us to obtain images of organ perfusion or blood flow, and positron emission tomography (PET) allows us to obtain images of metabolism or receptor binding. In other words, the various imaging techniques allow us to see inside the body in different ways—the "signal" is different in each case and can reveal information that the other methods cannot. Each of these different methods is a different imaging *modality*, and the "signals" that arise are intrinsically different. This harkens back to the opening question: What does the human body look like on the inside? The answer: It depends on the measured signal of interest.

In this book, we use a *signals and systems* approach to explain and analyze the most common imaging methods in radiology today. We want to answer the question: What do the images look like, and why? We will discover that medical imaging physics allows us to image certain parameters of the body's tissues, such as reflectivity in ultrasound imaging, linear attenuation coefficient in computed tomography, and hydrogen proton density in magnetic resonance imaging. These physical parameters, which we can think of as "signals" within the body, represent the input signal into an imaging system. In medical imaging, the "object" or "signal"

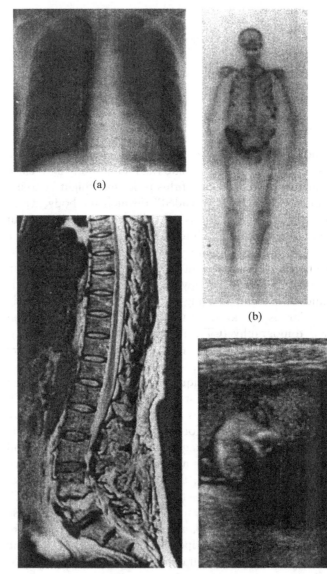

Figure I.1
The four main medical imaging signals discussed in this book: (a) x-ray transmission through the body, (b) gamma ray emission from within the body, (c) ultrasound echoes, and (d) nuclear magnetic resonance induction. The corresponding medical imaging modalities are projection radiography, planar scintigraphy, ultrasound imaging, and magnetic resonance imaging. ((a,b) Chen, et al., Basic Radiology, © 1996 McGraw Hill. This material is reproduced with permission of The McGraw Hill Companies. (c) From A. B. Wolbarst, Physics of Radiology, Second Edition (2005).)

arising from the patient depends on the physical processes governing a given imaging modality. Thus, a given patient represents an ensemble of different objects or signals. In considering a given medical image, it is thus important to start with the physics that underlie the creation of signals from the patient for that modality. Accordingly, each part of this book is organized so that the first chapter describes the relevant physics, and subsequent chapters describe those modalities based on the specific physical processes of that part.

The first output of any medical imaging system is based on physical measurements, which might be returning echoes in an ultrasound system, x-ray intensities in a CT system, or radio-frequency waves in an MRI system. The final output in this system is created through *image reconstruction*, the process of creating an image from measurements of signals. The overall quality of a medical image is determined by how well the image portrays the true spatial distribution of the physical parameter(s) of interest within the body. Resolution, noise, contrast, geometric distortion, and artifacts are important considerations in our study of image quality. Ultimately, the clinical utility of a medical image involves both the image's quality and the medical information contained in the parameters themselves.

Figure I.1 shows the four main medical imaging signals discussed in this book: (1) x-ray transmission through the body, (2) gamma ray emission from within the body, (3) ultrasound echoes, and (4) nuclear magnetic resonance induction. Part II covers modalities that use x-ray transmission signals; Part III covers modalities that use gamma ray emission; Part IV covers modalities that use ultrasound signals; and Part V covers magnetic resonance imaging, which uses signals that arise from nuclear magnetic resonance. Four specific medical imaging modalities are depicted in Figure I.1 (1) projection radiography, (2) planar scintigraphy, (3) ultrasound imaging, and (4) magnetic resonance imaging.

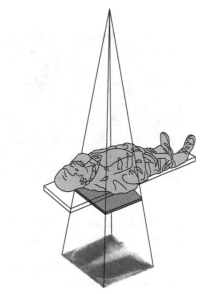

Figure I.2
The creation of a two-dimensional projection through the body. In this case, x-rays are transmitted through a patient, creating a radiograph. (Adapted from J. T. Bushberg, et al., *The Essential Physics of Medical Imaging*, 2d. ed., Lippincott, Williams, & Wilkins, 2002.)

In Figure I.1, parts (a) and (b) represent two-dimensional *projection* images of the three-dimensional human body. A projection is created as a two-dimensional "shadow" of the body, a process that is illustrated in Figure I.2. Figures I.1(c) and (d) are slices within the body. Figure I.3 depicts the three standard orientations of slice (or *tomographic*) images, *axial*, *coronal*, and *sagittal*. (Figure I.1(d) is a sagittal slice while Figure I.1(c) is an *oblique* slice, i.e., an orientation not corresponding to one of the standard slice orientations.)

Figure I.4 also shows slice images. In this case, each image is a transverse slice, oriented perpendicular to the head and body axis, through the brain. Each image is obtained from a different imaging modality: (1) computed tomography, (2) magnetic resonance imaging, and (3) positron emission tomography. Even though each image depicts (a slice through) the brain, the images are strikingly different, because the signals giving rise to each image are themselves strikingly different. In this part of the book, we study the common signal-processing concepts that relate to all imaging modalities, setting the groundwork for adding the physical differences that account for the distinct appearances of the imaging modalities, and hence their different uses in medicine.

Figure I.3
The three standard orientations of slice (or tomographic) images: (a) axial or transaxial or transverse, (b) coronal or frontal, and (c) sagittal.

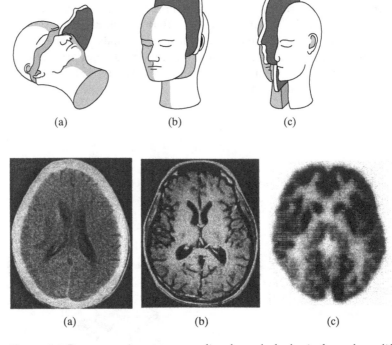

(a)   (b)   (c)

(a)   (b)   (c)

Figure I.4 Representative transverse slice through the brain from three difference imaging modalities: (a) computed tomography, (b) magnetic resonance imaging, and (c) positron emission tomography. ((a,b) From A. B. Wolbarst, Physics of Radiology, Second Edition (2005). (c) Source: M. E. Phelps, taken from Erich Krestel, Imaging Systems for Medical Diagnostics, Siemens 1990.)

Introduction

In this book, we take a signals and systems approach to the characterization of medical imaging. As discussed in the Overview, there are a variety of signals in which we are interested; ultimately, this interest stems from the biological and medical significance of these signals in patients with various diseases. In practice, these signals are transformed into images via medical imaging modalities. In this chapter, we begin to consider these modalities and their characteristics.

1.1 History of Medical Imaging

The first published medical image was a radiograph of the hand of Wilhelm Conrad Roentgen's wife in December 1895. Roentgen had been experimenting with a Crooke's tube (the forerunner of today's x-ray tube) and noticed that "a new kind of rays" (hence, *x-rays*) were emitted that could expose film even when optically shielded. It was immediately obvious to Roentgen that his discovery could have profound impact in medicine. Indeed, the first clinical use of x-rays occurred only two months later, in February 1896. The use of x-rays became widespread, and both static and dynamic (*fluoroscopic*) techniques were developed. Here, a *static* technique refers to an image taken at a single point in time, whereas a *dynamic* technique refers to a series of images acquired over time.

For many decades, these *planar* (i.e., two-dimensional projection) radiographs were the only medical images being produced. Ultimately, radiography was extended into transmission computed tomography, or *cross-sectional* imaging. Godfrey Hounsfield produced the first true CT scanner in 1972 at EMI in England. He used mathematical methods for image reconstruction developed a decade earlier by Allan Cormack of the United States. Hounsfield and Cormack shared the Nobel Prize in Medicine in 1979. Many radiologists consider CT scanning to be the most important development in medical imaging since Roentgen's original discovery.

As radiography arose from the discovery of x-rays, nuclear medicine arose from the discovery of radioactivity, by Antoine Henri Becquerel in 1896. Initially, radionuclides were used in cancer therapy rather than in medical imaging. The concept of using radioactive *tracers* to study physiology was introduced by George de Hevesy in 1923; de Hevesy is considered the father of nuclear medicine. A radiotracer is a radioactively labeled drug that mimics a biological compound of interest; the distribution of the radioactivity implies the distribution of the drug. Early studies with radiotracer used conventional nonimaging radiation detectors to roughly determine amounts of radioactivity in various body regions. In 1949, Benedict Cassen at UCLA started the development of the first imaging system in nuclear medicine, the *rectilinear scanner*. The modern *Anger scintillation camera* was developed by Hal Anger at UC Berkeley in 1952. The element of the most commonly used radionuclide in nuclear medicine, technetium-99m, was discovered in 1937 by Perrier and Emilio Segre; its first use in medicine was in 1961.

The interaction of acoustic waves with media was first described by Lord John Rayleigh over 100 years ago in the context of the propagation of sound in air. Modern ultrasound imaging had its roots in World War II Navy sonar technology. Initial medical applications focused on the brain. Ultrasound technology progressed through the 1960s from simple A-mode and B-mode scans to today's M-mode and Doppler two-dimensional (2-D) and even three-dimensional (3-D) systems.

The phenomenon of *nuclear magnetic resonance*, from which magnetic resonance imaging arises, was first described by Felix Bloch and Edward Purcell; the pair shared the 1952 Nobel Prize in Physics. This work was extended by Richard Ernst, who received the Nobel Prize in Chemistry in 1991. In 1971, Raymond Damadian published a paper suggesting the use of magnetic resonance in medical imaging; in 1973, a paper by Paul Lauterbur followed. Lauterbur received the Nobel Prize in Medicine in 2003, along with Peter Mansfield, who developed key methods in magnetic resonance imaging.

1.2 Physical Signals

In this book, we consider the detection of different physical signals arising from the patient, and their transformation into medical images. In practice, these signals arise from four processes:

- Transmission of x-rays through the body (in projection radiography and CT).
- Emission of gamma rays from radiotracers in the body (in nuclear medicine).
- Reflection of ultrasonic waves within the body (in ultrasound imaging).
- Precession of spin systems in a large magnetic field (in magnetic resonance imaging).

Radiography, CT scanning, nuclear medicine, and magnetic resonance imaging all make use of electromagnetic energy. Electromagnetic energy or waves consist of electric and magnetic waves traveling together at right angles. Wavelength and frequency are inversely related; frequency and energy are directly related. The electromagnetic spectrum spans the frequency range from DC to cosmic rays; only a relatively small portion of this spectrum is useful in medical imaging. At long wavelengths—e.g., longer than 1 angstrom—most electromagnetic energy is highly attenuated by the body, prohibiting its exit and external detection. At wavelengths shorter than about 10^{-2} angstroms, the corresponding energy is too high to be readily detected.

In this book, we express energy in units of *electron volts* (eV), where 1 eV is the amount of energy an electron gains when accelerated across 1 volt potential. We will concentrate on electromagnetic radiation whose wavelengths correspond to energies of roughly 25 keV to 500 keV.

Ultrasound imaging utilizes sound waves, and considerations of attenuation and detection are similar to those above. Image resolution is not adequate for wavelengths longer than a couple of millimeters, and attenuation is too high for very short wavelengths. An ideal frequency "window" for ultrasound in medical imaging is 1–20 MHz, where 1 Hz = 1 cycle/sec.

The signal in magnetic resonance imaging arises from the precession (like the motion of a child's top or dreidl) of nuclei of the hydrogen atom—i.e., protons. When placed on a large magnetic field, collections of protons, termed *spin systems*, can be set into motion by applying radio frequency (RF) currents through wire coils surrounding the patient. Although these spin systems precess at RF frequencies (64 MHz is typical), the primary signal source is not from radio waves, but from the Faraday induction of currents in the same or different wire coils.

1.3 Imaging Modalities

The medical imaging areas we consider in detail in this book are projection radiography, computed tomography, nuclear medicine, ultrasound imaging, and magnetic resonance imaging. An *imaging modality* is a particular imaging technique or system within one of these areas. In this section, we give a brief overview of these most common imaging modalities.

Projection radiography, computed tomography, and nuclear medicine all use ionizing radiation. The first two transmit x-rays through the body, using the fact that the body's tissues selectively *attenuate* (reduce) the x-ray intensities to form an image. These are termed *transmission* imaging modalities because they transmit energy through the body. In nuclear medicine, radioactive compounds are injected into the body. These compounds or *tracers* move selectively to different regions or organs within the body, emitting gamma rays with intensity proportional to the compound's local concentration. Nuclear medicine methods are *emission* imaging modalities because the radioactive sources emit radiation from within the body.

Ultrasound imaging fires high-frequency sound into the body and receives the echoes returning from structures within the body. This method is often called *reflection* imaging because it relies on acoustic reflections to create images. Finally, magnetic resonance imaging requires a combination of a high-strength magnetic field and radio waves to image properties of the proton nucleus of the hydrogen atom. This technique is called *magnetic resonance imaging* since it exploits the property of nuclear magnetic resonance.

1.4 Projection Radiography

Projection radiography includes the following modalities:

- *Routine diagnostic radiography*, including chest x-rays, fluoroscopy, mammography, and motion tomography (a form of tomography that is not *computed* tomography).
- *Digital radiography*, which includes all the scans in routine radiography, but with images that are recorded digitally instead of on film.
- *Angiography*, including universal angiography and angiocardiography, in which the systems are specialized for imaging the body's blood arteries and vessels.
- *Neuroradiology*, which includes specialized x-ray systems for precision studies of the skull and cervical spine.
- *Mobile x-ray systems*, which are small x-ray units designed for operating rooms or emergency vehicles.

All of these modalities are called "projection" radiography because they all represent the projection of a 3-D object or signal onto a 2-D image.

The common element in all of these systems is the *x-ray tube*. As we will see in Chapter 5, the x-ray tube generates an x-ray pulse in an approximately uniform "cone beam" (shaped like a cone) geometry. This pulse passes through the body and is attenuated by the intervening tissues. The x-ray intensity profile across the beam exiting from the body is no longer uniform—shadows have been created by dense objects (such as bone) in the body. This intensity distribution is revealed using a scintillator, which converts the x-rays to visible light. Finally, the light image on the scintillator is captured either on a large sheet of photographic film, a camera, or solid-state detectors.

The most common modality in projection radiography is the chest x-ray; a typical unit is shown in Figure 1.1(a). Here, the x-ray tube is located on the column on the right; the scintillator and film pack are located on the unit on the left. The

Figure 1.1
(a) A chest x-ray unit and (b) a chest x-ray image. (Used with permission of GE Healthcare.)

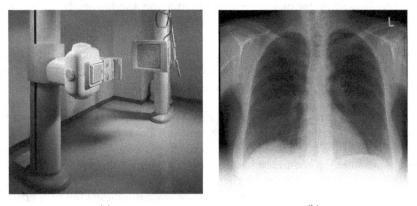

(a)   (b)

radiologic technologist stands at the console to the far left, protected by lead, but able to see through a window. A typical chest x-ray is shown in Figure 1.1(b). This image shows the spine, ribs, heart, lungs, and many other features radiologists are trained to identify and interpret. A key feature of this image is that structures located at different depths in the body are overlayed (or superimposed) on the 2-D image. For example, we can see both front and back ribs in the chest x-ray in Figure 1.1(b). This is a property of projection imaging, and it is common to all projection radiographic methods. True *tomography*, the imaging of a 2-D slice of the 3-D body, cannot be directly accomplished using any modality in projection radiography. More details about projection radiography are given in Chapter 5.

1.5   Computed Tomography

As in projection radiography, computed tomography (CT) uses x-rays. Instead of the x-rays traveling in a 3-D cone beam, they are collimated (i.e., restricted in their geometric spread) to travel within an approximate 2-D "fan beam." *Shadows* in the x-ray beam are created by the tissues in a 2-D cross-section of the body and detected upon exit by a large number of small detectors. These measurements, called *projections*, are collected for many angular orientations of the x-ray tube and detectors—they rotate around a stationary subject—and an image of the cross-section is *computed* from these projections.

If CT were considered to have separate modalities, it would be (standard) single-slice CT, *helical CT*, and *multislice* CT, which is currently emerging as a very important 3-D imaging technique. In helical CT, the x-ray tube and detectors continuously rotate around in a large circle, while the patient is moved in a continuous motion through the circle's center. From the patient's perspective, the x-ray tube carves out a helix; hence, the name helical CT. The importance of this technique is in its ability to rapidly acquire 3-D data, perhaps as a *whole body scan* in less than a minute. In multislice CT, there are several rows of detectors used to rapidly gather a *cone* of x-ray data, comprising a 2-D projection of the 3-D patient.

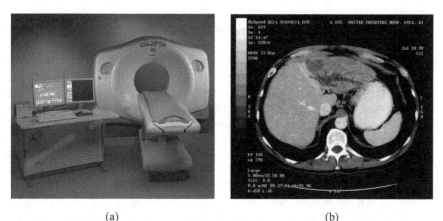

(a)   (b)

Figure 1.2
(a) A CT scanner and (b) a CT image of a slice through the liver. (Used with permission of GE Healthcare.)

With the rapid rotation of the x-ray source and detectors (1 to 2 revolutions per second), very rapid (near real-time) 3-D imaging is possible using these CT scanners.

A typical CT scanner is shown in Figure 1.2(a). In the center of the picture, we can see the tube in which the patient lies; a patient table is visible behind the left TV monitor. Around the tube is a housing containing both the x-ray tube and the detector array. The gantry holding these components is capable of spinning rapidly around the patient. The computer consoles and keyboard in the foreground are used for entering patient data and viewing images. Although CT images can be printed on paper or x-ray film, the images are completely digital in nature, since they are computed from the measured projections. The CT image shown in Figure 1.2(b) is a slice through the liver in which the data were acquired in 1 second. Computed tomography is described in more detail in Chapter 6.

1.6 Nuclear Medicine

Nuclear medicine imaging is distinguished from all other medical imaging modalities by the fact that images can only be made when appropriate radioactive substances are introduced into the body. These substances, which are either injected, ingested, or inhaled into the body, are trace amounts of biochemically active drugs whose molecules are labeled with radionuclides that emit gamma rays. These so-called *radiotracers* move within the body according to the body's natural uptake of the biological carrier molecule. For example, radioactive iodine can be used to study thyroid function. A nuclear medicine image reflects the local concentration of a radiotracer within the body. Since this concentration is tied to the physiological behavior within the body, nuclear medicine imaging is an example of a *functional imaging* method, whereas standard CT and MRI are *anatomical* or *structural imaging* methods.

There are three modalities within nuclear medicine: *conventional radionuclide imaging* or *scintigraphy*, *single-photon emission computed tomography* (SPECT), and *positron emission tomography* (PET). Conventional radionuclide imaging and SPECT utilize a special 2-D gamma ray scintillation detector called an *Anger camera*. This camera is designed to detect single x-rays or gamma rays, rather than simply detecting the intensity of a collective beam as in projection radiography and CT. In conventional radionuclide imaging, this array is conceptually analogous to the scintillator/film pack in projection radiography. A complication is that this procedure combines the effects of emission with the effects of attenuation of the rays by intervening body tissues, producing images that are 2-D projections of the 3-D distribution of radiotracers (which we wish to know) confounded by attenuation.

SPECT and PET produce images of slices within the body. SPECT does this by rotating the Anger camera around the body and using computed tomography methods to reconstruct images. Since the Anger camera is a 2-D imager, SPECT is fundamentally a 3-D imaging technique. In conventional radionuclide imaging and SPECT, a radioactive atom's decay produces a single gamma ray, which may intercept the Anger camera. In PET, however, a radionuclide decay produces a positron, which immediately annihilates (with an electron) to produce two gamma rays flying off in opposite directions. The PET scanner looks for coincident detections

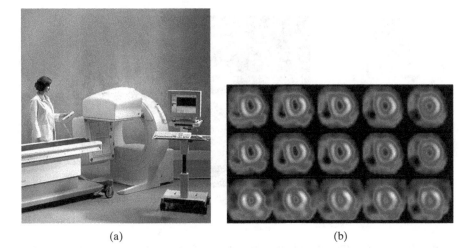

(a)   (b)

Figure 1.3
(a) A SPECT scanner and (b)
a series of SPECT heart
scans. (Used with permission
of GE Healthcare.)

from opposing detectors in its ring, thus determining the line that passes through the site where the annihilation occurred.

A SPECT scanner is shown in Figure 1.3(a). The Anger camera at the top is capable of rotating completely around the patient, who lies on the table. The table can move in coordination with the camera (for a helical tomographic scan), or the table can move with a stationary camera (for a whole-body standard projection scan). The collection of SPECT pictures shown in Figure 1.3(b) are cardiac scans taken at different spatial positions in the heart depicting blood flow to the heart muscle. Nuclear medicine imaging is described in more detail in Chapters 7 through 9.

1.7 Ultrasound Imaging

Ultrasound imaging uses electrical-to-acoustical *transducers* to generate repetitive bursts of high-frequency sound. These pulses travel into the soft tissue of the body and reflect back to the transducer. The time-of-return of these pulses gives information about the location (depth) of a reflector, and the intensity of these pulses gives information about the strength of a reflector. By rapidly moving or scanning the transducer or its acoustical beam, real-time images of cross-sections of soft tissue can be generated. Ultrasound imaging systems are comparatively inexpensive and completely noninvasive at the low imaging intensities typically used. Therefore, these systems are widespread and in common usage. They are designed primarily to image the anatomy, and although their image quality is relatively poor, they are real-time and highly adaptable to a wide variety of imaging goals.

Ultrasound imaging systems offer several imaging modalities:

- *A-mode imaging*, which generates a one-dimensional waveform, and as such does not really comprise an image. This mode, however, can provide very detailed information about rapid or subtle motion (of a heart valve, for example).

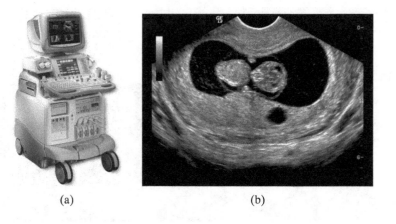

Figure 1.4
(a) An ultrasound scanner and (b) an ultrasound image of an 11-week-old human embryo. (Used with permission of GE Healthcare.)

(a)   (b)

- *B-mode imaging*, which is ordinary cross-sectional anatomical imaging. There are several arrangements of transducers that can give rise to images with different appearances.
- *M-mode imaging*, which generates a succession of A-mode signals, brightness modulated and displayed in time on a CRT. M-mode generates an image that is not anatomical but is important for measuring time-varying displacements of, for example, a heart valve.
- *Doppler imaging*, which uses the property of frequency or phase shift caused by moving objects (like a police siren that has a higher frequency when approaching and a lower frequency when departing) to generate images that are color coded by their motion. Doppler is most commonly used, however, in an audio mode. Running the frequency shifts through speakers allows an aural analysis of motion which is not possible with a visual display.

An ultrasound imaging system is shown in Figure 1.4(a). In addition to being inexpensive, ultrasound systems are small; most systems are on wheels, and can be rolled to the bedside or wherever needed. Two transducers are shown with their connecting cables; only one is used at any given time. Transducers with differing frequencies and geometries are usually available to serve the examination requirements. The ultrasound image shown in Figure 1.4(b) shows a human embryo at 11 weeks. The pear-shaped region is the amniotic fluid. The embryo is the fuzzy object within the sac. The textured appearance of ultrasound images is called "speckle," and it is a form of artifact. Ultrasound imaging is described in detail in Chapters 10 and 11.

1.8   Magnetic Resonance Imaging

Magnetic resonance (MR) scanners use the property of *nuclear magnetic resonance* to create images. In a strong magnetic field, the nucleus of the hydrogen atom—a proton—tends to align itself with the field. Given the vast numbers of hydrogen atoms in the body, this tendency results in a net magnetization of the body. It is

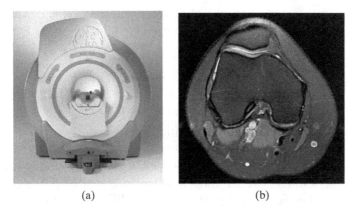

(a)   (b)

Figure 1.5
(a) A magnetic resonance scanner and (b) an MR image of a human knee. (Used with permission of GE Healthcare.)

then possible to selectively excite regions within the body, causing groups of these "little magnets" to tip away from the magnetic field direction. As the protons return back into alignment with the field, they precess like children's tops. Because protons are charged particles, this *precession* generates a radio-frequency electromagnetic signature, which can be sensed with an antenna.

There are many modes in which an MR scanner can operate; such scanners are incredibly flexible imaging devices. The most general categories of operation are the following:

- *Standard MRI*, which includes a whole host of *pulse sequences* (time-series of different excitation pulses).
- *Echo-planar imaging* (EPI), which utilizes specialized apparati to generate images in real time.
- *Magnetic resonance spectroscopic imaging*, which images other nuclei besides the hydrogen atom.
- *Functional MRI* (fMRI), which uses oxygenation-sensitive pulse sequences to image blood oxygenation in the brain. High regional blood oxygenation correlates to increased brain activity, and thus is a measure of brain function.

We discuss only standard MRI in this book.

An MR scanner is shown in Figure 1.5(a). Surrounding the tube in the center of the picture, within the housing, is a 2 tesla superconducting magnet. The field is very uniform within the bore, which is required for geometrically accurate imaging. The MR image in Figure 1.5(b) shows a cross-section of a human knee. Magnetic resonance imaging is discussed in detail in Chapters 12 and 13.

1.9 Summary and Key Concepts

In practice, radiologists look for specific patterns in medical images. These patterns depend on both the patient and the imaging modality. It is the job of the engineers and scientists who develop medical imaging systems to produce images that are as accurate and useful as possible; these systems depend on the physics of each modality. In this chapter, we presented the following key concepts that you should now understand:

1. Medical imaging relies on *noninvasive techniques* to image body structures and function.
2. Each technique or method is a different imaging *modality*.
3. The main imaging modalities are *projection radiography*, *computed tomography*, *nuclear medicine*, *ultrasound imaging*, and *magnetic resonance imaging*.
4. The *signal* of interest is defined by the modality and specific imaging parameters.
5. Radiologists are trained to look for *specific patterns*, defined by the modality, specific imaging parameters, and differences in the expected signal in health and disease.

Bibliography

Bushberg, J. T., Seibert, J. A., Leidholdt, E. M., and Boone, J. M. *The Essential Physics of Medical Imaging*, 2nd ed. Philadelphia: Lippincott Williams and Wilkins, 2002.

Cho, Z.-H., Jones, J. P., and Singh, M. *Foundations of Medical Imaging*. New York. Wiley, 1993.

Macovski, A. *Medical Imaging Systems*. Englewood Cliffs, NJ: Prentice Hall, 1983.

Shung, K. K., Smith, M. B., and Tsui, B. M. W. *Principles of Medical Imaging*. San Diego, CA: Academic Press, 1992.

Signals and Systems[1]

2.1 Introduction

Signals and *systems* are two fundamental concepts for modeling medical imaging systems. Signals are mathematical functions of one or more independent variables, capable of modeling a variety of physical processes. Systems respond to signals by producing new signals. They are useful for modeling how physical processes (signals) change in natural environments and how systems, such as medical imaging instruments, create new signals (i.e., images). This chapter provides an introduction to the theory of signals and systems, with a focus on those tools required for modeling medical imaging systems.

Signals can be classified into three categories: (1) continuous, (2) discrete, and (3) mixed. A *continuous signal* is a function of independent variables that range over a continuum of values. For example, in CT, the distribution of x-ray attenuation in a cross-section within the body can be mathematically modeled using a function $f(x, y)$ of two independent real-valued variables x and y, which represent two spatial dimensions. Physical processes are usually modeled using continuous signals. Some medical images, such as those obtained on x-ray film, are also modeled as continuous signals.

A *discrete signal* is a function of independent variables that range over discrete values. These signals can be used to model physical processes that naturally correspond to discrete values. For example, the times of arrival of photons in a radioactive decay process form a discrete sequence of arrival times—for example, a discrete signal. Discrete signals can also be used to represent continuous signals. For example, the x-ray distribution $f(x, y)$ described above might be represented in a computer as a function $f_d(m, n)$ of two independent discrete variables m and n.

Finally, a *mixed signal* is a function of some continuous and some discrete independent variables. Signals of this type are acquired by certain medical imaging instruments. For example, in computed tomography, a signal $g(\ell, \theta_k)$, $k = 1, 2, \ldots,$ is recorded as a result of a parallel or fan beam of x-rays passing through an object.

---

[1]The initial draft of this chapter was written by Dr. John I. Goutsias.

Variable ℓ is often continuous-valued, representing the distance of a particular ray from the center of the object, whereas θ_k is a discrete-valued variable, representing the beam's angle relative to a reference coordinate system (see Chapter 6).

Based on the previous signal classification, systems can also be classified into categories. For example, we may consider a *continuous-to-continuous system* that responds to a continuous signal by producing a continuous signal, or a *continuous-to-discrete system* that responds to a continuous signal by producing a discrete signal. For example, an analog-to-digital converter takes a continuous signal as input and produces a discrete signal as output through a process called *sampling*.

In this book, we consider one-dimensional (1-D), two-dimensional (2-D), and three-dimensional (3-D) signals and systems. In this chapter, we primarily develop 2-D continuous signals and systems; restriction to one dimension or extension to three dimensions is straightforward in most cases. In all cases, you can think of the *input signal* (to the imaging system) as coming from the patient and the *output signal* as coming from the imaging system. The challenge in medical imaging is to make the output signal a faithful representation of the input signal.

2.2 Signals

A *continuous signal f* is defined as a function

$$f(x, y), \qquad -\infty \leq x, \ y \leq \infty \tag{2.1}$$

of two independent real-valued variables x and y. This signal can be represented and visualized in two different ways. We may plot the signal as a function of the two independent variables x and y, or we may display it, by assigning an intensity or brightness proportional to its value at (x, y). This is illustrated in Figure 2.1. In the first case, we call $f(x, y)$ a *function* or a *signal* and (x, y) a *point*. In the second case, we call f an *image* and (x, y) a *pixel*, a word that is derived from "picture element." The 3-D analog of the pixel is the *voxel*, from "volume element." This is illustrated in Figure 2.2. The representation in Figure 2.1(a) is useful in mathematical calculations, while that in Figure 2.1(b) is most appropriate for human observers. Accordingly, in this book, we focus on two-dimensional pixel-based representations.

A number of special signals will frequently be used throughout this book. These include the point impulse, the comb and sampling functions, the line impulse, the rect and sinc functions, and exponential and sinusoidal signals. In addition, many signals will have properties that distinguish them from other signals and make their use in analysis easier. Separability and periodicity are two such common properties. We now study these special signals and properties.

2.2.1 Point Impulse

It is very useful in medical imaging to mathematically model the concept of a point source, which is used in the characterization of imaging system resolution. For example, if the source is very small, yet appears to be very large (blurred out) in

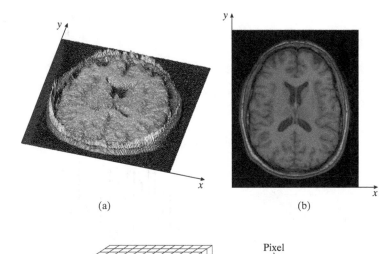

Figure 2.1
Two alternative signal visualizations: (a) a functional plot and (b) an image display.

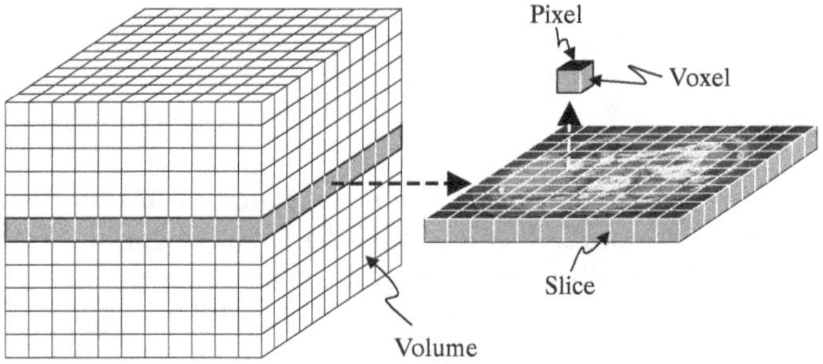

Figure 2.2
The representation of a 3-D object as a 2-D image. The 3-D object is represented by a collection of voxels; the 2-D image is comprised of pixels. In this example, the image is a slice through the 3-D object.

its radiological image, then we would say that the resolution of the system is poor. Precise definitions of resolution will be introduced in Chapter 3.

The concept of a point source in one dimension is known as the *1-D point impulse*, $\delta(x)$, which is defined by the following two properties:

$$\delta(x) = 0, \quad x \neq 0, \tag{2.2}$$

$$\int_{-\infty}^{\infty} f(x)\delta(x)\, dx = f(0). \tag{2.3}$$

In other references, the point impulse is also known as the *delta function*, the *Dirac function*, and the *impulse function*. The point impulse is not a function in the usual sense, but instead acts on other signals through integration, as we will see. It models the property of a point source by having infinitesimal width and unit area, which can be shown by taking $f(x) = 1$ in (2.3).

The *2-D point impulse*, $\delta(x, y)$ (also called the *2-D impulse function*, *2-D delta function*, or *2-D Dirac function*), is analogously characterized by

$$\delta(x, y) = 0, \quad (x, y) \neq (0, 0), \tag{2.4}$$

$$\int_{-\infty}^{\infty} \int_{-\infty}^{\infty} f(x, y)\delta(x, y)\, dx\, dy = f(0, 0). \tag{2.5}$$

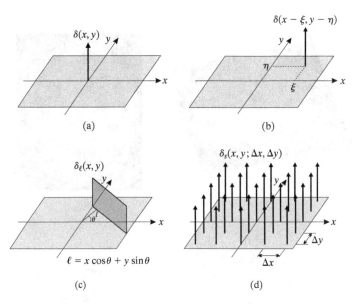

Figure 2.3
Signals derived from the point impulse: (a) point impulse $\delta(x, y)$, (b) shifted point impulse $\delta(x - \xi, y - \eta)$, (c) line impulse $\delta_\ell(x, y)$, and (d) sampling function $\delta_s(x, y; \Delta x, \Delta y)$.

The 2-D point impulse (also not a function in the usual sense) models the property of a point source located at $(0, 0)$ by having infinitesimal width and unit volume. Figure 2.3(a) illustrates this concept.

In (2.5), the point impulse "picks off" the value of $f(x, y)$ at the location $(0, 0)$ by multiplying $f(x, y)$ with the point impulse followed by integration over all space. Suppose we shift the delta function to position (ξ, η)—i.e., $\delta(x - \xi, y - \eta)$—as shown in Figure 2.3(b). Using a change of variables in (2.5), it can be shown that

$$f(\xi, \eta) = \int_{-\infty}^{\infty} \int_{-\infty}^{\infty} f(x, y)\delta(x - \xi, y - \eta)\, dx\, dy. \tag{2.6}$$

This important property of the point impulse is known as the *sifting property*.

EXAMPLE 2.1

Consider the point impulse shifted to position (ξ, η)—i.e., $\delta(x - \xi, y - \eta)$.

Question   What is the nature of the "function" that is defined as the product of $f(x, y)$ with the shifted point impulse?

Answer   Let $g(x, y) = f(x, y)\delta(x - \xi, y - \eta)$. Whenever $(x, y) \neq (\xi, \eta)$, the function $\delta(x - \xi, y - \eta)$ is zero, which means that

$$g(x, y) = 0, \qquad (x, y) \neq (\xi, \eta) .$$

When $(x, y) = (\xi, \eta)$, the shifted delta function is undefined (because its value is infinity), so $g(\xi, \eta)$ is also undefined. However, we know from (2.6) that the integral of $g(x, y)$—i.e., its volume—is $f(\xi, \eta)$. Therefore, we can conclude that $g(x, y)$ is a point impulse function located at (ξ, η) having volume $f(\xi, \eta)$, or

$$g(x, y) = f(\xi, \eta)\delta(x - \xi, y - \eta) . \tag{2.7}$$

This result tells us that we can interpret the product of a function with a point impulse as another point impulse whose volume is equal to the value of the function at the location of the point impulse. ∎

Two other properties of the point impulse will be used throughout the book. The *scaling property* of the point impulse is given by

$$\delta(ax, by) = \frac{1}{|ab|}\,\delta(x, y)\,. \tag{2.8}$$

Finally, we will occasionally need to use the fact that the point impulse is an *even function*; that is,

$$\delta(-x, -y) = \delta(x, y)\,, \tag{2.9}$$

which follows from (2.8) by setting $a = b = -1$.

2.2.2 Line Impulse

When calibrating medical imaging equipment, it is sometimes easier to use a linelike rather than a pointlike object. For example, it may be easier to position a wire than a small bead to assess the resolution of a projection radiography system. For this reason, we would like a mathematical model for a line source.

Consider the set of points defined by

$$L(\ell, \theta) = \{(x, y) \mid x\cos\theta + y\sin\theta = \ell\}\,. \tag{2.10}$$

It can be shown that $L(\ell, \theta)$ is a line whose unit normal is oriented at an angle θ relative to the x-axis and is at distance ℓ from the origin in the direction of the unit normal. This geometry can be seen in Figure 2.3(c).

The *line impulse* $\delta_\ell(x, y)$ associated with line $L(\ell, \theta)$ is given by

$$\delta_\ell(x, y) = \delta(x\cos\theta + y\sin\theta - \ell)\,. \tag{2.11}$$

This signal, illustrated in Figure 2.3(c), is effectively a 1-D impulse function "spread out" on the line $L(\ell, \theta)$.

2.2.3 Comb and Sampling Functions

We saw above that through the use of a shifted point impulse [within a double integral; (2.6)], we could "pick off" the value of a function at a single point. In nearly all medical imaging modalities, we need to "pick off" values not just at a single point, but on a grid or matrix of points, a process that is called *sampling*. Collectively, this matrix of values represents the medical image that is observed. For example, a CT image is typically a 1024×1024 matrix of *CT numbers*, representing a physical parameter called the *linear attenuation coefficient* of a cross-section of the human body. An MR image is typically a 256×256 matrix of values, representing one of several possible nuclear magnetic resonance properties of tissues.

As a first step toward characterizing sampling mathematically, we introduce the *comb function*, given by

$$\text{comb}(x) = \sum_{n=-\infty}^{\infty} \delta(x-n). \tag{2.12}$$

The comb function is also known as the *shah function*. It is called the comb function because the set of shifted point impulses comprising it resembles the teeth of a comb. In 2-D, the comb function is given by

$$\text{comb}(x,y) = \sum_{m=-\infty}^{\infty} \sum_{n=-\infty}^{\infty} \delta(x-m, y-n). \tag{2.13}$$

It is useful in describing signal sampling (see Section 2.8) to space the point impulses in the comb function by amounts Δx in the x-direction, and Δy in the y-direction. This yields the *sampling function* $\delta_s(x, y; \Delta x, \Delta y)$, defined by

$$\delta_s(x, y; \Delta x, \Delta y) = \sum_{m=-\infty}^{\infty} \sum_{n=-\infty}^{\infty} \delta(x-m\Delta x, y-n\Delta y). \tag{2.14}$$

It can be shown using the scaling property (2.8) of the point impulse, that the sampling function is related to the comb function as follows:

$$\delta_s(x, y; \Delta x, \Delta y) = \frac{1}{\Delta x \Delta y} \text{comb}\left(\frac{x}{\Delta x}, \frac{y}{\Delta y}\right). \tag{2.15}$$

The sampling function is a sequence of point impulses located at points $(m\Delta x, n\Delta y)$, $-\infty < m, n < \infty$, of the plane, as illustrated in Figure 2.3(d). The concept of the sampling function is critical for understanding *discretization*—the process of going from a continuous signal to a discrete signal.

2.2.4 Rect and Sinc Functions

Two signals that are frequently used in the study of medical imaging systems are the rect and sinc functions. The *rect function* is given by

$$\text{rect}(x,y) = \begin{cases} 1, & \text{for } |x| < \dfrac{1}{2} \text{ and } |y| < \dfrac{1}{2} \\[2mm] 0, & \text{for } |x| > \dfrac{1}{2} \text{ or } |y| > \dfrac{1}{2} \end{cases}, \tag{2.16}$$

where the value at $|x| = 1/2, |y| \leq 1/2$ or $|x| \leq 1/2, |y| = 1/2$ is immaterial. If this value is desired, we usually set it to $1/2$. The rect function is a finite energy signal, with unit total energy, that models signal concentration over a unit square centered around point $(0, 0)$. We can use the product

$$f(x, y) \, \text{rect}\left(\frac{x - \xi}{X}, \frac{y - \eta}{Y}\right) \tag{2.17}$$

to select that part of signal $f(x, y)$ centered at a point (ξ, η) of the plane with width X and height Y, and set the rest to zero.

The rect function can be written as the product of two 1-D rect functions,

$$\text{rect}(x, y) = \text{rect}(x) \, \text{rect}(y), \tag{2.18}$$

where

$$\text{rect}(x) = \begin{cases} 1, & \text{for } |x| < \frac{1}{2} \\[2mm] 0, & \text{for } |x| > \frac{1}{2} \end{cases} . \tag{2.19}$$

The 1-D rect function is plotted in Figure 2.4(a).

The *sinc function* is given by

$$\text{sinc}(x, y) = \begin{cases} 1, & \text{for } x = y = 0 \\[2mm] \dfrac{\sin(\pi x) \sin(\pi y)}{\pi^2 x y}, & \text{otherwise.} \end{cases} \tag{2.20}$$

The sinc function is a finite energy signal with unit total energy. Its maximum value equals 1, and is attained at point $(0, 0)$. The sinc function can be written as the product of two 1-D sinc functions,

$$\text{sinc}(x, y) = \text{sinc}(x) \, \text{sinc}(y), \tag{2.21}$$

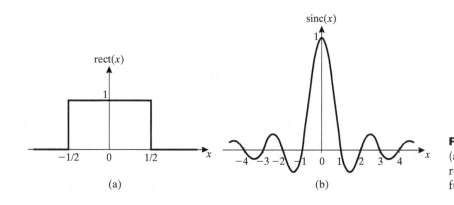

Figure 2.4
(a) The 1-D rect function rect(x), and (b) the 1-D sinc function sinc(x).

where

$$\text{sinc}(x) = \frac{\sin(\pi x)}{\pi x}. \tag{2.22}$$

The 1-D sinc function is plotted in Figure 2.4(b). This function consists of a main lobe and several side lobes, which eventually diminish to zero for large values of x. It alternates between positive and negative values, by passing through zero at points $x = \pm 1, \pm 2, \dots$.

The rect and sinc functions are useful in understanding the influence of the finite width of pixels in portraying an originally continuous signal as a discrete image.

2.2.5 Exponential and Sinusoidal Signals

Another continuous signal is the *complex exponential signal*, given by

$$e(x, y) = e^{j2\pi(u_0 x + v_0 y)}, \tag{2.23}$$

where u_0 and v_0 are two real-valued parameters and $j^2 = -1$. These parameters are usually referred to as *fundamental frequencies*, and have units that are the inverse of the units of x and y. For example, if x and y are given in mm, then u_0 and v_0 have the unit mm^{-1}.

The complex exponential signal can be decomposed into real and imaginary parts using two *sinusoidal signals*:

$$s(x, y) = \sin[2\pi(u_0 x + v_0 y)] \qquad \text{and} \qquad c(x, y) = \cos[2\pi(u_0 x + v_0 y)]. \tag{2.24}$$

Indeed,

$$
\begin{aligned}
e(x, y) &= e^{j2\pi(u_0 x + v_0 y)} \\
&= \cos[2\pi(u_0 x + v_0 y)] + j\sin[2\pi(u_0 x + v_0 y)] \\
&= c(x, y) + js(x, y).
\end{aligned} \tag{2.25}
$$

On the other hand, a sinusoidal signal can be written in terms of two complex exponential signals, since

$$s(x, y) = \sin[2\pi(u_0 x + v_0 y)] = \frac{1}{2j}e^{j2\pi(u_0 x + v_0 y)} - \frac{1}{2j}e^{-j2\pi(u_0 x + v_0 y)}, \tag{2.26}$$

$$c(x, y) = \cos[2\pi(u_0 x + v_0 y)] = \frac{1}{2}e^{j2\pi(u_0 x + v_0 y)} + \frac{1}{2}e^{-j2\pi(u_0 x + v_0 y)}. \tag{2.27}$$

The fundamental frequencies u_0, v_0 affect the oscillating behavior of the sinusoidal signals in the x and y directions, respectively. For example, small values of u_0 result in slow oscillations in the x-direction, whereas large values result in fast oscillations. This is illustrated in Figure 2.5.

The concepts of exponential signals are particularly useful in Fourier analysis, which is used in image reconstruction and in understanding MRI .

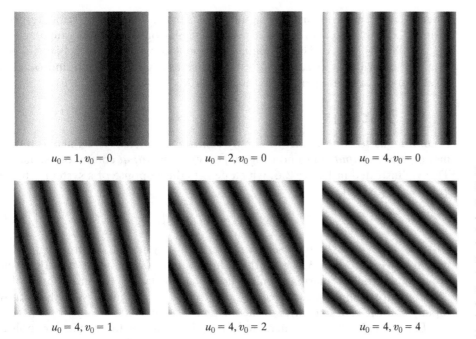

$u_0 = 1, v_0 = 0$ $u_0 = 2, v_0 = 0$ $u_0 = 4, v_0 = 0$

$u_0 = 4, v_0 = 1$ $u_0 = 4, v_0 = 2$ $u_0 = 4, v_0 = 4$

Figure 2.5
Six instances of the sinusoidal signal $s(x, y) = \sin[2\pi(u_0 x + v_0 y)]$, $0 \le x, y \le 1$, for various values of the fundamental frequencies u_0, v_0. Notice that small values result in slow oscillations in the corresponding direction, whereas large values result in fast oscillations.

2.2.6 Separable Signals

Separable signals form another class of continuous signals. A signal $f(x, y)$ is a *separable signal* if there exist two 1-D signals $f_1(x)$ and $f_2(y)$ such that

$$f(x, y) = f_1(x)f_2(y). \tag{2.28}$$

A 2-D separable signal that is a function of two independent variables x and y can be separated into a product of two 1-D signals, one of which is only a function of x and the other only of y. Notice that the point impulse is a separable signal, since $\delta(x, y) = \delta(x)\delta(y)$. From (2.18) and (2.21), the rect and sinc functions are separable signals as well.

Separable signals are limited, in the sense that they can only model signal variations independently in the x- and y-directions. However, there are instances in which use of separable signals is appropriate. Of major importance is the fact that operating on separable signals is much simpler than operating on purely 2-D signals, since, for separable signals, 2-D operations reduce to simpler consecutive 1-D operations.

2.2.7 Periodic Signals

A signal $f(x, y)$ is a *periodic signal* if there exist two positive constants X and Y such that

$$f(x, y) = f(x + X, y) = f(x, y + Y), \tag{2.29}$$

where X and Y are known as the signal *periods* in the x- and y-direction, respectively. From (2.29), we need to know a periodic signal only in the rectangular window $0 \le x < X$, $0 \le y < Y$. The sampling function $\delta_s(x, y; \Delta x, \Delta y)$ [(2.14)] is a periodic signal with periods $X = \Delta x$ and $Y = \Delta y$, whereas exponential and sinusoidal signals [(2.23) and (2.24)] are periodic with periods $X = 1/u_0$ and $Y = 1/v_0$.

2.3 Systems

A *continuous-to-continuous* (or simply *continuous*) *system* is defined as a transformation δ of an *input* continuous signal $f(x, y)$ to an *output* continuous signal $g(x, y)$. This is illustrated in Figure 2.6, which depicts the response of a system δ to a point impulse. In general,

$$g(x, y) = \delta[f(x, y)], \qquad (2.30)$$

which is known as the *input-output equation* of system δ. With a slight abuse of notation, we denote by $\delta[f(x, y)]$ the value, at point (x, y), of the signal obtained by applying the transformation δ on the *entire* signal $f(x, y)$. The importance of (2.30) is that it implies that we can predict the output of an imaging system if we know the input f and the characteristics of the imaging system δ.

Equation (2.30) is too general to be useful in practice. In a particular application, there might be many choices for δ, and the question is which one is more appropriate. For a choice of δ to be useful, it should be mathematically tractable. It is therefore necessary to limit our choices, by considering systems obtained using transformations δ that satisfy certain simplifying assumptions. Of course, any choice of δ should accurately portray the actual situation.

2.3.1 Linear Systems

A simplifying assumption is that of linearity. A system δ is a *linear system* if, when the input consists of a weighted summation of several signals, the output will also be a weighted summation of the responses of the system to each individual input signal. More precisely, for any collection $\{f_k(x, y), k = 1, 2, \ldots, K\}$ of input signals, and for any collection $\{w_k, k = 1, 2, \ldots, K\}$ of weights, we have

$$\delta\left[\sum_{k=1}^{K} w_k f_k(x, y)\right] = \sum_{k=1}^{K} w_k \delta\left[f_k(x, y)\right]. \qquad (2.31)$$

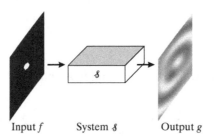

Figure 2.6
The response of a continuous system to a point impulse.

Input f System δ Output g

The linearity assumption leads to a class of relatively simple systems, which are tractable and enjoy a number of useful properties. Linearity is thus a critical assumption used in this book.

How reasonable is the assumption of linearity in practice? Are linear systems good mathematical approximations of the physical phenomena underlying medical imaging? It turns out that many medical imaging systems are approximately linear. In practice, this means that the image of an ensemble of signals is (nearly) identical to the sum of separate images of each signal, which is considered next.

2.3.2 Impulse Response

We now investigate how a linear system responds to an arbitrary input signal $f(x, y)$, and demonstrate the fact that linearity leads to tractable systems. However, before we pursue this investigation, it will be beneficial to first consider the output of a system $\mathcal{S}$ to a point impulse. Let us denote by $h(x, y; \xi, \eta)$ the output of system $\mathcal{S}$ to input $\delta_{\xi\eta}(x, y) = \delta(x - \xi, y - \eta)$ (i.e., a point impulse located at (ξ, η)). In this case,

$$h(x, y; \xi, \eta) = \mathcal{S}[\delta_{\xi\eta}(x, y)] . \qquad (2.32)$$

In general, this output depends on four independent variables; it is therefore a four-dimensional signal. The response $h(x, y; \xi, \eta)$ is known as the *point spread function* (PSF) of system $\mathcal{S}$, or, equivalently, the *impulse response function*.

Let us now assume that $\mathcal{S}$ is a linear system. If $f(x, y)$ is an arbitrary input to $\mathcal{S}$, then the output $g(x, y)$ will be given by

$$g(x, y) = \mathcal{S}[f(x, y)] \text{ [from (2.30)]}$$

$$= \mathcal{S}\left[\int_{-\infty}^{\infty} \int_{-\infty}^{\infty} f(\xi, \eta)\delta(x - \xi, y - \eta) \, d\xi \, d\eta \right] \qquad \text{[from (2.6)]},$$

$$= \int_{-\infty}^{\infty} \int_{-\infty}^{\infty} \mathcal{S}\left[f(\xi, \eta)\delta_{\xi\eta}(x, y) \right] d\xi \, d\eta \qquad \text{[from (2.31)]},$$

$$= \int_{-\infty}^{\infty} \int_{-\infty}^{\infty} f(\xi, \eta)\mathcal{S}[\delta_{\xi\eta}(x, y)] \, d\xi \, d\eta \qquad \text{[from (2.31)]},$$

$$= \int_{-\infty}^{\infty} \int_{-\infty}^{\infty} f(\xi, \eta)h(x, y; \xi, \eta) \, d\xi \, d\eta \qquad \text{[from (2.32)]} . \qquad (2.33)$$

Therefore, for any linear system $\mathcal{S}$ with PSF $h(x, y; \xi, \eta)$, the input-output equation is given by

$$g(x, y) = \int_{-\infty}^{\infty} \int_{-\infty}^{\infty} f(\xi, \eta)h(x, y; \xi, \eta) \, d\xi \, d\eta . \qquad (2.34)$$

The integral in (2.34) is known as the *superposition integral*. This equation shows that the PSF $h(x, y; \xi, \eta)$ uniquely characterizes a linear system; i.e., we only need to know the PSF of a linear system in order to calculate its output in response to a given input.

In fact, (2.34) facilitates computation of the output $g(x, y)$ from any arbitrary-but-known input $f(x, y)$. If the PSF $h(x, y; \xi, \eta)$ is known for every (x, y, ξ, η), then this computation reduces to calculating the double integral in (2.34). In practice, however, knowing the PSF for every (x, y, ξ, η) is a formidable task, since this either requires that $h(x, y; \xi, \eta)$ is known analytically, which is true only in limited circumstances of usually no practical interest, or that $h(x, y; \xi, \eta)$ is experimentally measured and stored in numerical form, which is usually not practical due to the four-dimensional nature of $h(x, y; \xi, \eta)$. Therefore, additional simplification is needed at this point, which may be achieved by imposing additional conditions on system $\mathscr{S}$.

2.3.3 Shift Invariance

An additional simplifying assumption is *shift invariance*. A system $\mathscr{S}$ is shift-invariant if an arbitrary translation of the input results in an identical translation in the output. In mathematical terms, if

$$f_{x_0 y_0}(x, y) = f(x - x_0, y - y_0) \tag{2.35}$$

is a translated version of an input signal $f(x, y)$ shifted at a point (x_0, y_0), then

$$g(x - x_0, y - y_0) = \mathscr{S}[f_{x_0 y_0}(x, y)], \tag{2.36}$$

where $g(x, y)$ is given by (2.30). Therefore, the response of a shift-invariant system to a translated input equals the response of the system to the actual input translated by the same amount.

Shift invariance does not require or imply linearity: a system may be shift-invariant but not linear and vice versa. However, if a linear system $\mathscr{S}$ is shift-invariant and

$$h(x, y) = \mathscr{S}[\delta(x, y)], \tag{2.37}$$

then

$$\mathscr{S}[\delta_{\xi\eta}(x, y)] = h(x - \xi, y - \eta). \tag{2.38}$$

In this case, the PSF is a 2-D signal. Therefore, by imposing shift invariance on a linear system, we are able to reduce the dimensionality of the PSF by a factor of two.

In practice, this means that the PSF is the same throughout the field-of-view of a shift-invariant imaging system. If we measure the PSF at one position, we can assume the same PSF at all other positions within the imaging field. Such a measurement is accomplished by using a (very!) small point object (to mimic a point impulse) and (2.37), which indicates that the image of this point object *is* the PSF.

A system δ that is both linear and shift-invariant is known as a *linear shift-invariant* (LSI) *system*. If δ is an LSI system, then [see (2.32), (2.34), and (2.38)]

$$g(x, y) = \int_{-\infty}^{\infty} \int_{-\infty}^{\infty} f(\xi, \eta) h(x - \xi, y - \eta) \, d\xi \, d\eta, \qquad (2.39)$$

or, simply,

$$g(x, y) = h(x, y) * f(x, y). \qquad (2.40)$$

Equations (2.39) and (2.40) are known as the *convolution integral* and the *convolution equation*, respectively. Convolution plays a fundamental role in the theory of signals and systems. In this book, we almost exclusively deal with LSI systems. Treating imaging systems as LSI systems significantly simplifies our analysis of these systems, and in most cases is accurate enough for practical use.

EXAMPLE 2.2
Consider a continuous system with input-output equation

$$g(x, y) = 2f(x, y). \qquad (2.41)$$

Question   Is this system linear and shift-invariant?

Answer   If $g'(x, y)$ is the response of the system to input $\sum_{k=1}^{K} w_k f_k(x, y)$, then

$$g'(x, y) = 2 \left(\sum_{k=1}^{K} w_k f_k(x, y) \right),$$

$$= \sum_{k=1}^{K} w_k 2 f_k(x, y),$$

$$= \sum_{k=1}^{K} w_k g_k(x, y), \qquad (2.42)$$

where $g_k(x, y)$ is the response of the system to input $f_k(x, y)$. Therefore, the system is linear. On the other hand, if $g'(x, y)$ is the response of the system to input $f(x - x_0, y - y_0)$, then

$$g'(x, y) = 2f(x - x_0, y - y_0) = g(x - x_0, y - y_0) \qquad (2.43)$$

and the system is also shift-invariant. ■

EXAMPLE 2.3
Consider a continuous system with input-output equation

$$g(x, y) = xyf(x, y). \qquad (2.44)$$

Question   Is this system linear and shift-invariant?

Answer   If $g'(x, y)$ is the response of the system to input $\sum_{k=1}^{K} w_k f_k(x, y)$, then

$$g'(x, y) = xy \left(\sum_{k=1}^{K} w_k f_k(x, y) \right),$$

$$= \sum_{k=1}^{K} w_k xy f_k(x, y),$$

$$g'(x, y) = \sum_{k=1}^{K} w_k g_k(x, y), \tag{2.45}$$

where $g_k(x, y)$ is the response of the system to input $f_k(x, y)$. Therefore, the system is linear. On the other hand, if $g'(x, y)$ is the response of the system to input $f(x - x_0, y - y_0)$ where $x_0 \neq 0$ and $y_0 \neq 0$, then

$$g'(x, y) = xyf(x - x_0, y - y_0),$$

$$\neq (x - x_0)(y - y_0)f(x - x_0, y - y_0).$$

Thus,

$$g'(x, y) \neq g(x - x_0, y - y_0), \tag{2.46}$$

and the system is not shift-invariant.   ∎

2.3.4 Connections of LSI Systems

LSI systems can be stand-alone or connected with other LSI systems. Two types of connections are usually considered: (1) *cascade* or *serial connections*; and (2) *parallel connections*.

Figure 2.7 illustrates two equivalent cascade connections of two LSI systems with PSFs $h_1(x, y)$ and $h_2(x, y)$, as well as the equivalent "single" LSI system. In this case, the following is true:

$$g(x, y) = h_2(x, y) * [h_1(x, y) * f(x, y)],$$

$$= h_1(x, y) * [h_2(x, y) * f(x, y)],$$

$$= [h_1(x, y) * h_2(x, y)] * f(x, y), \tag{2.47}$$

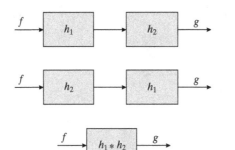

Figure 2.7
Cascade connection of two
LSI systems.

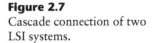

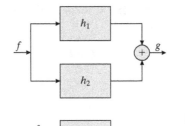

Figure 2.8
Parallel connection of two
LSI systems.

and

$$h_1(x, y) * h_2(x, y) = h_2(x, y) * h_1(x, y) .\tag{2.48}$$

Equations (2.47) and (2.48) are two properties of convolution, known as *associativity* and *commutativity*, respectively. From the commutativity property, we see that $g(x, y) = h(x, y) * f(x, y) = f(x, y) * h(x, y)$, which leads to

$$g(x, y) = \int_{-\infty}^{\infty} \int_{-\infty}^{\infty} f(\xi, \eta) h(x - \xi, y - \eta)\, d\xi\, d\eta ,$$

$$= \int_{-\infty}^{\infty} \int_{-\infty}^{\infty} h(\xi, \eta) f(x - \xi, y - \eta)\, d\xi\, d\eta .\tag{2.49}$$

Figure 2.8 illustrates a parallel connection of two LSI systems with PSFs $h_1(x, y)$ and $h_2(x, y)$, as well as the resulting equivalent "single" LSI system. In this case,

$$g(x, y) = h_1(x, y) * f(x, y) + h_2(x, y) * f(x, y) ,$$

$$= [h_1(x, y) + h_2(x, y)] * f(x, y) ,\tag{2.50}$$

which is another property of convolution, known as *distributivity*.

The properties of associativity, commutativity, and distributivity associated with LSI systems significantly simplify our analysis. (We imagine you're getting the idea why we like treating medical imaging systems as LSI so much!)

EXAMPLE 2.4
Consider two LSI systems connected in cascade, with Gaussian PSFs of the form

$$h_1(x, y) = \frac{1}{2\pi\sigma_1^2} e^{-(x^2+y^2)/2\sigma_1^2} \quad \text{and} \quad h_2(x, y) = \frac{1}{2\pi\sigma_2^2} e^{-(x^2+y^2)/2\sigma_2^2} ,\tag{2.51}$$

where σ_1 and σ_2 are two positive constants.

Question What is the point spread function of this system?

Answer   This connection is equivalent to a single LSI system with impulse response $h(x, y)$, given by

$$h(x, y) = h_1(x, y) * h_2(x, y),$$

$$= \int_{-\infty}^{\infty} \int_{-\infty}^{\infty} h_2(\xi, \eta) h_1(x - \xi, y - \eta) \, d\xi \, d\eta,$$

$$= \frac{1}{4\pi^2 \sigma_1^2 \sigma_2^2} \int_{-\infty}^{\infty} \int_{-\infty}^{\infty} e^{-(\xi^2 + \eta^2)/2\sigma_2^2} e^{-[(x-\xi)^2 + (y-\eta)^2]/2\sigma_1^2} \, d\xi \, d\eta,$$

$$= \frac{1}{4\pi^2 \sigma_1^2 \sigma_2^2} \int_{-\infty}^{\infty} e^{-\xi^2/2\sigma_2^2 - (x-\xi)^2/2\sigma_1^2} \, d\xi$$

$$\int_{-\infty}^{\infty} e^{-\eta^2/2\sigma_2^2 - (y-\eta)^2/2\sigma_1^2} \, d\eta. \tag{2.52}$$

However,

$$\int_{-\infty}^{\infty} e^{-\xi^2/2\sigma_2^2 - (x-\xi)^2/2\sigma_1^2} \, d\xi = e^{-x^2/2(\sigma_1^2 + \sigma_2^2)} \int_{-\infty}^{\infty} e^{-\frac{\sigma_1^2 + \sigma_2^2}{2\sigma_1^2 \sigma_2^2} \left\{ \xi - [\sigma_2^2/(\sigma_1^2 + \sigma_2^2)]x \right\}^2} \, d\xi,$$

$$= e^{-x^2/2(\sigma_1^2 + \sigma_2^2)} \int_{-\infty}^{\infty} e^{-\frac{\sigma_1^2 + \sigma_2^2}{2\sigma_1^2 \sigma_2^2} \tau^2} \, d\tau,$$

$$= \frac{\sqrt{2\pi} \sigma_1 \sigma_2}{\sqrt{\sigma_1^2 + \sigma_2^2}} e^{-x^2/2(\sigma_1^2 + \sigma_2^2)}, \tag{2.53}$$

in which case

$$h(x, y) = \frac{1}{2\pi (\sigma_1^2 + \sigma_2^2)} \exp \left\{ \frac{-(x^2 + y^2)}{2(\sigma_1^2 + \sigma_2^2)} \right\}. \tag{2.54}$$

To obtain the second equality in (2.53), we have substituted the variable $\xi - [\sigma_2^2/(\sigma_1^2 + \sigma_2^2)]x$ with τ; whereas, to obtain the third equality, we have used the fact that

$$\int_{-\infty}^{\infty} e^{-a^2\tau^2} \, d\tau = \frac{\sqrt{\pi}}{a}, \quad \text{for } a \neq 0. \tag{2.55}$$

The resulting PSF is Gaussian as well.   ∎

2.3.5 Separable Systems

Separable systems form an important class of LSI systems. As with separable signals, a 2-D LSI system with PSF $h(x, y)$ is a *separable system* if there exist two 1-D systems with PSFs $h_1(x)$ and $h_2(y)$, such that

$$h(x, y) = h_1(x) h_2(y). \tag{2.56}$$

A 2-D separable system consists of a cascade of two 1-D systems, one applied in the x-direction and the other one in the y-direction. Therefore, for a 2-D separable system, the 2-D convolution integral (2.39) can be calculated using two simpler 1-D convolution integrals, as follows:

- Compute $w(x, y) = \int_{-\infty}^{\infty} f(\xi, y) h_1(x - \xi) \, d\xi$, for every y.
- Compute $g(x, y) = \int_{-\infty}^{\infty} w(x, \eta) h_2(y - \eta) \, d\eta$, for every x.

Indeed,

$$\int_{-\infty}^{\infty} w(x, \eta) h_2(y - \eta) \, d\eta = \int_{-\infty}^{\infty} \left[\int_{-\infty}^{\infty} f(\xi, \eta) h_1(x - \xi) d\xi \right] h_2(y - \eta) \, d\eta,$$

$$= \int_{-\infty}^{\infty} \int_{-\infty}^{\infty} f(\xi, \eta) h_1(x - \xi) h_2(y - \eta) \, d\xi \, d\eta,$$

$$= \int_{-\infty}^{\infty} \int_{-\infty}^{\infty} f(\xi, \eta) h(x - \xi, y - \eta) \, d\xi \, d\eta \quad \text{[from (2.56)]},$$

$$= h(x, y) * f(x, y) = g(x, y). \tag{2.57}$$

The first step keeps y fixed, treats image $f(x, y)$ as a 1-D signal, and convolves it with the PSF $h_1(x)$ to obtain $w(x, y)$, for every y. The second step keeps x fixed, treats image $w(x, y)$ as a 1-D signal, and convolves it with the PSF $h_2(y)$. This is illustrated in Figure 2.9, which depicts the result of convolving an image with a 2-D Gaussian PSF of the form

$$h(x, y) = \frac{1}{2\pi\sigma^2} e^{-(x^2 + y^2)/2\sigma^2}, \tag{2.58}$$

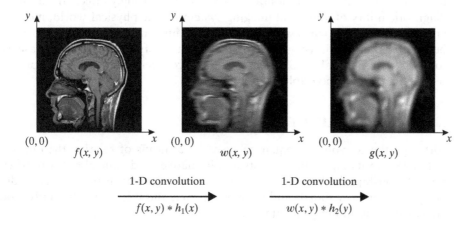

Figure 2.9
Calculation of the output of a 2-D separable system by using two 1-D steps in cascade.

with $\sigma = 4$, by using two 1-D steps in cascade. This PSF is separable, with

$$h_1(x) = \frac{1}{\sqrt{2\pi}\sigma}e^{-x^2/2\sigma^2} \qquad \text{and} \qquad h_2(y) = \frac{1}{\sqrt{2\pi}\sigma}e^{-y^2/2\sigma^2}. \qquad (2.59)$$

In practice, it is often faster (and easier) to execute two consecutive 1-D operations than a single 2-D operation, so the concept of separability is quite useful (when applicable).

2.3.6 Stable Systems

Stability is an important property of medical imaging systems. Informally, a medical imaging system is *stable* if small inputs lead to outputs that do not diverge. Although there are many ways to characterize system stability, we only consider BIBO stability here. A system is a *bounded-input bounded-output* (BIBO) *stable system* if, when the input is a *bounded signal*—i.e., when

$$|f(x, y)| \leq B < \infty, \qquad \text{for every } (x, y), \qquad (2.60)$$

for some finite B—there exists a finite B' such that

$$|g(x, y)| = |h(x, y) * f(x, y)| \leq B' < \infty, \qquad \text{for every } (x, y), \qquad (2.61)$$

in which case the output will also be a bounded signal. It can be shown that an LSI system is a BIBO stable system if and only if its PSF is absolutely integrable, in which case

$$\int_{-\infty}^{\infty} \int_{-\infty}^{\infty} |h(x, y)| \, dx \, dy < \infty. \qquad (2.62)$$

In practice, it is very important to design medical imaging systems that are stable. The output of an unstable system may grow out of bound, even if the input is small. This is an undesirable problem that strongly affects the accuracy and diagnostic utility of a medical imaging system. In the physical world, most systems are stable. However, when we design a medical imaging system, we may occasionally be tempted to model it using a PSF that does not satisfy (2.62), in which case we need to find alternative stable PSFs. This will be encountered in Chapter 6, for example, where computed tomography is discussed.

2.4   The Fourier Transform

Although the convolution equation provides a means of relating the output of an LSI system to its input, there exists an alternative (and equivalent) way of viewing this relationship: the Fourier transform. The Fourier transform provides a different perspective on how signals and systems interact, and it leads to alternative tools for system analysis and implementation.

Recall that the convolution equation [(2.40)] was obtained by decomposing a signal into point impulses [see (2.6) and (2.33)]. An alternative way to decompose a signal is in terms of complex exponential signals. It can be shown that, if

$$F(u, v) = \int_{-\infty}^{\infty} \int_{-\infty}^{\infty} f(x, y) e^{-j2\pi(ux+vy)} \, dx \, dy, \qquad (2.63)$$

then

$$f(x, y) = \int_{-\infty}^{\infty} \int_{-\infty}^{\infty} F(u, v) e^{j2\pi(ux+vy)} \, du dv. \qquad (2.64)$$

The signal $F(u, v)$ in (2.63) is known as the (2-D) *Fourier transform* of $f(x, y)$, whereas the signal decomposition in (2.64) is known as the (2-D) *inverse Fourier transform*. We use the notation

$$F(u, v) = \mathcal{F}_{2D}(f)(u, v) \quad \text{and} \quad f(x, y) = \mathcal{F}_{2D}^{-1}(F)(x, y) \qquad (2.65)$$

to denote the Fourier transform and its inverse, respectively. The subscript "2D" is used to denote that the Fourier transform is 2-D.

Before we can use the Fourier transform and its inverse, we need to make sure that the integrals in (2.63) and (2.64) exist. A sufficient condition for the existence of such integrals is that the signal $f(x, y)$ is continuous, or has a finite number of discontinuities, and that it is absolutely integrable. These conditions are almost always satisfied in practice.

Equation (2.64) shows that the Fourier transform produces a decomposition of a signal $f(x, y)$ into complex exponentials $e^{j2\pi(ux+vy)}$ with strength $F(u, v)$. The variables u and v are known as the x and y (spatial) *frequencies* of signal $f(x, y)$, respectively. The Fourier transform $F(u, v)$ is frequently referred to as the *spectrum* of $f(x, y)$. Since

$$e^{j2\pi(u_0 x + v_0 y)} = \cos[2\pi(u_0 x + v_0 y)] + j \sin[2\pi(u_0 x + v_0 y)], \qquad (2.66)$$

the Fourier transform provides information on the sinusoidal composition of a signal $f(x, y)$ at different frequencies. Notice the relationship between the Fourier transform and the exponential and sinusoidal signals discussed in Section 2.2.5.

In practice, the Fourier transform allows us to *separately* consider the action of an LSI system on each sinusoidal frequency. This general characterization of an LSI system then carries over to the action on arbitrary signals by considering the signals' Fourier transforms.

In general, the Fourier transform is a complex-valued signal, even if $f(x, y)$ is real-valued. It is quite common then to separately consider the Fourier transform's *magnitude* and *phase*, given by

$$|F(u, v)| = \sqrt{F_R^2(u, v) + F_I^2(u, v)}, \qquad (2.67)$$

and

$$\angle F(u, v) = \tan^{-1}\left(\frac{F_I(u, v)}{F_R(u, v)}\right), \tag{2.68}$$

where $F_R(u, v)$ and $F_I(u, v)$ are the real and imaginary parts of $F(u, v)$, respectively [i.e., $F(u, v) = F_R(u, v) + jF_I(u, v)$], in which case

$$F(u, v) = |F(u, v)| e^{j\angle F(u,v)}. \tag{2.69}$$

The magnitude $|F(u, v)|$ and phase $\angle F(u, v)$ are usually referred to as the *magnitude spectrum* and the *phase spectrum*, respectively. Moreover, the square $|F(u, v)|^2$ of the magnitude spectrum is usually referred to as the *power spectrum*. Notice that both the magnitude and phase spectra are required to uniquely determine $f(x, y)$.

We now provide a few examples of signals and their corresponding Fourier transforms. Some basic Fourier transform pairs are summarized in Table 2.1.

EXAMPLE 2.5
Consider the point impulse $\delta(x, y)$.

Question What is its Fourier transform?

Answer From (2.63), we have

$$\mathcal{F}_{2D}(\delta)(u, v) = \int_{-\infty}^{\infty}\int_{-\infty}^{\infty} \delta(x, y)e^{-j2\pi(ux+vy)}\,dx\,dy,$$

$$= \int_{-\infty}^{\infty}\int_{-\infty}^{\infty} \delta(x, y)e^{-j2\pi(u0+v0)}\,dx\,dy \quad \text{[from (2.7)]},$$

$$= \int_{-\infty}^{\infty}\int_{-\infty}^{\infty} \delta(x, y)\,dx\,dy = 1 \quad \text{[from (2.5)]}. \tag{2.70}$$

TABLE 2.1

Basic Fourier transform pairs	
Signal	Fourier Transform
1	$\delta(u, v)$
$\delta(x, y)$	1
$\delta(x - x_0, y - y_0)$	$e^{-j2\pi(ux_0+vy_0)}$
$\delta_s(x, y; \Delta x, \Delta y)$	$\text{comb}(u\Delta x, v\Delta y)$
$e^{j2\pi(u_0x+v_0y)}$	$\delta(u - u_0, v - v_0)$
$\sin[2\pi(u_0x + v_0y)]$	$\frac{1}{2j}[\delta(u - u_0, v - v_0) - \delta(u + u_0, v + v_0)]$
$\cos[2\pi(u_0x + v_0y)]$	$\frac{1}{2}[\delta(u - u_0, v - v_0) + \delta(u + u_0, v + v_0)]$
$\text{rect}(x, y)$	$\text{sinc}(u, v)$
$\text{sinc}(x, y)$	$\text{rect}(u, v)$
$\text{comb}(x, y)$	$\text{comb}(u, v)$
$e^{-\pi(x^2+y^2)}$	$e^{-\pi(u^2+v^2)}$

In this case, the magnitude spectrum is 1, whereas the phase spectrum is 0. Therefore, the Fourier transform of a point impulse is constant over all frequencies, with strength equal to 1.   ∎

EXAMPLE 2.6

Consider the complex exponential signal

$$f(x, y) = e^{j2\pi(u_0 x + v_0 y)}. \tag{2.71}$$

Question   What is its Fourier transform?

Answer   From (2.63), we have

$$\mathcal{F}_{2D}(f)(u, v) = \int_{-\infty}^{\infty} \int_{-\infty}^{\infty} f(x, y) e^{-j2\pi(ux+vy)} \, dx \, dy,$$

$$= \int_{-\infty}^{\infty} \int_{-\infty}^{\infty} e^{j2\pi(u_0 x + v_0 y)} e^{-j2\pi(ux+vy)} \, dx \, dy,$$

$$= \int_{-\infty}^{\infty} \int_{-\infty}^{\infty} e^{-j2\pi[(u-u_0)x+(v-v_0)y]} \, dx \, dy,$$

$$= \delta(u - u_0, v - v_0), \tag{2.72}$$

where the last step is based on the fact that

$$\int_{-\infty}^{\infty} \int_{-\infty}^{\infty} e^{-j2\pi(ux+vy)} \, dx \, dy = \delta(u, v), \tag{2.73}$$

which relates the complex exponential to the point impulse. In this case, the magnitude spectrum is a point impulse located at frequency (u_0, v_0), whereas the phase spectrum is 0. When $u_0 = v_0 = 0$, the complex exponential becomes the unit-valued signal $f(x, y) = 1$, for every (x, y), in which case (2.72) implies that the Fourier transform of a unit-valued signal is a point impulse at frequency $(0, 0)$.   ∎

A point impulse, which has an extremely nonuniform profile across space, results in a uniform frequency content (i.e., a magnitude spectrum with constant value). On the other hand, a constant signal, which does not vary in space, has a spectrum that is all concentrated at frequency $(0, 0)$. This extreme behavior reveals a more general property. Slow signal variation in space produces a spectral content that is primarily concentrated at low frequencies, whereas fast signal variation results in spectral content at high frequencies. This is illustrated in Figure 2.10, where three images with decreasing spatial variation (from left to right) produce spectra with decreasing high frequency content. Notice that, in order to reduce the dynamic range of values in the magnitude spectrum $|F(u, v)|$, when it exceeds the capabilities of a display device, we calculate and display the 2-D function $\log(1 + |F(u, v)|)$ instead.

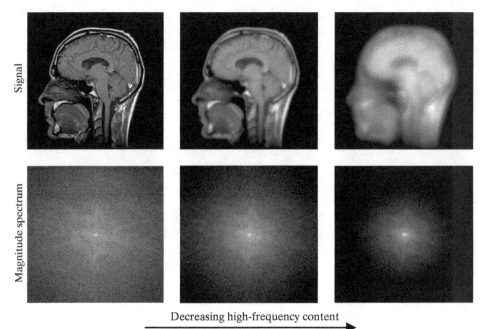

Figure 2.10
Three images of decreasing spatial variation (from left to right) and the associated magnitude spectra [depicted as $\log(1 + |F(u,v)|)$].

Decreasing high-frequency content

The Fourier transform of a 1-D signal and its inverse can be easily obtained from (2.63) and (2.64) by simply eliminating one of the two dimensions. In fact, if $f(x)$, $-\infty \leq x \leq \infty$, is a 1-D signal, then

$$F(u) = \mathcal{F}_{1D}(f)(u) = \int_{-\infty}^{\infty} f(x)e^{-j2\pi ux}\, dx, \tag{2.74}$$

and

$$f(x) = \mathcal{F}_{1D}^{-1}(F)(x) = \int_{-\infty}^{\infty} F(u)e^{j2\pi ux}\, du, \tag{2.75}$$

where the subscript "$1D$" denotes that the Fourier transform is one-dimensional.

EXAMPLE 2.7
Consider the 1-D rect function

$$\mathrm{rect}(x) = \begin{cases} 1, & \text{for } |x| < \frac{1}{2} \\ 0, & \text{for } |x| > \frac{1}{2} \end{cases}. \tag{2.76}$$

Question What is its Fourier transform?

Answer   From (2.74), we have

$$\mathcal{F}_{1D}(\text{rect})(u) = \int_{-\infty}^{\infty} \text{rect}(x)e^{-j2\pi ux}\,dx,$$

$$= \int_{-1/2}^{1/2} e^{-j2\pi ux}\,dx,$$

$$= \frac{1}{j2\pi u}\, e^{-j2\pi ux}\Big|_{-1/2}^{1/2},$$

$$= \frac{1}{\pi u}\frac{e^{j\pi u} - e^{-j\pi u}}{2j},$$

$$= \frac{\sin(\pi u)}{\pi u} = \text{sinc}(u). \qquad (2.77)$$

To derive (2.77), we have used the fact that, if $f(x)$ is a single-valued, bounded, and integrable function on $[a, b]$, and there exists a function $F(x)$ such that $dF(x)/dx = f(x)$, for $a \le x \le b$, then

$$\int_a^x f(\xi)\,d\xi = F(\xi)\Big|_a^x = F(x) - F(a), \qquad (2.78)$$

for $a \le x \le b$. Therefore, the 1-D Fourier transform of the rect function is the sinc function (and vice versa).   ∎

2.5   Properties of the Fourier Transform

The Fourier transform satisfies a number of useful properties. Most of them are used in both theory and application to simplify calculations. The following sections discuss the most important ones. Refer to Table 2.2 for a summary.

2.5.1 Linearity

If the Fourier transforms of two signals $f(x, y)$ and $g(x, y)$ are $F(u, v)$ and $G(u, v)$, respectively, then

$$\mathcal{F}_{2D}(a_1 f + a_2 g)(u, v) = a_1 F(u, v) + a_2 G(u, v), \qquad (2.79)$$

where a_1 and a_2 are two constants. This property can be extended to a linear combination of an arbitrary number of signals.

2.5.2 Translation

If $F(u, v)$ is the Fourier transform of a signal $f(x, y)$, and if

$$f_{x_0 y_0}(x, y) = f(x - x_0, y - y_0), \qquad (2.80)$$

TABLE 2.2

Properties of the Fourier transform		
Property	Signal	Fourier Transform
	$f(x, y)$	$F(u, v)$
	$g(x, y)$	$G(u, v)$
Linearity	$a_1 f(x, y) + a_2 g(x, y)$	$a_1 F(u, v) + a_2 G(u, v)$
Translation	$f(x - x_0, y - y_0)$	$F(u, v) e^{-j2\pi(ux_0 + vy_0)}$
Conjugation	$f^*(x, y)$	$F^*(-u, -v)$
Conjugate symmetry	$f(x, y)$ is real-valued	$F(u, v) = F^*(-u, -v)$
		$F_R(u, v) = F_R(-u, -v)$
		$F_I(u, v) = -F_I(-u, -v)$
		$|F(u, v)| = |F(-u, -v)|$
		$\angle F(u, v) = -\angle F(-u, -v)$
Signal reversing	$f(-x, -y)$	$F(-u, -v)$
Scaling	$f(ax, by)$	$\dfrac{1}{|ab|} F\left(\dfrac{u}{a}, \dfrac{v}{b}\right)$
Rotation	$f(x\cos\theta - y\sin\theta, x\sin\theta + y\cos\theta)$	$F(u\cos\theta - v\sin\theta, u\sin\theta + v\cos\theta)$
Circular symmetry	$f(x, y)$ is circularly symmetric	$F(u, v)$ is circularly symmetric
		$|F(u, v)| = F(u, v)$
		$\angle F(u, v) = 0$
Convolution	$f(x, y) * g(x, y)$	$F(u, v) G(u, v)$
Product	$f(x, y) g(x, y)$	$F(u, v) * G(u, v)$
Separable product	$f(x) g(y)$	$F(u) G(v)$
Parseval's theorem	$\int_{-\infty}^{\infty} \int_{-\infty}^{\infty} |f(x, y)|^2 \, dx \, dy = \int_{-\infty}^{\infty} \int_{-\infty}^{\infty} |F(u, v)|^2 \, du \, dv.$	

then

$$\mathcal{F}_{2D}(f_{x_0 y_0})(u, v) = F(u, v) e^{-j2\pi(ux_0 + vy_0)}. \tag{2.81}$$

Notice that, in this case,

$$\left|\mathcal{F}_{2D}(f_{x_0 y_0})(u, v)\right| = |F(u, v)|, \tag{2.82}$$

and

$$\angle \mathcal{F}_{2D}(f_{x_0 y_0})(u, v) = \angle F(u, v) - 2\pi(ux_0 + vy_0). \tag{2.83}$$

Therefore, translating a signal $f(x, y)$ does not affect its magnitude spectrum but subtracts a constant phase of $2\pi(ux_0 + vy_0)$ at each frequency (u, v).

2.5.3 Conjugation and Conjugate Symmetry

If $F(u, v)$ is the Fourier transform of a complex-valued signal $f(x, y)$, then

$$\mathcal{F}_{2D}(f^*)(u, v) = F^*(-u, -v), \qquad (2.84)$$

where $*$ denotes the complex conjugate. This is known as the *conjugation property* of the Fourier transform. When f is real-valued, its Fourier transform exhibits *conjugate symmetry*, which is defined by

$$F(u, v) = F^*(-u, -v). \qquad (2.85)$$

In this case, the real part $F_R(u, v)$ of $F(u, v)$ and the magnitude spectrum $|F(u, v)|$ are *symmetric functions*, whereas the imaginary part $F_I(u, v)$ and phase spectrum $\angle F(u, v)$ are *antisymmetric functions*:

$$F_R(u, v) = F_R(-u, -v) \qquad \text{and} \qquad F_I(u, v) = -F_I(-u, -v), \qquad (2.86)$$

$$|F(u, v)| = |F(-u, -v)| \qquad \text{and} \qquad \angle F(u, v) = -\angle F(-u, -v). \qquad (2.87)$$

Notice that the three magnitude spectra depicted in Figure 2.10 are symmetric around the origin.

2.5.4 Scaling

If $F(u, v)$ is the Fourier transform of a signal $f(x, y)$, and if

$$f_{ab}(x, y) = f(ax, by), \qquad (2.88)$$

where a and b are two nonzero constants, then

$$\mathcal{F}_{2D}(f_{ab})(u, v) = \frac{1}{|ab|} F\left(\frac{u}{a}, \frac{v}{b}\right). \qquad (2.89)$$

If we set $a = b = -1$, then the Fourier transform of signal $f(-x, -y)$ will be $F(-u, -v)$. Thus, reversing a signal in space also reverses its Fourier transform.

EXAMPLE 2.8

Detectors of many medical imaging systems can be modeled as rect functions of different sizes and locations.

Question Compute the Fourier transform of the following scaled and translated rect function:

$$f(x, y) = \text{rect}\left(\frac{x - x_0}{\Delta x}, \frac{y - y_0}{\Delta y}\right)$$

Answer   The Fourier transform of the rect function is the sinc function:

$$\mathcal{F}_{2D}(\text{rect})(u, v) = \text{sinc}(u, v).$$

We see that $f(x, y)$ is the rect function scaled by factors Δx and Δy in each direction and then translated to (x_0, y_0). By using the scaling property of the Fourier transform, we have

$$\mathcal{F}_{2D}\left\{\text{rect}\left(\frac{x}{\Delta x}, \frac{y}{\Delta y}\right)\right\} = \Delta x \Delta y \, \text{sinc}(\Delta x u, \Delta y v).$$

By using the translation property, we have

$$\mathcal{F}_{2D}(f)(u, v) = \Delta x \Delta y \, \text{sinc}(\Delta x u, \Delta y v) e^{-j2\pi(ux_0 + vy_0)}. \qquad \blacksquare$$

2.5.5 Rotation

Let us denote by $f_\theta(x, y)$ the signal

$$f_\theta(x, y) = f(x \cos\theta - y \sin\theta, x \sin\theta + y \cos\theta). \qquad (2.90)$$

We note that $f_\theta(x, y)$ is a rotated version of $f(x, y)$, rotated by an angle θ around the origin $(0, 0)$. If $F(u, v)$ is the Fourier transform of $f(x, y)$, then

$$\mathcal{F}_{2D}(f_\theta)(u, v) = F(u \cos\theta - v \sin\theta, u \sin\theta + v \cos\theta). \qquad (2.91)$$

If $f(x, y)$ is rotated by an angle θ, then its Fourier transform is rotated by angle θ as well.

2.5.6 Convolution

If $F(u, v)$ and $G(u, v)$ are the Fourier transforms of two signals $f(x, y)$ and $g(x, y)$, then the Fourier transform of the convolution $f(x, y) * g(x, y)$ equals the product of the individual Fourier transforms:

$$\mathcal{F}_{2D}(f * g)(u, v) = F(u, v)G(u, v). \qquad (2.92)$$

This property, known as the *convolution theorem*, provides a fundamental and very useful link between the space (f) and frequency (F) domains.

EXAMPLE 2.9

The convolution theorem provides a practical tool for computing the convolution of two signals, which may be difficult to conduct in spatial domain.

Question   Consider the two signals $f(x, y)$ and $g(x, y)$ given by

$$f(x, y) = \text{sinc}(Ux, Vy), \qquad \text{and}$$

$$g(x, y) = \text{sinc}(Vx, Uy),$$

where $0 < V \le U$. What is the convolution $f(x, y) * g(x, y)$?

Answer   From Table 2.1, we know that

$$\mathcal{F}_{2D}\{\text{sinc}(x, y)\} = \text{rect}(u, v).$$

By using the scaling property of the Fourier transform, we have

$$F(u, v) = \mathcal{F}_{2D}(f)(u, v) = \frac{1}{UV} \text{rect}\left(\frac{u}{U}, \frac{v}{V}\right)$$

and

$$G(u, v) = \mathcal{F}_{2D}(g)(u, v) = \frac{1}{UV} \text{rect}\left(\frac{u}{V}, \frac{v}{U}\right)$$

The convolution of $f(x, y)$ and $g(x, y)$ is

$$
\begin{aligned}
f(x, y) * g(x, y) &= \mathcal{F}_{2D}^{-1}\{F(u, v)G(u, v)\} \\
&= \mathcal{F}_{2D}^{-1}\left\{\frac{1}{(UV)^2} \text{rect}\left(\frac{u}{U}, \frac{v}{V}\right) \text{rect}\left(\frac{u}{V}, \frac{v}{U}\right)\right\} \\
&= \mathcal{F}_{2D}^{-1}\left\{\frac{1}{(UV)^2} \text{rect}\left(\frac{u}{V}, \frac{v}{V}\right)\right\} \\
&= \frac{1}{U^2} \mathcal{F}_{2D}^{-1}\left\{\frac{1}{V^2} \text{rect}\left(\frac{u}{V}, \frac{v}{V}\right)\right\} \\
&= \frac{1}{U^2} \text{sinc}(Vx, Vy) \qquad\qquad \blacksquare
\end{aligned}
$$

2.5.7 Product

If $F(u, v)$ and $G(u, v)$ are the Fourier transforms of two signals $f(x, y)$ and $g(x, y)$, the Fourier transform of the product $f(x, y)g(x, y)$ equals the convolution of the two Fourier transforms:

$$
\begin{aligned}
\mathcal{F}_{2D}(fg)(u, v) &= F(u, v) * G(u, v), \\
&= \int_{-\infty}^{\infty} \int_{-\infty}^{\infty} G(\xi, \eta) F(u - \xi, v - \eta) \, d\xi \, d\eta.
\end{aligned}
\qquad (2.93)
$$

2.5.8 Separable Product

If f is a separable signal, such that

$$f(x, y) = f_1(x)f_2(y), \qquad\qquad (2.94)$$

then

$$\mathcal{F}_{2D}(f)(u, v) = F_1(u)F_2(v), \qquad\qquad (2.95)$$

where

$$F_1(u) = \mathcal{F}_{1D}(f_1)(u) \qquad \text{and} \qquad F_2(v) = \mathcal{F}_{1D}(f_2)(v) \qquad (2.96)$$

are the 1-D Fourier transforms of $f_1(x)$ and $f_2(y)$, respectively. Therefore, the Fourier transform of a separable signal is also separable. In this case, the Fourier transform of a separable 2-D signal $f(x, y)$ can be computed by independently calculating the two 1-D Fourier transforms of $f_1(x)$ and $f_2(y)$ and then multiplying the results.

2.5.9 Parseval's Theorem

If $F(u, v)$ is the Fourier transform of a signal $f(x, y)$, then

$$\int_{-\infty}^{\infty} \int_{-\infty}^{\infty} |f(x, y)|^2 \, dx \, dy = \int_{-\infty}^{\infty} \int_{-\infty}^{\infty} |F(u, v)|^2 \, du \, dv, \qquad (2.97)$$

which is known as *Parseval's theorem*. This relationship says that the total energy of a signal $f(x, y)$ in the spatial domain equals its total energy in the frequency domain. In other words, the Fourier transform (as well as its inverse) are energy-preserving ("unit gain") transformations.

2.5.10 Separability

The Fourier transform $F(u, v)$ of a 2-D signal $f(x, y)$ can be calculated using two simpler 1-D Fourier transforms, as follows:

- Compute $r(u, y) = \int_{-\infty}^{\infty} f(x, y)e^{-j2\pi ux} \, dx$, for every y.
- Compute $F(u, v) = \int_{-\infty}^{\infty} r(u, y)e^{-j2\pi vy} \, dy$, for every u.

Indeed,

$$\int_{-\infty}^{\infty} r(u, y)e^{-j2\pi vy} \, dy = \int_{-\infty}^{\infty} \left[\int_{-\infty}^{\infty} f(x, y)e^{-j2\pi ux} \, dx \right] e^{-j2\pi vy} \, dy,$$

$$= \int_{-\infty}^{\infty} \int_{-\infty}^{\infty} f(x, y)e^{-j2\pi (ux+vy)} \, dx \, dy,$$

$$= F(u, v). \qquad (2.98)$$

The first step fixes y, treats image $f(x, y)$ as a 1-D signal, and calculates its 1-D Fourier transform $r(u, y)$. The second step fixes u, treats $r(u, y)$ as a 1-D signal, and calculates its 1-D Fourier transform. This is illustrated in Figure 2.11. Notice that calculation of 1-D Fourier transforms is simpler than their 2-D counterparts. Therefore, this technique is strongly recommended for calculating 2-D Fourier transforms (and their inverses).

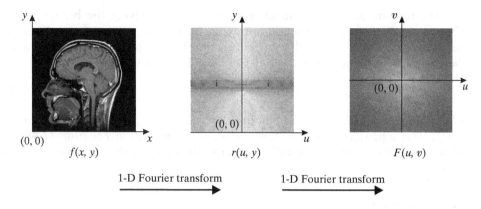

$f(x, y)$ $r(u, y)$ $F(u, v)$

1-D Fourier transform   1-D Fourier transform

Figure 2.11
Calculation of the 2-D
Fourier transform by using
two 1-D Fourier transforms
in cascade.

2.6 Transfer Function

Our discussion of LSI systems and their response to a point impulse makes it clear that in order to determine the PSF $h(x, y)$ of an LSI system $\mathcal{S}$, we observe the output of the system to a point impulse [see (2.37)]. We now replace the point impulse with one of the complex exponential signals $e^{j2\pi(ux+vy)}$ in decomposition (2.64), for some (u, v). As a direct consequence of the convolution integral (2.39), we have

$$g(x, y) = \int_{-\infty}^{\infty} \int_{-\infty}^{\infty} e^{j2\pi(u\xi+v\eta)} h(x - \xi, y - \eta) \, d\xi \, d\eta,$$

$$= \int_{-\infty}^{\infty} \int_{-\infty}^{\infty} h(\xi, \eta) e^{j2\pi[u(x-\xi)+v(y-\eta)]} \, d\xi \, d\eta,$$

$$= \left[\int_{-\infty}^{\infty} \int_{-\infty}^{\infty} h(\xi, \eta) e^{-j2\pi(u\xi+v\eta)} \, d\xi \, d\eta \right] e^{j2\pi(ux+vy)},$$

$$= H(u, v) e^{j2\pi(ux+vy)}, \tag{2.99}$$

where

$$H(u, v) = \int_{-\infty}^{\infty} \int_{-\infty}^{\infty} h(\xi, \eta) e^{-j2\pi(u\xi+v\eta)} \, d\xi \, d\eta, \tag{2.100}$$

and we have used (2.49) in the second line of (2.99). Notice that, in (2.99), the output equals the input multiplied by $H(u, v)$.

From (2.63) and (2.100), it is clear that $H(u, v)$ is the Fourier transform of the PSF $h(x, y)$. $H(u, v)$ is known as the *transfer function* of the LSI system $\mathcal{S}$. $H(u, v)$ is also known as the *frequency response* or the *optical transfer function* (OTF) of system $\mathcal{S}$. The transfer function uniquely characterizes an LSI system since, given $H(u, v)$, the PSF is uniquely determined using the inverse Fourier transform:

$$h(x, y) = \int_{-\infty}^{\infty} \int_{-\infty}^{\infty} H(u, v) e^{j2\pi(ux+vy)} \, du \, dv. \tag{2.101}$$

The transfer function is an important tool for studying the behavior of LSI systems because of the convolution property of the Fourier transform. If an input $f(x, y)$ to an LSI system $\mathcal{S}$ with transfer function $H(u, v)$ produces an output $g(x, y)$, then, as a direct consequence of (2.40) and (2.92), we have

$$G(u, v) = H(u, v)F(u, v), \tag{2.102}$$

where $F(u, v)$ and $G(u, v)$ are the Fourier transforms of $f(x, y)$ and $g(x, y)$, respectively. This equation is much simpler than the convolution equation (2.40), and allows an alternative (and simpler) method of studying the behavior of an LSI system.

EXAMPLE 2.10

Consider an idealized system whose PSF is $h(x, y) = \delta(x - x_0, y - y_0)$.

Question   What is the transfer function $H(u, v)$ of the system, and what is the output $g(x, y)$ of the system to an input signal $f(x, y)$?

Answer   The transfer function of the system is

$$H(u, v) = \mathcal{F}\{h(x, y)\}$$

$$= \mathcal{F}\{\delta(x - x_0, y - y_0)\}$$

$$= e^{-j2\pi(ux_0 + vy_0)}$$

For an input signal $f(x, y)$, the output is its translated version $f(x - x_0, y - y_0)$. This conclusion can be drawn from the sifting property of the δ function. Now, let's prove this using the transfer function.

The Fourier transform of the input signal $f(x, y)$ is $F(u, v) = \mathcal{F}(f)(u, v)$. The Fourier transform of the output signal $g(x, y)$ is

$$G(u, v) = F(u, v)H(u, v)$$

$$= F(u, v)e^{-j2\pi(ux_0 + vy_0)}$$

By applying the translation property of the Fourier transform, we have $g(x, y) = \mathcal{F}^{-1}(G)(x, y)$, which is the input signal $f(x, y)$ translated to (x_0, y_0).   ∎

Remember that in Fourier or frequency space we are dealing with the representation of an object as sine waves of different frequencies. In medical images, these are frequencies in space—*spatial frequencies*. Low spatial frequencies represent signals that vary slowly across the image; high spatial frequencies represent signals that vary locally (e.g., at the edges of structures within an image).

As an example, consider an LSI system $\mathcal{S}$ whose transfer function $H(u, v)$ is given by

$$H(u, v) = \begin{cases} 1, & \text{for } \sqrt{u^2 + v^2} \leq c \\ 0, & \text{for } \sqrt{u^2 + v^2} > c \end{cases}. \tag{2.103}$$

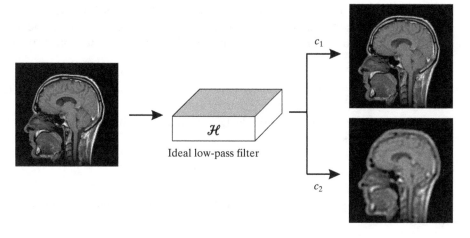

Figure 2.12
The response of an ideal low-pass filter for two values of the cutoff frequency c ($c_1 > c_2$).

This is known as an *ideal low-pass filter* with *cutoff frequency c*. From (2.102) and (2.103), we have

$$G(u, v) = \begin{cases} F(u, v), & \text{for } \sqrt{u^2 + v^2} \leq c \\ 0, & \text{for } \sqrt{u^2 + v^2} > c \end{cases}. \qquad (2.104)$$

Therefore, the input spectrum will be eliminated at high spatial frequencies (u, v) given by $\sqrt{u^2 + v^2} > c$. Applying such a system to an input signal will result in *signal smoothing*, with fine signal details (details that are mainly responsible for high-frequency spectral content) being eliminated. The amount of smoothing depends directly on the value of c, with smaller values of c producing more smoothing. This is illustrated in Figure 2.12. Most imaging systems can be modeled as some sort of low-pass filter.

We should point out here that, usually, the numerical implementation of the convolution equation (2.40) is not done in the space domain [i.e., by calculating the double integral in (2.39)], but in the frequency domain, using (2.102). The main reason is the existence of an efficient algorithm (the *fast Fourier transform*, or FFT) that allows a fast computer implementation of (2.102).

2.7 Circular Symmetry and the Hankel Transform

Often, the performance of a medical imaging system does not depend on the orientation of the patient with respect to the system. This orientation independence arises from the circular symmetry of certain signals, especially the PSF. A 2-D signal $f(x, y)$ is defined to be *circularly symmetric* if it satisfies

$$f_\theta(x, y) = f(x, y), \qquad \text{for every } \theta, \qquad (2.105)$$

where $f_\theta(x, y)$ is a rotated version of $f(x, y)$ [see (2.90)]. It follows from (2.91) that $\mathcal{F}_{2D}(f_\theta)(u, v)$ is also circularly symmetric. If $f(x, y)$ is circularly symmetric, then it is even in both x and y, and $F(u, v)$ is even in both u and v. In addition, $F(u, v)$ is real; therefore,

$$|F(u, v)| = F(u, v) \quad \text{and} \quad \angle F(u, v) = 0. \tag{2.106}$$

Functions $f(x, y)$ that are circularly symmetric can be written as a function of radius only; that is,

$$f(x, y) = f(r), \tag{2.107}$$

where $r = \sqrt{x^2 + y^2}$. Since in this case $F(u, v)$ is also circularly symmetric, $F(u, v)$ must satisfy

$$F(u, v) = F(q), \tag{2.108}$$

where $q = \sqrt{u^2 + v^2}$. The functions $f(r)$ and $F(q)$ are 1-D functions representing 2-D functions that are related by the Fourier transform. It can be shown (see Problem 2.19) that $f(r)$ and $F(q)$ are related by

$$F(q) = 2\pi \int_0^\infty f(r) J_0(2\pi q r)\, r\, dr, \tag{2.109}$$

where $J_0(r)$ is the zero-order Bessel function of the first kind. The nth order Bessel function of the first kind is given by

$$J_n(r) = \frac{1}{\pi} \int_0^\pi \cos(nr - r\sin\phi)\, d\phi, \quad n = 0, 1, 2, \ldots, \tag{2.110}$$

and therefore

$$J_0(r) = \frac{1}{\pi} \int_0^\pi \cos(r\sin\phi)\, d\phi. \tag{2.111}$$

The relation (2.109) is called the *Hankel transform*, and we will use the notation

$$F(q) = \mathcal{H}\{f(r)\}. \tag{2.112}$$

The inverse Hankel transform is identical to the forward transform,

$$f(r) = 2\pi \int_0^\infty F(q) J_0(2\pi q r)\, q\, dq. \tag{2.113}$$

If a 2-D signal is circularly symmetric, the Fourier transform of the signal can be found using the Hankel transform. Some Hankel transform pairs are given in Table 2.3.

TABLE 2.3

Selected Hankel Transform Pairs	
Signal	Hankel Transform
$\exp\{-\pi r^2\}$	$\exp\{-\pi q^2\}$
1	$\delta(q)/\pi q = \delta(u, v)$
$\delta(r - a)$	$2\pi a J_0(2\pi a q)$
$\text{rect}(r)$	$\frac{J_1(\pi q)}{2q}$
$\text{sinc}(r)$	$\frac{2\,\text{rect}(q)}{\pi\sqrt{1-4q^2}}$
$\frac{1}{r}$	$\frac{1}{q}$

Table 2.1 presented the Fourier transform of the 2-D Gaussian function as

$$\mathcal{F}\left\{e^{-\pi(x^2+y^2)}\right\} = e^{-\pi(u^2+v^2)}, \tag{2.114}$$

which itself is a 2-D Gaussian function. These Gaussian functions are circularly symmetric, and by direct substitution of $r^2 = x^2 + y^2$ and $q^2 = u^2 + v^2$, we find the Hankel transform pair

$$\mathcal{H}\left\{e^{-\pi r^2}\right\} = e^{-\pi q^2}. \tag{2.115}$$

The Gaussian function is a good model for the blurring inherent in medical imaging systems, and we will have many occasions throughout the text to use it and its Fourier (or Hankel) transform.

The *unit disk*, defined as

$$f(r) = \text{rect}(r), \tag{2.116}$$

is another example of a circularly symmetric function. It can be shown (see Problem 2.20) that the Hankel transform of the unit disk is given by

$$\mathcal{H}\left\{\text{rect}(r)\right\} = \frac{J_1(\pi q)}{2q}, \tag{2.117}$$

where J_1 is the first-order Bessel function of the first kind [see (2.110)]. Forming an analogy to the special relationship between the rect and sinc functions in Fourier transforms, the so-called jinc function is defined as

$$\text{jinc}(q) = \frac{J_1(\pi q)}{2q}. \tag{2.118}$$

Therefore, the rect and jinc form a Hankel transform pair.

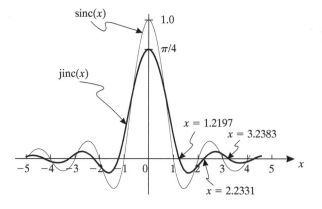

Figure 2.13
The 1-D jinc and sinc
functions.

A plot of the jinc(x) function, shown with the sinc(x) function for comparison, is shown in Figure 2.13. Like the sinc, the jinc function has a central lobe with its maximum value at the origin (see Figure 2.13),

$$\text{jinc}(0) = \frac{\pi}{4}, \tag{2.119}$$

and has a set of zeros as $x \to \infty$. The first few zeros are at the following locations:

$$\text{jinc}(1.2197) = 0 \tag{2.120a}$$

$$\text{jinc}(2.2331) = 0 \tag{2.120b}$$

$$\text{jinc}(3.2383) = 0. \tag{2.120c}$$

The jinc function takes on half its peak value at $x = 0.70576$.

The scaling theorem for the Hankel transform is derived from that of the Fourier transform. First, it is recognized that the scaling parameters, a and b, must be equal, $a = b$, in order for circular symmetry to be preserved. From (2.89) it follows that

$$\mathcal{H}\left\{f(ar)\right\} = \frac{1}{a^2}F(q/a). \tag{2.121}$$

EXAMPLE 2.11

In some medical imaging systems, only spatial frequencies smaller than ϱ_0 can be imaged.

Question   What is the function having uniform spatial frequencies within the disk of radius ϱ_0 and what is its inverse Fourier transform?

Answer   The disk of radius ϱ_0 is a circularly symmetric function, a scaled version of rect(q) given by

$$F(q) = \text{rect}\left(\frac{q}{2\varrho_0}\right).$$

Since the inverse Hankel transform is the same as the forward Hankel transform, we have

$$\text{jinc}(r) = \mathcal{H}^{-1}\left\{\text{rect}(q)\right\}.$$

The scaling theorem for the Hankel transform, (2.121), is then used to yield

$$f(r) = \mathcal{H}^{-1}\left\{\text{rect}\left(\frac{q}{2\varrho_0}\right)\right\}$$

$$= 4\varrho_0^2 \text{jinc}(2\varrho_0 r).$$ ∎

2.8 Sampling

To electronically sense, store, and process continuous signals using computers, we must transform them into collections of numbers. This transformation, called *discretization* or *sampling*, means that we only retain representative signal values and discard the rest. There are many ways to do this. In this book, we focus our attention on the so-called *rectangular sampling* scheme. According to this scheme, a 2-D continuous signal is replaced by a discrete signal whose values are the values of the continuous signal at the vertices of a 2-D rectangular grid. More precisely, given a 2-D continuous signal $f(x, y)$, rectangular sampling generates a 2-D discrete signal $f_d(m, n)$, such that

$$f_d(m, n) = f(m\Delta x, n\Delta y), \qquad \text{for } m, n = 0, 1, \ldots. \tag{2.122}$$

In (2.122), Δx and Δy are the *sampling periods* in the x and y directions, respectively. This is illustrated in Figure 2.14. Notice that $f_d(m, n)$ forms an array of numbers that contains the values of the 2-D continuous signal $f(x, y)$ at the discrete points $(m\Delta x, n\Delta y)$. The inverses $1/\Delta x$ and $1/\Delta y$ of Δx and Δy are referred to as the *sampling frequencies* in the x- and y-direction, respectively. As shown below, these sampling frequencies are related to the comb and sampling functions presented earlier.

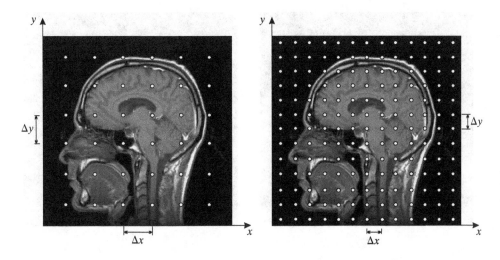

Figure 2.14
A coarse and a fine rectangular sampling scheme. Although coarse sampling results in fewer samples, it may not allow reconstruction of the original continuous signal from these samples.

Ideally, we would like to keep as few samples as possible. This strategy minimizes the number of detectors (or the scanning time needed by a single detector) and reduces signal storage and subsequent processing requirements. So we ask: Given a 2-D continuous signal $f(x, y)$, what are the maximum possible values for Δx and Δy such that $f(x, y)$ can be reconstructed from the 2-D discrete signal $f_d(m, n)$, obtained by (2.122)? As illustrated in Figure 2.14, we may be tempted to perform *coarse sampling* as opposed to *fine sampling*. However, if we give in to this temptation, we will end up (in the extreme case) with only one sample, which is clearly not enough to represent a continuous signal in general (unless the signal is constant).

Sampling a continuous signal with too few samples results in a type of signal corruption called *aliasing*, in which higher frequencies "take the alias of" lower frequencies. A signal sampled with too few samples cannot be reconstructed from its discrete representation; the best possible continuous reconstruction will be corrupted by aliasing. The visual appearance of this artifact depends on the signal's spectrum, but generally it appears as new patterns where none should exist. The spatial frequency of these patterns is always lower than it should be but may be relatively high compared with other spatial information in the image. An example of this phenomenon is depicted in Figure 2.15. Obviously, it is vital that medical imaging systems sample without aliasing.

2.8.1 Sampling Signal Model

It is intuitive to believe that a slowly varying signal could be reconstructed from fewer samples than a rapidly varying signal, which may require finer sampling at regions of rapid signal variation. Since signal variation in space is directly associated with frequency content, we suspect that there must be a direct relationship between the appropriate values of sampling periods Δx and Δy and the frequency content of the signal under consideration. Here, we prove this intuition mathematically, and derive a recipe for choosing appropriate values for Δx and Δy. We do this by using

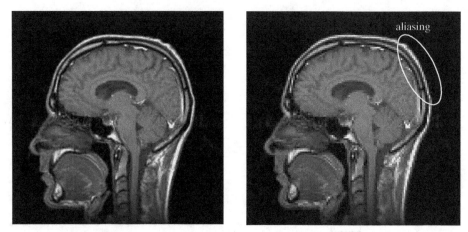

Figure 2.15
(a) An original image, and (b) a sampled version that suffers from aliasing.

(a)   (b)

the comb function comb(x, y), given by (2.13), and its close relative, the sampling function $\delta_s(x, y; \Delta x, \Delta y)$, given by (2.14).

Consider multiplying a continuous signal $f(x, y)$ by the sampling function. We have

$$f_s(x, y) = f(x, y)\delta_s(x, y; \Delta x, \Delta y),$$

$$= \sum_{m=-\infty}^{\infty} \sum_{n=-\infty}^{\infty} f(x, y)\delta(x - m\Delta x, y - n\Delta y) \quad [\text{from } (2.14)],$$

$$= \sum_{m=-\infty}^{\infty} \sum_{n=-\infty}^{\infty} f(m\Delta x, n\Delta y)\delta(x - m\Delta x, y - n\Delta y) \quad [\text{from } (2.7)],$$

$$= \sum_{m=-\infty}^{\infty} \sum_{n=-\infty}^{\infty} f_d(m, n)\delta(x - m\Delta x, y - n\Delta y), \tag{2.123}$$

where we have used (2.122) for the final step. It follows that, given the discrete signal $f_d(m, n) = f(m\Delta x, n\Delta y)$, we can calculate the continuous signal $f_s(x, y)$, regardless of the sampling periods Δx and Δy. Thus, if we can reconstruct $f(x, y)$ from $f_s(x, y)$, then $f(x, y)$ can also be reconstructed from $f_d(m, n)$. The most important point here is that, in order to understand the effects of sampling, we need only look at the *continuous* signal $f_s(x, y)$ and its relationship to $f(x, y)$.

Since $f_s(x, y)$ is the product of two functions, its Fourier transform is the convolution of the Fourier transforms of the two functions. This is a consequence of the product property (2.93) of the Fourier transform. Therefore,

$$F_s(u, v) = F(u, v) * \text{comb}(u\Delta x, v\Delta y)$$

$$= F(u, v) * \sum_{m=-\infty}^{\infty} \sum_{n=-\infty}^{\infty} \delta(u\Delta x - m, v\Delta y - n) \quad [\text{from } (2.13)],$$

$$= \frac{1}{\Delta x \Delta y} F(u, v) * \sum_{m=-\infty}^{\infty} \sum_{n=-\infty}^{\infty} \delta(u - m/\Delta x, v - n/\Delta y) \quad [\text{from } (2.8)],$$

$$= \frac{1}{\Delta x \Delta y} \sum_{m=-\infty}^{\infty} \sum_{n=-\infty}^{\infty} F(u, v) * \delta(u - m/\Delta x, v - n/\Delta y),$$

$$= \frac{1}{\Delta x \Delta y} \sum_{m=-\infty}^{\infty} \sum_{n=-\infty}^{\infty} \left[\int_{-\infty}^{\infty} \int_{-\infty}^{\infty} F(\xi, \eta)\delta(u - m/\Delta x - \xi, v - n/\Delta y - \eta) \, d\xi \, d\eta \right],$$

$$= \frac{1}{\Delta x \Delta y} \sum_{m=-\infty}^{\infty} \sum_{n=-\infty}^{\infty} F(u - m/\Delta x, v - n/\Delta y) \quad [\text{from } (2.6)], \tag{2.124}$$

where we have used the fact that the Fourier transform of the sampling function is given by

$$\mathcal{F}_{2D}(\delta_s(x, y; \Delta x, \Delta y)) = \text{comb}(u\Delta x, v\Delta y), \qquad (2.125)$$

as indicated in Table 2.1.

From (2.124), it is clear that the spectrum $F_s(u, v)$ of $f_s(x, y)$ is calculated by shifting the spectrum $F(u, v)$ of $f(x, y)$ to locations $(m/\Delta x, n/\Delta y)$, for all m and n, adding all shifted spectra and dividing the result by $\Delta x \Delta y$. This is illustrated in Figure 2.16. If the shifted spectra in $F_s(u, v)$ do not overlap, then the original spectrum $F(u, v)$ of $f(x, y)$, and thus $f(x, y)$ itself, can be recovered by filtering $f_s(x, y)$ to "pick" one of the (equivalent) spectra. It is customary to use a low-pass filter [see (2.103)] to capture the spectrum centered at the origin, like the one outlined with a dark box in Figure 2.16(b).

2.8.2 Nyquist Sampling Theorem

In order that the spectra in $F_s(u, v)$ do not overlap, it is first necessary that the spectrum of $f(x, y)$ be zero outside a rectangle in frequency space. Such signals are called *band-limited*. If the highest frequencies present in $f(x, y)$ in the x and y directions are U and V, respectively, then if

$$\Delta x \leq \frac{1}{2U} \qquad \text{and} \qquad \Delta y \leq \frac{1}{2V}, \qquad (2.126)$$

Figure 2.16
(a) The spectrum $F(u, v)$ of a band-limited continuous signal $f(x, y)$ with cutoff frequencies U and V, and (b) the spectrum $F_s(u, v)$ of signal $f_s(x, y)$ obtained by sampling $f(x, y)$ with sampling periods $\Delta x < 1/2U$ and $\Delta y < 1/2V$. In this case, $F(u, v)$ can be perfectly reconstructed from $F_s(u, v)$. The spectrum $F(u, v)$ takes value one within the light gray area and zero outside.

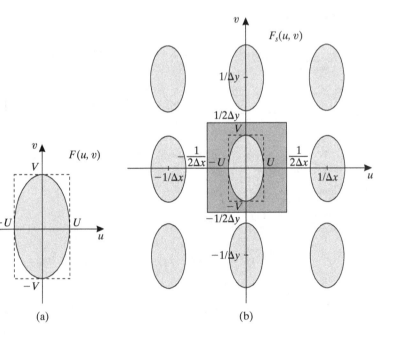

(a)   (b)

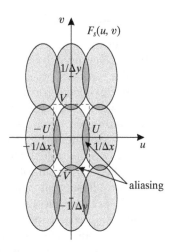

Figure 2.17
The spectrum $F_s(u, v)$ of $f_s(x, y)$ obtained by sampling the 2-D continuous signal $f(x, y)$ whose spectrum is depicted in Figure 2.16(a), with sampling periods $\Delta x > 1/2U$ and $\Delta y > 1/2V$. In this case, $f_s(x, y)$ experiences aliasing.

the spectrum $F(u, v)$ can be reconstructed from the spectrum $F_s(u, v)$, in which case $f(x, y)$ can be reconstructed from $f_s(x, y)$ and thus from the samples $f_d(m, n)$.

If $\Delta x > 1/2U$ or $\Delta y > 1/2V$, there is overlap of "high" frequencies of $F(u, v)$ in $F_s(u, v)$, producing aliasing. Aliasing is illustrated in Figure 2.17. In this case, the spectrum $F(u, v)$ cannot be recovered from the spectrum $F_s(u, v)$ and, therefore, $f(x, y)$ cannot be reconstructed from its samples $f_d(m, n)$.

In summary, we have the following important result: A 2-D continuous band-limited signal $f(x, y)$, with cutoff frequencies U and V, can be uniquely determined from its samples $f_d(m, n) = f(m\Delta x, n\Delta y)$, if and only if the sampling periods Δx and Δy satisfy

$$\Delta x \leq \frac{1}{2U} \qquad \text{and} \qquad \Delta y \leq \frac{1}{2V} \, .$$

This is known as the *sampling theorem* (or the *Nyquist sampling theorem*, after its discoverer). To avoid aliasing, the maximum allowed values for Δx and Δy are given by

$$(\Delta x)_{\text{max}} = \frac{1}{2U} \qquad \text{and} \qquad (\Delta y)_{\text{max}} = \frac{1}{2V}, \qquad (2.127)$$

and are known as the *Nyquist sampling periods*.

From our previous discussion, aliasing-free sampling requires band-limited continuous signals. From (2.127), the minimum number of samples required for aliasing-free sampling is directly proportional to the cutoff frequencies U and V. This verifies the fact that slowly varying signals, which are characterized by small values of U and V, require fewer samples than rapidly varying signals, which are characterized by large values of U and V. From the preceding discussion, we can see why undersampled signals tend to show high-frequency artifacts: the overlap of high-frequency spectra in the aliased Fourier transform artificially boosts high-frequency content.

2.8.3 **Anti-Aliasing Filters**

In medical imaging systems, there is an inherent trade-off between the number of samples acquired by a detector and image quality. Acquiring a large number of samples typically produces the highest image resolution but is almost always either expensive or time-consuming. Reducing the number of samples may lead to aliasing, which will introduce unwanted artifacts into the image. An alternative is to first filter the continuous signal using a low-pass filter, and then sample with fewer samples. In this case, the image is degraded by blurring rather than aliasing artifacts, which is usually preferable. Such a low-pass filter is called an *anti-aliasing filter*, and must be applied before sampling.

Anti-aliasing filtering is often an inherent part of a medical imaging instrument. This is because most medical imaging systems are integrators rather than point samplers. That is, most imaging instruments do not sample a signal at an array of points, but integrate the continuous signal locally. Thus, modeling sampling by multiplying a continuous signal $f(x, y)$ with the sampling function $\delta_s(x, y; \Delta x, \Delta y)$ is not usually accurate. Local integration can be modeled by convolving $f(x, y)$ with the PSF $h(x, y)$ of the detector, in which case the sampling process can be mathematically modeled by

$$f_d(m, n) = f_s(m\Delta x, n\Delta y), \tag{2.128}$$

$$f_s(x, y) = [h(x, y) * f(x, y)] \, \delta_s(x, y; \Delta x, \Delta y). \tag{2.129}$$

Since integration is a low-pass process, this resembles the anti-aliasing filtering approach described above: The continuous signal $f(x, y)$ is first filtered by a low-pass filter with PSF $h(x, y)$, from which $f_d(m, n)$ is obtained by sampling.

It should also be noted that a pixel often represents an area rather than a point sample in a discrete image. In such a case, discretization itself must be modeled as the sampling of the original signal [(2.14)] convolved with a rect function [(2.16)].

EXAMPLE 2.12

Consider a medical imaging system with sample periods Δ in both the x- and y-directions.

Question   What is the highest frequency allowed in the images so that the sampling is free of aliasing? If an anti-aliasing filter, whose PSF is modeled as a rect function, is used and we ignore all the side lobes of its transfer function, what are the widths of the rect function?

Answer   The sampling periods are Δ in both directions. From Nyquist sampling theorem, we know that if the image contains frequency higher than $1/2\Delta$, then there will be aliasing. So the highest allowed frequency is $1/2\Delta$.

If an anti-aliasing filter is used, the highest frequency of the image being sampled is the cutoff frequency of the filter (here, we assume the transfer function of the filter has value 0 outside the cutoff frequency). In this example, the anti-aliasing filter is modeled as a rect function. Its transfer function is given by the sinc function (we ignore the magnitude of the rect function, since it does not change the cutoff frequency)

$$H(u, v) = \text{sinc}(\Delta_x u, \Delta_y v),$$

where Δ_x and Δ_y are the widths of the filter in the x- and y-directions, respectively.

If we ignore the side lobes of $H(u, v)$, the cutoff frequencies of the filter are the first zeros of the sinc function, which are $1/\Delta_x$ and $1/\Delta_y$. For the given sampling periods Δ, we must have

$$\frac{1}{\Delta_x} \leq \frac{1}{2\Delta} \quad \text{and} \quad \frac{1}{\Delta_y} \leq \frac{1}{2\Delta}$$

which is equivalent to

$$\Delta_x \geq 2\Delta \text{ and } \Delta_y \geq 2\Delta. \qquad \blacksquare$$

2.9 Summary and Key Concepts

In this book, we take a signals and systems approach to the understanding and analysis of medical imaging systems. Signals model physical processes; systems model how medical imaging systems create new signals (images, in our case) from these original signals. In this chapter, we presented the following key concepts that you should now understand:

1. *Signals* are mathematical functions, and they can be continuous, discrete, or mixed.
2. The *point impulse* or *impulse function* is used as the input signal to characterize the response of a system, which is the impulse response.
3. *Comb* and *sampling functions*, *rect* and *sinc functions*, and *exponential* and *sinusoidal signals* are used to help describe and characterize imaging systems.
4. A system is *linear* if, when the input consists of a collection of signals, the output is a summation of the responses of the system to each individual input signal.
5. A system is *shift-invariant* if an arbitrary translation of the input signal results in an identical translation of the output.
6. A signal is *separable* if it can be represented as the product of 1-D signals.
7. The *Fourier transform* represents signals as a sum of sinusoids of different frequencies, with associated magnitude and phase.
8. The Fourier space analog of the impulse response is the *transfer function*.
9. The *output* of a linear, shift-invariant system is the convolution of the input with the impulse response; in Fourier space, it is the product of the Fourier transform of the input and the transfer function.
10. Continuous signals are transformed by *sampling* or *discretization* into discrete signals in order to be digitally represented.
11. *Aliasing* is caused by improper sampling of a continuous signal, yielding artifacts in the resulting digital image.

Bibliography

Bracewell, R. N. *Two-Dimensional Imaging*. Englewood Cliffs, NJ: Prentice Hall, 1995.

Gonzalez, R. C., and Woods, R. E. *Digital Image Processing*, 2nd ed. Upper Saddle River, NJ: Prentice Hall, 2002.

Macovski, A. *Medical Imaging Systems*. Englewood Cliffs, NJ: Prentice Hall, 1983.

Oppenheim, A. V., Willsky, A. S., and Nawad, S. H. *Signals and Systems*, 2nd ed. Upper Saddle River, NJ: Prentice Hall, 1997.

Problems

Signals and Their Properties

2.1 Determine whether the following signals are separable. Fully justify your answers.

(a) $\delta_s(x, y) = \sum_{m=-\infty}^{\infty} \sum_{n=-\infty}^{\infty} \delta(x - m, y - n)$.

(b) $\delta_l(x, y) = \delta(x \cos \theta + y \sin \theta - l)$.

(c) $e(x, y) = \exp\{j2\pi(u_0 x + v_0 y)\}$.

(d) $s(x, y) = \sin[2\pi(u_0 x + v_0 y)]$.

2.2 Determine whether the following signals are periodic. If they are, find the smallest periods.

(a) $\delta(x, y)$.

(b) $\text{comb}(x, y)$.

(c) $f(x, y) = \sin(2\pi x) \cos(4\pi y)$.

(d) $f(x, y) = \sin(2\pi(x + y))$.

(e) $f(x, y) = \sin(2\pi(x^2 + y^2))$.

(f) $f_d(m, n) = \sin\left(\frac{\pi}{5} m\right) \cos\left(\frac{\pi}{5} n\right)$.

(g) $f_d(m, n) = \sin\left(\frac{1}{5} m\right) \cos\left(\frac{1}{5} n\right)$.

2.3 We define the *energy* of a signal f in the finite rectangular window $-X \leq x \leq X$, $-Y \leq y \leq Y$; $X, Y < \infty$, by

$$E_{XY} = \int_{-X}^{X} \int_{-Y}^{Y} |f(x, y)|^2 \, dx \, dy,$$

where f might be complex-valued, in which case $|\cdot|$ denotes complex magnitude (or modulus). The *total energy* is defined as

$$E_\infty = \lim_{X \to \infty} \lim_{Y \to \infty} E_{XY}.$$

The *power* of signal f is defined as

$$P_{XY} = \frac{1}{4XY} \int_{-X}^{X} \int_{-Y}^{Y} |f(x, y)|^2 \, dx \, dy = \frac{E_{XY}}{4XY}$$

and the *total power* as

$$P_\infty = \lim_{X\to\infty} \lim_{Y\to\infty} P_{XY}.$$

For each of the signals in Problem 2.1, determine the values of E_∞ and P_∞.

Systems and Their Properties

2.4 Two LSI systems are connected in cascade. Show that the overall system is an LSI system and prove equations (2.47) and (2.48).

2.5 Show that an LSI system is BIBO stable if and only if its PSF is absolutely integrable.

2.6 Determine whether the system $g(x, y) = f(x, -1) + f(0, y)$ is
  (a) linear.
  (b) shift-invariant.

2.7 For each system with the following input-output equation, determine whether the system is (1) linear and/or (2) shift-invariant.
  (a) $g(x, y) = f(x, y)f(x - x_0, y)$.
  (b) $g(x, y) = \int_{-\infty}^{\infty} f(x, \eta)d\eta$.

2.8 For each system with the following PSF, determine whether the system is stable.
  (a) $h(x, y) = x^2 + y^2$.
  (b) $h(x, y) = \exp\{-(x^2 + y^2)\}$.
  (c) $h(x, y) = x^2\exp\{-y^2\}$.

2.9 Consider the 1D system whose input-output equation is given by

$$g(x) = f(x) * f(x),$$

where $*$ denotes convolution.
  (a) Write an integral expression that gives $g(x)$ as a function of $f(x)$.
  (b) Determine whether the system is linear.
  (c) Determine whether the system is shift-invariant.

Convolution of Signals

2.10 Given a continuous signal $f(x, y) = x + y^2$, evaluate the following:
  (a) $f(x, y)\delta(x - 1, y - 2)$.
  (b) $f(x, y) * \delta(x - 1, y - 2)$.
  (c) $\int_{-\infty}^{\infty} \int_{-\infty}^{\infty} \delta(x - 1, y - 2)f(x, 3)dx\, dy$;
  (d) $\delta(x - 1, y - 2) * f(x + 1, y + 2)$.

2.11 Consider two continuous signals $f(x, y)$ and $g(x, y)$ that are separable—i.e., $f(x, y) = f_1(x)f_2(y)$ and $g(x, y) = g_1(x)g_2(y)$.

      (a) Show that their convolution is also separable.

      (b) Express the convolution in terms of $f_1(x)$, $f_2(y)$, $g_1(x)$, and $g_2(y)$.

2.12 By using two 1-D convolution integrals, calculate the 2-D convolution of signal $f(x, y) = x + y$ with the exponential PSF $h(x, y) = \exp\{-(x^2 + y^2)\}$.

Fourier Transforms and Their Properties

2.13 Find the Fourier transforms of the following continuous signals:

      (a) $\delta_s(x, y) = \sum_{m=-\infty}^{\infty} \sum_{n=-\infty}^{\infty} \delta(x - m, y - n)$.

      (b) $\delta_s(x, y; \Delta x, \Delta y) = \sum_{m=-\infty}^{\infty} \sum_{n=-\infty}^{\infty} \delta(x - m\Delta x, y - n\Delta y)$.

      (c) $s(x, y) = \sin[2\pi(u_0 x + v_0 y)]$.

      (d) $c(x, y) = \cos[2\pi(u_0 x + v_0 y)]$.

      (e) $f(x, y) = \frac{1}{2\pi\sigma^2}\exp\{-(x^2 + y^2)/2\sigma^2\}$.

2.14 Suppose $F(u)$ is the Fourier transform of a 1-D real signal $f(x)$, $F(u) = \mathcal{F}[f(x)]$. Prove the following, where * is complex conjugate:

      (a) If $f(x) = f(-x)$, then $F^*(u) = F(u)$.

      (b) If $f(x) = -f(-x)$, then $F^*(u) = -F(u)$.

2.15 In the above problem, if $f(x)$ is not a real signal, can we still arrive at the conclusions? If not, what kind of symmetric property does $F(u) = \mathcal{F}[f(x)]$ have when $f(x) = f(-x)$?

2.16 Prove the following properties of the Fourier transform:

      (a) Conjugate and conjugate symmetry.

      (b) Scaling.

      (c) Convolution.

      (d) Product.

2.17 Show that the 2-D Fourier transform of the rect function is the sinc function, whereas the 2-D Fourier transform of the sinc function is the rect function. Determine the values of E_∞ and P_∞ for the sinc function (see Problem 2.3).

2.18 Find the Fourier transform of the separable continuous signal $f(x, y) = \sin(2\pi ax)\cos(2\pi by)$.

2.19 Prove that a circularly symmetric function and its Fourier transform are related by the zero-order Bessel function of the first kind, as expressed by (2.109).

2.20 Show that the Hankel transform of a unit disk is a jinc function.

Transfer Function

2.21 A new imaging system with which you are experimenting has anisotropic properties. You measure the impulse response function as $h(x, y) = e^{-\pi(x^2 + y^2)/4}$.

      (a) Sketch the impulse response function.

      (b) What is the transfer function?

2.22 A medical imaging system has the following line spread function, where $\alpha = 2$ radians/cm:

$$l(x) = \begin{cases} \cos(\alpha x) & |\alpha x| \le \pi/2 \\ 0 & \text{otherwise} \end{cases}.$$

(a) Suppose a bar phantom is imaged, where the bars have width w, separation w, and unity height. Assume $\pi/2\alpha \le w \le \pi/\alpha$. What are the responses of the system at the center of a bar and halfway between two bars?
(b) From the line spread function, can we tell whether the system is isotropic?
(c) Assume the system is separable with $h(x, y) = h_{1D}(x)h_{1D}(y)$, compute its transfer function.

Sampling Theory

2.23 A signal $f(t)$ is defined as

$$f(t) = \begin{cases} \sin(2\pi t/T), & 0 \le t \le T \\ 0, & \text{otherwise} \end{cases}.$$

We sample the signal with a sampling period of $\Delta T = 0.25T$.
(a) What is $f_s(t)$, and what is $f_d(m)$?
(b) Define a new signal $f_h(t)$ as

$$f_h(t) = f_d(k), \qquad \text{for} \qquad k\Delta T \le t < (k+1)\Delta T.$$

Sketch $f_h(t)$ and find its Fourier transform.
(c) Repeat the above using a sampling period of $\Delta T = 0.5T$.

2.24 The Nyquist sampling periods for 1-D band-limited signals $f(x)$ and $g(x)$ are Δ_f and Δ_g, respectively. Find the Nyquist sampling periods for the following signals:
(a) $f(x - x_0)$, where x_0 is a given constant.
(b) $f(x) + g(x)$.
(c) $f(x) * f(x)$.
(d) $f(x)g(x)$.
(e) $|f(x)|$.

2.25 We want to sample the 2-D continuous signal $f(x, y) = \exp\{-\pi(x^2 + y^2)\}$ by means of a rectangular sampling scheme to obtain 1.5 samples per millimeter. Determine the PSF $h(x, y)$ of an ideal low-pass anti-aliasing filter with the maximum possible frequency content. What percentage of the spectrum energy (see Problem 2.3) of $f(x, y)$ is preserved by this filter? In practice, can we sample $f(x, y)$ alias-free without using an anti-aliasing filter?

Applications, Extensions, and Advanced Topics

2.26 The point spread function of a medical imaging system is given by

$$h(x, y) = e^{-(|x|+|y|)}$$

(x and y are in millimeters.)
(a) Is the system separable? Explain.
(b) Is this system circularly symmetric? Explain.
(c) What is the response of the system to the line impulse $f(x, y) = \delta(x)$?
(d) What is the response of the system to the line impulse $f(x, y) = \delta(x - y)$?

2.27 Consider a one-dimensional linear imaging system whose point spread function is given by

$$h(x; \xi) = e^{\frac{-(x-\xi)^2}{2}},$$

which represents the response to the shifted impulse $\delta(x - \xi)$.
(a) Is this imaging system shift-invariant? Explain.
(b) Write the output image $g(x)$ of this system when the input signal is $f(x) = \delta(x + 1) + \delta(x) + \delta(x - 1)$?

2.28 Consider a 1-D ideal *high-pass filter* whose spectrum is shown in Figure P2.1.

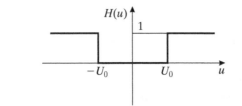

Figure P2.1
An ideal high-pass filter
$H(u)$.

(a) Compute the impulse response of the filter.
(b) Using the linearity of convolution, compute the system response to (1) a constant function $f(t) = c$; and (2) a unit step $f(t) = \begin{cases} 1, & t \geq 0 \\ 0, & t < 0 \end{cases}$.

2.29 The impulse response of a 1-D LTI system is

$$h(t) = \frac{1}{T}\left[-\operatorname{rect}\left(\frac{t + 0.75T}{0.5T}\right) + \operatorname{rect}\left(\frac{t}{T}\right) - \operatorname{rect}\left(\frac{t - 0.75T}{0.5T}\right) \right].$$

(a) Plot $h(t)$ with labeled axes. Is the system stable? Is it causal?
(b) Find and plot the response of the system to a constant signal $f(t) = c$.

(c) Find and plot the response of the system to a unit step signal

$$f(t) = \begin{cases} 1, & t \geq 0 \\ 0, & t < 0 \end{cases}.$$

(d) Compute the Fourier transform of $h(t)$.

(e) Plot the magnitude spectrum $\|H(u)\|$ for $T = 0.25$, $T = 0.1$, and $T = 0.05$.

(f) What kind of filter is this LTI system?

2.30 In CT image reconstruction, a ramp filter is used in *filtered backprojection*. The transfer function of the ramp filter is defined as

$$H(\varrho) = |\varrho|.$$

For practical reasons, a windowed ramp filter $\hat{H}(\varrho) = W(\varrho)|\varrho|$ is used instead of $H(\varrho)$.

(a) Assume that $W(\varrho)$ is a rectangular window defined as

$$W(\varrho) = \begin{cases} 1, & |\varrho| \leq \varrho_0 \\ 0, & \text{otherwise} \end{cases},$$

where ϱ_0 is the cutoff frequency. Find $\hat{h}(r) = \mathcal{F}^{-1}\{\hat{H}(\varrho)\}$.

(b) What are the responses of a ramp filter to (1) a constant function $f(r) = c$; (2) a sinusoid function $f(r) = \sin(\omega r)$.

2.31 For the imaging system depicted in Figure P2.2, show that the output is a scaled and inverted replica of the input.

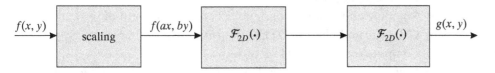

Figure P2.2
A cascade of subsystems representing the imaging system in Problem 2.31.

2.32 A continuous signal

$$f(x, y) = \begin{cases} 1, & \text{for } x = y = 0 \\ \dfrac{\sin(ax)\sin(by)}{\pi^2 xy}, & \text{otherwise} \end{cases}$$

is corrupted by an additive sinusoidal signal

$$\eta(x, y) = \cos\left[2\pi(Ax + By)\right], \quad A, B \geq 0,$$

and a set of measurements

$$g(x, y) = f(x, y) + \eta(x, y)$$

is obtained. Design an ideal low-pass filter (an LSI system whose transfer function equals 1, for $0 \leq |u| \leq U, 0 \leq |v| \leq V$, and 0 otherwise) with impulse response $h(x, y)$, such that

$$f(x, y) = h(x, y) * g(x, y).$$

Given a and b, determine all possible values of A and B for which this is possible. Explain your answers in full detail. You might use the formula

$$\int_{-\infty}^{\infty} \frac{\sin(\alpha t)}{\pi t} e^{-j\tau t} dt = \begin{cases} 1, & \text{for } |\tau| \leq \alpha \\ 0, & \text{for } |\tau| > \alpha \end{cases}.$$

2.33 The *discrete-time Fourier transform* (DTFT) of a 1-D aperiodic discrete time signal with finite energy, $f(m)$, is defined as

$$F(e^{j\omega}) = \sum_{m=-\infty}^{+\infty} f(m)e^{-j\omega m} \tag{P2.1}$$

and the inverse transform is

$$f(m) = \frac{1}{2\pi} \int_{-\pi}^{\pi} F(e^{j\omega})e^{j\omega m} d\omega. \tag{P2.2}$$

A continuous signal $g(x)$ is sampled with a given sampling period Δx_1 to give a discrete signal $g_1(m)$.

(a) Compute and plot the continuous Fourier transform of $g(x)$.
(b) Calculate and plot the DTFT of $g_1(m)$, $G_1(\omega)$.
(c) Compute the inverse DTFT of $G_1(\omega)$, $\hat{g}_1(m)$.
(d) If $g(x)$ is sampled at period $\Delta x_2 > \Delta x_1$ (aliasing) to get $g_2(m)$, compute and plot the DTFT of $g_2(m)$, $G_2(\omega)$.
(e) Compute the inverse DTFT of $G_2(\omega)$, $\hat{g}_2(m)$.
(f) Use zeroth order holding to reconstruct continuous signal from its discrete samples.
(g) Conclude that when sampling period is larger than the inverse of the Nyquist frequency, it is not possible to reconstruct the continuous signal from its discrete samples.

Image Quality[1]

3.1 Introduction

The primary purpose of a medical imaging system is to create images of the internal structures and functions of the human body that can be used by medical professionals to diagnose abnormal conditions, determine the underlying mechanisms that produce and control these conditions, guide therapeutic procedures, and monitor the effectiveness of treatment. The ability of medical professionals to successfully accomplish these tasks strongly depends on the *quality* of the images acquired by the medical imaging system at hand, where by "quality" we mean the degree to which an image allows medical professionals to accomplish their goals.

Image quality depends on the particular imaging modality used. With each modality, the range of image quality may be considerable, depending on the characteristics and setup of the particular medical imaging system, the skill of the operator handling the system, and several other factors, such as patient characteristics and imaging time. Studying how these factors affect image quality is an important and complicated task. This task is simplified by focusing on the following six important factors: (1) contrast, (2) resolution, (3) noise, (4) artifacts, (5) distortion, and (6) accuracy.

The ability of medical professionals to discriminate among anatomical or functional features in a given image strongly depends on *contrast*. Contrast quantifies the difference between image characteristics (e.g., shades of gray or color) of an object (or feature within an object) and surrounding objects or background. High contrast allows easier identification of individual objects in an image, whereas low contrast makes this task difficult. In the overview to Part I, we presented images from different modalities that have different contrast. For example, the brain structures in the PET image in Figure I.4(c) are of higher contrast than the same structures in Figure I.4(a) or (b), because the actual signal coming from the patient is of intrinsically higher contrast in PET.

---

[1]The initial draft of this chapter was written by Dr. John I. Goutsias.

Sometimes, medical images are blurry and lack detail. The ability of a medical imaging system to depict details is known as *resolution*. High resolution systems create images of high diagnostic quality. Low resolution systems create images that lack fine detail. For example, while the PET image in Figure I.4(c) has higher contrast than the CT image in Figure I.4(a) or the MRI image in Figure I.4(b), it has poorer spatial resolution.

A medical image may be corrupted by random fluctuations in image intensity that do not contribute to image quality. This is known as *noise*. The source, amount, and type of noise depends on the particular imaging modality used. Object visibility is often reduced by the presence of noise, because the noise masks image features. Nuclear medicine images tend to have the highest noise, as reflected in the PET image in Figure I.4(c).

Most medical imaging systems can create image features that do not represent a valid object within, or characteristic of, the patient. These features are known as *artifacts*, and can frequently obscure important features, or be falsely interpreted as abnormal findings. Medical images should not only make desired features visible but should also give an accurate impression of their shape, size, position, and other geometric characteristics. Unfortunately, for many reasons to be explained later in this book, medical imaging systems frequently introduce *distortion* of these important factors. Distortion in medical images should be corrected in order to improve the diagnostic quality of these images.

Ultimately, the quality of medical images should be judged on their utility in the context of a specific clinical application. For example, medical images that increase the chance of tumor detectability in nuclear medicine should be preferable to images with poor tumor detectability. Fundamentally, we are interested in the *accuracy* of medical images in the context of a clinical application, where "accuracy" means both conformity to truth and clinical utility.

The user of a medical imaging system is very interested in adjusting the system to produce images of the highest possible quality, while maintaining a safe environment for the patient. In order to achieve this, methods must be developed for evaluating image quality. Since image quality depends primarily on the six previously discussed factors, we need to mathematically quantify these factors and systematically study their influence on image quality. The purpose of this chapter is to provide a fundamental exposition on contrast, resolution, noise, artifacts, distortion, and accuracy. More detailed discussions on these factors, and how they affect image quality for a specific imaging modality, will be provided later.

3.2 Contrast

Contrast refers to differences between the image intensity of an object and surrounding objects or background. This difference, or *image contrast*, is itself the result of the inherent *object contrast* within the patient. In general, the goal of a medical imaging system is to accurately portray or preserve the true object contrast in the image. Particularly for detection of abnormalities, a medical imaging system that produces high contrast images is preferable to a system that produces low contrast images, since anatomical and functional features are easier to identify in high contrast images.

3.2.1 Modulation

Use of a periodic signal and its modulation is an effective way to quantify contrast. The *modulation* m_f of a periodic signal $f(x, y)$, with maximum and minimum values $f_{\max}$ and $f_{\min}$, is defined by

$$m_f = \frac{f_{\max} - f_{\min}}{f_{\max} + f_{\min}}. \tag{3.1}$$

Modulation quantifies the relative amount by which the amplitude (or difference) $(f_{\max} - f_{\min})/2$ of $f(x, y)$ stands out from the average value (or background) $(f_{\max} + f_{\min})/2$. In general, m_f refers to the *contrast* of the periodic signal $f(x, y)$ relative to its average value. We assume here that $f(x, y)$ has nonnegative values, in which case $0 \le m_f \le 1$. Of importance, $m_f = 1$ only when $f_{\min} = 0$. Thus, in practice, the usual presence of a nonzero "background" intensity in a medical image reduces image contrast. If $m_f = 0$ (in which case, $f_{\min} = f_{\max}$), we say that $f(x, y)$ has no contrast. If $f(x, y)$ and $g(x, y)$ are two periodic signals with the same average value, we say that $f(x, y)$ has more contrast than $g(x, y)$ if $m_f > m_g$.

3.2.2 Modulation Transfer Function

The way a medical imaging system affects contrast can be investigated by imaging a *sinusoidal object* $f(x, y)$ of the form

$$f(x, y) = A + B \sin(2\pi u_0 x), \tag{3.2}$$

where A and B are two nonnegative constants such that $A \ge B$. This is a sinusoidal object that varies only in the x-direction with spatial frequency u_0. Notice that $f_{\max} = A + B$ and $f_{\min} = A - B$, so the modulation of $f(x, y)$ is given by

$$m_f = \frac{B}{A}. \tag{3.3}$$

Figure 3.1 depicts four instances of $f(x, y)$, for the cases when $m_f = 0, 0.2, 0.5, 1$. Notice that as modulation increases, it becomes much easier to distinguish differences in shades of gray in the image of $f(x, y)$; in other words, contrast increases.

We are now interested in determining how an LSI imaging system with PSF $h(x, y)$ affects the modulation of $f(x, y)$; i.e., we are interested in mathematically relating the modulation m_g of the output $g(x, y)$ to the modulation m_f of the input $f(x, y)$. In order to simplify our discussion, we assume that $h(x, y)$ is circularly symmetric [see (2.105)]. Since

$$f(x, y) = A + B \sin(2\pi u_0 x) = A + \frac{B}{2j} \left[e^{j2\pi u_0 x} - e^{-j2\pi u_0 x} \right], \tag{3.4}$$

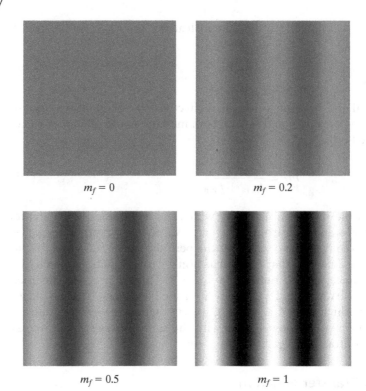

$m_f = 0$ $m_f = 0.2$

$m_f = 0.5$ $m_f = 1$

Figure 3.1
Four instances of a sinusoidal signal $f(x,y) = A + B\sin(2\pi u_0 x)$, with $m_f = B/A = 0$, 0.2, 0.5, 1.

the output $g(x,y)$ of the system is given by

$$g(x,y) = AH(0,0) + B\,|H(u_0,0)|\sin(2\pi u_0 x)\,. \tag{3.5}$$

Notice that the recorded image $g(x,y)$ of the sinusoidal object $f(x,y)$ is also sinusoidal with (the same) frequency u_0. From (3.5), $g_{\max} = AH(0,0) + B|H(u_0,0)|$ and $g_{\min} = AH(0,0) - B|H(u_0,0)|$, in which case the modulation of $g(x,y)$ is given by

$$m_g = \frac{B|H(u_0,0)|}{AH(0,0)} = m_f\frac{|H(u_0,0)|}{H(0,0)}\,. \tag{3.6}$$

The modulation m_g of $g(x,y)$ depends on the spatial frequency u_0.

The way an LSI medical imaging system affects modulation, and therefore contrast, is illustrated in Figure 3.2. The output modulation m_g is a scaled version of the input modulation m_f, the scaling factor being the magnitude spectrum $|H(u_0,0)|$ of the medical imaging system under consideration. If $H(0,0) = 1$ and $|H(u_0,0)| < 1$, then $m_g < m_f$, and since both signals $f(x,y)$ and $g(x,y)$ have the same average value, the output $g(x,y)$ will have less contrast than the input $f(x,y)$.

The ratio of the output modulation to the input modulation, as a function of spatial frequency, is called the *modulation transfer function* (MTF), and

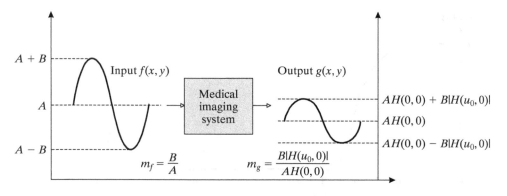

Figure 3.2
Basic principles for determining the modulation of the output of an LSI medical imaging system from the modulation of the input, when the input object is sinusoidal.

[by rearranging (3.6)] is given by

$$\text{MTF}(u) = \frac{m_g}{m_f} = \frac{|H(u,0)|}{H(0,0)}. \tag{3.7}$$

This shows that the MTF of a medical imaging system is, in essence, the "frequency response" of the system, and it can be directly obtained from the Fourier transform of the PSF of the system (remember that $H(u,v) = \mathcal{F}_{2D}\{h(x,y)\}$). Since $|H(u,0)| = |H(-u,0)|$, the MTF is usually considered only at nonnegative frequencies. Since the MTF, which characterizes contrast, can be mathematically related to the PSF, which characterizes blurring (or resolution), we can assume that blurring reduces contrast. We will have more to say about this in Section 3.3.3.

The MTF quantifies degradation of contrast as a function of spatial frequency. For most medical imaging systems,

$$0 \le \text{MTF}(u) \le \text{MTF}(0) = 1, \qquad \text{for every } u, \tag{3.8}$$

with the MTF becoming significantly less than unity (or even zero) at high spatial frequencies.

EXAMPLE 3.1

A typical MTF is depicted in Figure 3.3. Note that $0 \le \text{MTF}(u) \le 1$, with the maximum value attained at $u = 0$. Note also that the MTF monotonically decreases to zero with increasing frequency, becoming zero at spatial frequencies larger than 0.8 mm^{-1}.

Question What can we learn about the contrast behavior of an imaging system with this MTF?

Answer At spatial frequency 0.6 mm^{-1}, the MTF takes a value of 0.5. This means that the contrast of a sinusoidal object at spatial frequency 0.6 mm^{-1} is reduced by half when imaged through this system. Moreover, since the MTF is zero at all spatial frequencies larger than 0.8 mm^{-1}, any sinusoidal input with frequency larger than 0.8 mm^{-1} will be imaged as a constant output of zero contrast. ∎

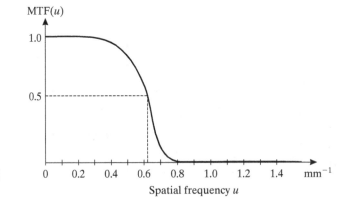

Figure 3.3
A typical MTF of a medical imaging system.

As indicated above, we can think of loss of contrast as the result of the blurring action of a medical imaging system. This is illustrated in Figure 3.4, which depicts the outputs of three radiographic imaging systems with increasingly poorer MTFs. Here, a poorer MTF is one that drops to zero at lower spatial frequencies. A poorer MTF results in less contrast.

It should be noted here that the PSF of a medical imaging system need not be *isotropic*—i.e., equivalent in all (2-D or 3-D) directions. In a nonisotropic

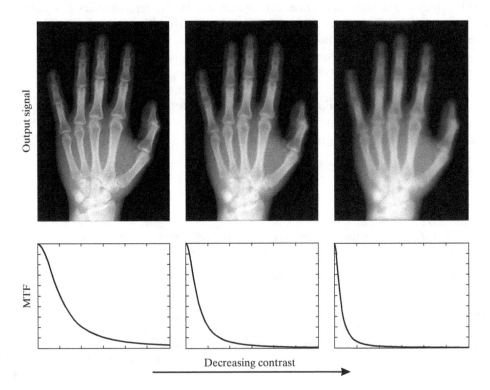

Figure 3.4
Impact of the MTF of a medical imaging system on contrast. Clearly, a poorer MTF results in less contrast. (Used with the permission of GE Healthcare.)

Decreasing contrast

system, the profile through the PSF changes with orientation; thus, the system has an orientation-dependent response. In the nonisotropic (2-D) case, the MTF is rotationally dependent, and (3.7) can be generalized to

$$\text{MTF}(u, v) = \frac{m_g}{m_f} = \frac{|H(u, v)|}{H(0, 0)}, \tag{3.9}$$

in which case

$$m_g = m_f \frac{|H(u, v)|}{H(0, 0)}. \tag{3.10}$$

For a typical nonisotropic medical imaging system,

$$0 \le \text{MTF}(u, v) = \frac{|H(u, v)|}{H(0, 0)} \le \text{MTF}(0, 0) = 1, \qquad \text{for every } u, v. \tag{3.11}$$

3.2.3 Local Contrast

The identification of some specific object or feature within an image is only possible if its value differs from that of surrounding areas. The definition of modulation or contrast for sinusoidal signals can be adapted for use in this situation as well. It is common in many imaging modalities (e.g., nuclear medicine) to consider an object of interest (e.g., a tumor in the liver) that we refer to as the *target*, as illustrated in Figure 3.5. Suppose that the target has a nominal image intensity of f_t. The target (i.e., the tumor) is surrounded by other tissues (i.e., the liver tissue), called the *background*, which may obscure our ability to see or detect the target. Suppose that the background has a nominal image intensity of f_b. The difference between the target and its background is captured by the *local contrast*, defined as

$$C = \frac{f_t - f_b}{f_b}. \tag{3.12}$$

The definition of local contrast in (3.12) differs from the definition of modulation in (3.1) for sinusoidal signals in that the intensities f_t and f_b may be selected locally—e.g., within the liver—and they need not be the maximum and minimum intensities within the image as a whole. For example, f_t could be taken to be the average image intensity within the tumor, whereas f_b could be taken as the average

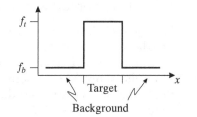

Figure 3.5
Local contrast scenario.

image intensity within the liver. The tumor intensity could be larger than f_b or less than f_b. If f_t is less than f_b, then C is negative and, in certain circumstances, it may be convenient to report its absolute value.

EXAMPLE 3.2

Consider an image showing an organ with intensity I_o and a tumor with intensity $I_t > I_o$.

Question   What is the local contrast of the tumor? If we add a constant intensity $I_c > 0$ to the image, what is the local contrast? Is the local contrast improved?

Answer   By definition, the local contrast of the tumor is

$$C = \frac{I_t - I_o}{I_o}.$$

If we add a constant intensity I_c to the image, the intensities of the background and the target become $f_b = I_o + I_c$, and $f_t = I_t + I_c$. The local contrast of the processed image is

$$C' = \frac{(I_t + I_c) - (I_o + I_c)}{I_o + I_c} = \frac{I_t - I_o}{I_o + I_c} = C\frac{I_o}{I_o + I_c} < C.$$

So the local contrast is worse if we add a constant intensity I_c to the image. ∎

3.3   Resolution

A basic measure of image quality is *resolution*. For our purposes, resolution can be thought of as the ability of a medical imaging system to accurately depict two distinct events in space, time, or frequency as separate. In this case, we talk about *spatial*, *temporal*, or *spectral resolution*, respectively. Resolution can also be thought of as the degree of smearing, or blurring, a medical imaging system introduces to a single event in space, time, or frequency. These two ways of looking at resolution are related, because the less smearing a system introduces, the closer in space, time, or frequency two events can be and still be distinguished as separate. Therefore, a high resolution medical imaging system is characterized by low smearing, whereas a low resolution system is characterized by high smearing. In this section, we mostly focus on spatial resolution, although a brief discussion of temporal and spectral resolution is also provided. Therefore, when we talk about "resolution," we mean spatial resolution, unless otherwise specified.

3.3.1 Line Spread Function

As suggested above, we can think of resolution as the degree of smearing, or blurring, a medical imaging system introduces to a single event (e.g., a point) in space. This is the traditional PSF described in Chapter 2, and the response of an imaging system to a point source (i.e., a point impulse) is often used to characterize resolution. As an alternative, we can consider the response of a medical imaging system to a line impulse (see Section 2.2.2).

Consider an LSI medical imaging system with isotropic PSF $h(x, y)$ that is normalized to 1. Suppose that a line source, passing through the origin of the spatial domain, is imaged through the system. We will mathematically represent this line source by the *line impulse* $f(x, y) = \delta_\ell(x, y)$ [see (2.11)]. Since the system is isotropic, it is sufficient to consider the response to a vertical line through the origin; from (2.11), we see that for this case, $f(x, y) = \delta(x)$. Then, the output $g(x, y)$ of the system will be given by

$$g(x, y) = \int_{-\infty}^{\infty} \int_{-\infty}^{\infty} h(\xi, \eta) f(x - \xi, y - \eta) \, d\xi \, d\eta,$$

$$= \int_{-\infty}^{\infty} \left[\int_{-\infty}^{\infty} h(\xi, \eta) \delta(x - \xi) \, d\xi \right] d\eta,$$

$$= \int_{-\infty}^{\infty} h(x, \eta) \, d\eta, \tag{3.13}$$

where, in the third equality, we have used the 1-D analog of (2.6).

The resulting image $g(x, y)$ is only a function of x, say $l(x)$. This is known as the *line spread function* (LSF) of the system under consideration, and it can be used to quantify resolution. This function is directly related to the PSF $h(x, y)$, since, from (3.13), we have

$$l(x) = \int_{-\infty}^{\infty} h(x, \eta) \, d\eta. \tag{3.14}$$

Notice that, since the PSF $h(x, y)$ is assumed to be isotropic, $l(x)$ is symmetric [i.e., $l(x) = l(-x)$], and since the PSF is normalized to 1,

$$\int_{-\infty}^{\infty} l(x) \, dx = 1. \tag{3.15}$$

Moreover, the 1-D Fourier transform $L(u)$ of the LSF $l(x)$ is related to the transfer function $H(u, v)$ of the system, since

$$L(u) = \mathcal{F}_{1D}[l](u),$$

$$= \int_{-\infty}^{\infty} l(x) e^{-j2\pi ux} \, dx,$$

$$= \int_{-\infty}^{\infty} \int_{-\infty}^{\infty} h(x, \eta) e^{-j2\pi ux} \, dx \, d\eta,$$

$$= H(u, 0). \tag{3.16}$$

Therefore, the values of the transfer function along the horizontal line that passes through the origin of the frequency domain are adequate for determining the 1-D Fourier transform of the LSF, and hence the LSF itself. Since the PSF $h(x, y)$ is assumed to be isotropic, the transfer function is isotropic as well. Thus, the LSF is adequate for determining the PSF of the system. Indeed, from the LSF $l(x)$, we can calculate the 1-D Fourier transform $L(u)$, and $H(u, 0) = L(u)$. However, since the transfer function $H(u, v)$ is isotropic, the values of $H(u, v)$ along any line that passes through the origin of the frequency domain will be the same as the values of $H(u, 0)$.

3.3.2 Full Width at Half Maximum

Given the LSF (or the PSF) of a medical imaging system, its resolution can be quantified using a measure called the *full width at half maximum* (FWHM). This is the (full) width of the LSF (or the PSF) at one-half its maximum value. The FWHM is usually expressed in millimeters. Provided there is no geometric scaling, the FWHM equals the minimum distance that two lines (or points) must be separated in space in order to appear as separate in the recorded image. This is depicted in Figure 3.6, where we see the profiles of two points through a 1-D imaging system with PSF $h(x)$. The points move closer to each other as we go from (a) to (d). In a linear system, the observed profile is the sum of the individual profiles from the two points. In practice, a residual "dip" between the points must occur in the (summed) profile for the two points to be visualized or resolved as separate. The profile depicted in Figure 3.6(c) shows the separation distance at which the two points are just distinguishable. Notice that, in this case, the two points are separated by the FWHM. Therefore, a decrease in the FWHM indicates an improvement in resolution.

3.3.3 Resolution and Modulation Transfer Function

Another way to quantify the resolution of a medical imaging system is as the smallest separation (in mm) between two adjacent maxima (or minima) in a sinusoidal input that can be resolved in the image. Consider the case when the input of a medical imaging system is sinusoidal of the form $f(x, y) = B \sin(2\pi ux)$, with amplitude B and frequency u. From (3.5) and (3.7), the output of the system is given by

$$g(x, y) = \text{MTF}(u)H(0, 0)B \sin(2\pi ux). \tag{3.17}$$

Notice that the separation between two adjacent maxima (or minima) of the sinusoidal input $f(x, y)$ is $1/u$. The recorded image $g(x, y)$ is also sinusoidal, with $1/u$ being the separation between two adjacent maxima (or minima) as well. However, the amplitude of the output image equals the amplitude of the input multiplied by the MTF at spatial frequency u. In practice, $\text{MTF}(u) \neq 0$, for every $u \leq u_c$, and $\text{MTF}(u) = 0$, for every $u > u_c$, for some spatial cutoff frequency u_c, in which case, $g(x, y) = 0$, for every $u > u_c$. In this case, the resolution of the system will be $1/u_c$.

EXAMPLE 3.3

The MTF depicted in Figure 3.3 becomes zero at spatial frequencies larger than 0.8 mm^{-1}.

Question   What is the resolution of this system?

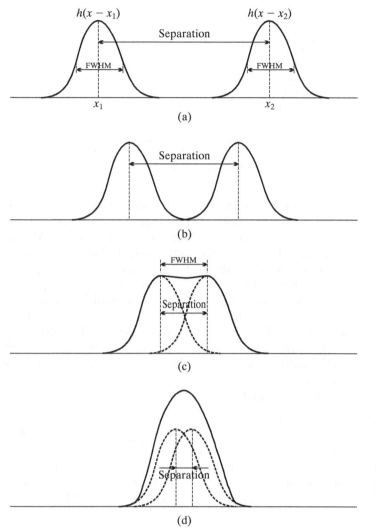

Figure 3.6
An example of the effect of system resolution on the ability to differentiate two points. The FWHM equals the minimum distance that the two points must be separated in order to be distinguishable.

Answer The resolution of a system with such an MTF is $1/(0.8 \text{ mm}^{-1}) = 1.25$ mm. Fine structures of an object to be imaged by such a system, with spatial frequencies larger than 0.8 mm^{-1}, cannot be seen at the output of the system. ∎

From our discussion, we find that the MTF can be effectively used to compare two competing medical imaging systems in terms of their contrast and resolution. If the MTFs of the two systems under consideration are of a similar shape but have a different cutoff frequency u_c, we can conclude that the system with higher MTF values will be better in terms of contrast and resolution. For example, the radiographic imaging system depicted in the first panel in Figure 3.4 is better in terms of contrast and resolution than the system depicted in the third panel.

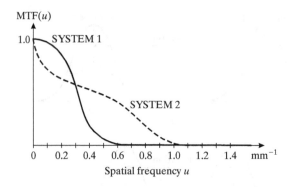

Figure 3.7
MTF curves of two
competing medical imaging
systems.

If the MTF curves are of different shapes, the situation is more complicated. Figure 3.7 depicts two MTF curves that correspond to two competing medical imaging systems: SYSTEM 1 and SYSTEM 2. SYSTEM 1 has better low-frequency contrast, and is thus better for imaging coarse details, while SYSTEM 2 has a better high frequency contrast, and is thus better for imaging fine details. Since contrast, as quantified by the MTF, is a function of spatial frequency, we can make this sort of frequency-by-frequency comparison. Spatial resolution, as described here, is not frequency-dependent, so it becomes harder to directly compare MTFs in the context of "better resolution." As described above, the FWHM of the PSF or LSF is the most direct metric of resolution. Further understanding of a system's resolution from its MTF usually comes from MTF values at higher spatial frequencies, and the cutoff frequency u_c.

The MTF can be directly obtained from the LSF. Indeed, from (3.7) and (3.16),

$$\mathrm{MTF}(u) = \frac{|L(u)|}{L(0)}, \qquad \text{for every } u. \tag{3.18}$$

Therefore, the MTF equals the magnitude of the 1-D Fourier transform of the LSF. The next example shows that this relationship can be used to determine the FWHM of a medical imaging system directly from its MTF.

EXAMPLE 3.4
Sometimes, the PSF, LSF, or MTF can be described by an analytical function. Such a function can arise by either fitting observed data or by making simplifying assumptions about its shape. Assume that the MTF of a medical imaging system is given by

$$\mathrm{MTF}(u) = e^{-\pi u^2}. \tag{3.19}$$

Question   What is the FWHM of this system?

Answer   By using the 1-D inverse Fourier transform and the fact that $|L(u)| = \mathrm{MTF}(u)$, we have $l(x) = e^{-\pi x^2}$. The FWHM will then be given by $\mathrm{FWHM} = 2x_0$, where x_0 is such that

$$e^{-\pi x_0^2} = \frac{1}{2}. \tag{3.20}$$

This results in $\mathrm{FWHM} = 2\sqrt{\ln 2/\pi}$.   ∎

3.3.4 Subsystem Cascade

Medical imaging systems are often modeled as a cascade of LSI subsystems, as introduced in Section 2.3.4. Accordingly, the recorded image $g(x, y)$ can be modeled as the convolution of the input object $f(x, y)$ with the PSF of the first subsystem, followed by the convolution with the PSF of the second subsystem, etc. (see Section 2.3.4). For example, in the case of K subsystems with PSFs $h_1(x, y), h_2(x, y), \ldots, h_K(x, y)$,

$$g(x, y) = h_K(x, y) * \cdots * (h_2(x, y) * (h_1(x, y) * f(x, y))) . \tag{3.21}$$

The quality of the overall system, in terms of contrast and resolution, can be predicted by considering the quality of each subsystem.

If resolution is quantified using the FWHM, then the FWHM of the overall system can be determined approximately from the FWHMs $R_1, R_2, \ldots, R_K$ of the individual subsystems, by

$$R = \sqrt{R_1^2 + R_2^2 + \cdots + R_K^2} . \tag{3.22}$$

Notice that the overall FWHM R is dominated by the largest (i.e., the poorest resolution) term. Thus, small improvements in any given subsystem's resolution do not often yield improvements in overall system resolution.

The following example shows that, when a medical imaging system is composed of subsystems with Gaussian PSFs, then (3.22) gives the exact value of the FWHM of the overall system.

EXAMPLE 3.5

Consider a 1-D medical imaging system with PSF $h(x)$ composed of two subsystems with Gaussian PSFs of the form

$$h_1(x) = \frac{1}{\sqrt{2\pi}\sigma_1} \exp\left\{\frac{-x^2}{2\sigma_1^2}\right\} \quad \text{and} \quad h_2(x) = \frac{1}{\sqrt{2\pi}\sigma_2} \exp\left\{\frac{-x^2}{2\sigma_2^2}\right\} . \tag{3.23}$$

Question What is the FWHM of this system?

Answer The FWHMs R_1 and R_2 associated with the two subsystems h_1 and h_2, respectively, are given by $R_1 = 2x_1$ and $R_2 = 2x_2$, where x_1 and x_2 are such that

$$h_1(x_1) = \frac{1}{\sqrt{2\pi}\sigma_1} \exp\left\{\frac{-x_1^2}{2\sigma_1^2}\right\} = \frac{1}{2\sqrt{2\pi}\sigma_1}, \tag{3.24}$$

and

$$h_2(x_2) = \frac{1}{\sqrt{2\pi}\sigma_2} \exp\left\{\frac{-x_2^2}{2\sigma_2^2}\right\} = \frac{1}{2\sqrt{2\pi}\sigma_2} . \tag{3.25}$$

After algebraic manipulation of (3.24) and (3.25),

$$R_1 = 2\sigma_1\sqrt{2\ln 2} \quad \text{and} \quad R_2 = 2\sigma_2\sqrt{2\ln 2} . \tag{3.26}$$

Following Example 2.4, it can be shown that the PSF $h(x)$ of the overall system is given by

$$h(x) = h_1(x) * h_2(x) = \frac{1}{\sqrt{2\pi(\sigma_1^2 + \sigma_2^2)}} \exp\left\{\frac{-x^2}{2(\sigma_1^2 + \sigma_2^2)}\right\}. \tag{3.27}$$

The FWHM R of this system is given by $R = 2x_0$, where x_0 is such that

$$h(x_0) = \exp\left\{\frac{-x_0^2}{2(\sigma_1^2 + \sigma_2^2)}\right\} = 0.5, \tag{3.28}$$

from which we obtain

$$R = 2\sqrt{\sigma_1^2 + \sigma_2^2}\sqrt{2\ln 2}. \tag{3.29}$$

$\blacksquare$

If contrast and resolution are quantified using the MTF, then the MTF of the overall system will be given by

$$\text{MTF}(u, v) = \text{MTF}_1(u, v)\text{MTF}_2(u, v)\cdots\text{MTF}_K(u, v), \tag{3.30}$$

in terms of the MTFs $\text{MTF}_k(u, v)$, $k = 1, 2, \ldots, K$, of the individual subsystems. This is a direct consequence of (3.9) and the fact that the frequency response $H(u, v)$ of the overall system is given by

$$H(u, v) = H_1(u, v)H_2(u, v)\cdots H_K(u, v), \tag{3.31}$$

in terms of the frequency responses $H_k(u, v)$, $k = 1, 2, \ldots, K$, of the individual subsystems.

Figure 3.8 depicts the MTF curves of three subsystems of a medical imaging system, together with the MTF curve of the overall system. If one subsystem has a small value of the MTF at some spatial frequency, then the MTF of the overall system will be small at that frequency as well. In other words, the MTF of the

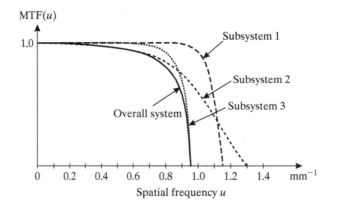

Figure 3.8
MTF curves of three
subsystems of a medical
imaging system and the MTF
curve of the overall system.

overall system will always be less than the MTF of each subsystem. This follows from (3.30), which suggests that, for every $k = 1, 2, \ldots, K$,

$$\text{MTF}(u, v) \leq \text{MTF}_k(u, v), \qquad \text{for every } u, v, \qquad (3.32)$$

provided that $\text{MTF}_k(u, v) \leq 1$, for every $k = 1, 2, \ldots, K$. Therefore, the overall quality of a medical imaging system, in terms of contrast and resolution, will be inferior to the quality of each subsystem.

From the previous discussion, the resolution of a medical imaging system can be specified by either its PSF or its LSF. In some medical imaging systems, we are able to create an object consisting of a very small "point" or "line" of some material. For example, we may create a very small and highly radioactive point or line source for measuring the resolution of a nuclear medicine camera. When a medical imaging system is mathematically modeled using an LSI system, its output to these objects will be the PSF and the LSF, respectively. This provides a practical way for calculating these two important response functions.

From our discussion, we can see that spatial resolution and image contrast are tightly linked, since the Fourier transforms of the PSF and the LSF, which are measures of resolution, yield the MTF, which is a measure of contrast. Indeed, spatial resolution can be thought of as the ability of an imaging system to preserve object contrast in the image, since blurring, due to poor resolution, is what actually reduces contrast.

In a nonisotropic system, the profile through the PSF changes with orientation; thus, the system has an orientation-dependent resolution. A good example of this situation is in ultrasound imaging systems, in which case the *range resolution* (along the transducer axis) is usually substantially better than the *lateral resolution* (orthogonal to the transducer axis). (See Chapter 10 for more details.) In the nonisotropic case, the MTF is given by (3.9).

It is also possible for a medical imaging system to be linear but not shift-invariant. Then, the resolution is spatially dependent. For example, this situation exists in ultrasound systems because the acoustic energy spreads out with increasing distance from the transducer. This makes the FWHM increase with depth, which corresponds to a degradation of lateral resolution. Magnification in pinhole collimators, or focused collimators in nuclear medicine, also leads to spatially dependent resolution, in which case the image plane resolution is related, in a precise geometric way, to actual resolution within the object.

3.3.5 Resolution Tool

Images with better resolution are preferable to images with poorer resolution, since they contain more image details. Resolution can be easily quantified in terms of the ability of a system to image details of a given test pattern. For example, one common way to measure resolution for a particular system is to image the so-called *resolution tool* or *bar phantom*, such as the one depicted in Figure 3.9. This tool is composed of groups of parallel lines of a certain width, separated by gaps having the same width as the line width (i.e., an overall duty cycle of 50 percent). Each group

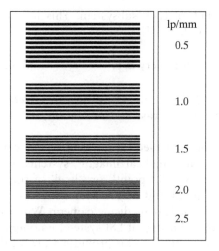

Figure 3.9
A resolution tool or bar phantom.

is characterized by the density of such lines, measured in *line pairs per millimeter* (lp/mm). The tool is imaged through the system under consideration, and system resolution is reported as the frequency (in lp/mm) of the finest line group that can be resolved at the output. For example, the resolution might be 6–8 lp/mm for a projection radiography system and 2 lp/mm for a CT scanner.

3.3.6 Temporal and Spectral Resolution

The preceding discussion on spatial resolution and the concepts embodied in the PSF, LSF, and FWHM apply equally well to temporal and spectral resolution. *Temporal resolution* is the ability to distinguish two events in time as being separate. *Spectral resolution* is the ability to distinguish two different frequencies (or, equivalently, energies). Conceptually, we could create a frequency distribution histogram of the number of observed events as a function of time or energy from a single-time or single-energy process (i.e., a point impulse in time or energy) to yield the equivalent of a PSF. As with the PSF, the ideal system response would be a delta function, and the actual FWHM (in time or energy) would quantify the resolution. In other words, the concept of PSF applies equally well to events in time or frequency as it does to events in space.

3.4 Noise

An unwanted characteristic of medical imaging systems is *noise*. Noise is a generic term that refers to any type of random fluctuation in an image, and it can have a dramatic impact on image quality; image quality decreases as noise increases. The source and amount of noise depend on the imaging method used and the particular medical imaging system at hand. For example, in projection radiography, x-rays arrive at the detector in discrete packets of energy, called *quanta* or *photons*. The discrete nature of their arrival leads to random fluctuations, called *quantum mottle*,

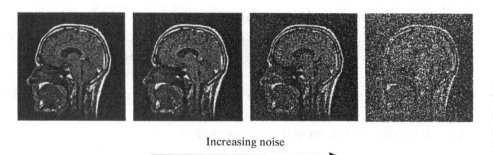

Increasing noise

Figure 3.10
The effect of noise on image quality: image quality decreases rapidly with increasing noise contamination.

which give an x-ray image a textured or grainy appearance. On the other hand, in magnetic resonance imaging, radio-frequency pulses generated by nuclear spin systems are sensed by antennas connected to amplifiers. Since these signals have very low power, they compete with signals being generated in the antenna from natural thermal vibrations. Thermal vibrations are unpredictable—i.e., random—and therefore comprise one source of noise in magnetic resonance images. The effect of increasing noise is shown in Figure 3.10.

The source of noise in a medical imaging system depends on the physics and instrumentation of the particular modality, subjects that will be developed in depth later in this book. The main objective of this section is to provide an introductory exposition of the tools used to mathematically characterize noise. A general way to characterize noise is to consider it as the numerical outcome of a random event or experiment. In nuclear medicine, for example, certain radioactive sources emit gamma ray photons that are received and recorded by a detector. Although governed by fixed physical properties (such as photon energy and decay rate), the specific nature of radioactive decay is random: photons are emitted at random times in random directions. We think of the noise as the deviation from a nominal value that would be predicted from purely deterministic arguments. This deviation, arising from the random nature of radioactive emissions, accounts for the noise that is present in all nuclear medicine images.

3.4.1 Random Variables

The numerical quantity associated with a random event or *experiment* is called a *random variable*. Different repetitions of the experiment may produce different observed values—i.e., the experiment has a random outcome. A random variable is mathematically described by $P_N(\eta)$, its *probability distribution function* (PDF), given by

$$P_N(\eta) = \Pr[N \leq \eta], \tag{3.33}$$

where $\Pr[\cdot]$ denotes *probability*. The PDF gives the probability that random variable N will take on a value less than or equal to η. Notice that $0 \leq P_N(\eta) \leq 1$, $P_N(-\infty) = 0$, $P_N(\infty) = 1$, and $P_N(\eta_1) \leq P_N(\eta_2)$, for $\eta_1 \leq \eta_2$.

3.4.2 **Continuous Random Variables**

If $P_N(\eta)$ is a continuous function of η, then N is a *continuous* random variable. This random variable is uniquely specified by its *probability density function* (pdf),[1]

$$p_N(\eta) = \frac{dP_N(\eta)}{d\eta} \,. \tag{3.34}$$

Any pdf satisfies the following three properties:

$$p_N(\eta) \geq 0 \,, \tag{3.35}$$

$$\int_{-\infty}^{\infty} p_N(\eta) \, d\eta = 1 \,, \tag{3.36}$$

$$P_N(\eta) = \int_{-\infty}^{\eta} p_N(u) \, du \,. \tag{3.37}$$

In practice, the pdf of a random variable may not be known. Instead, a random variable is often characterized by its *expected value*

$$\mu_N = \mathrm{E}[N] = \int_{-\infty}^{\infty} \eta p_N(\eta) \, d\eta \,, \tag{3.38}$$

also called its *mean*, and its *variance*

$$\sigma_N^2 = \mathrm{Var}[N] = \mathrm{E}[(N - \mu_N)^2] = \int_{-\infty}^{\infty} (\eta - \mu_N)^2 p_N(\eta) \, d\eta \,, \tag{3.39}$$

where $\mathrm{E}[\cdot]$ and $\mathrm{Var}[\cdot]$ denote expectation and variance, respectively. The square root σ_N of the variance is called the *standard deviation* of N.

The mean can be thought of as the average value of the random variable, whereas the standard deviation can be thought of as the "average" variation of the values of the random variable about its mean. The larger the standard deviation, the "more random" the random variable. As the standard deviation approaches zero, the observed values of the random variable more tightly cluster around the mean, and, in the limit, the random variable becomes a constant that equals μ_N.

Uniform Random Variable   A random variable N is said to be *uniform* over the interval $[a, b]$ if its pdf is of the form

$$p_N(\eta) = \begin{cases} \dfrac{1}{(b - a)}, & \text{for } a \leq \eta < b \\[2mm] 0, & \text{otherwise} \end{cases} \,. \tag{3.40}$$

---

[1]By convention, the abbreviation for probability density function is (lower case) pdf, and the abbreviation for probability distribution function is (upper case) PDF.

In this case, the distribution function is given by

$$P_N(\eta) = \begin{cases} 0, & \text{for } \eta < a \\ \dfrac{\eta - a}{b - a}, & \text{for } a \leq \eta \leq b, \\ 1, & \text{for } \eta > b \end{cases} \tag{3.41}$$

whereas the expected value and variance are given by

$$\mu_N = \frac{a+b}{2} \quad \text{and} \quad \sigma_N^2 = \frac{(b-a)^2}{12}, \tag{3.42}$$

respectively.

Gaussian Random Variable   If the pdf of a random variable N is given by

$$p_N(\eta) = \frac{1}{\sqrt{2\pi\sigma^2}} e^{-(\eta-\mu)^2/2\sigma^2}, \tag{3.43}$$

then N is a *Gaussian* random variable. In this case, the distribution function is given by

$$P_N(\eta) = \frac{1}{2} + \text{erf}\left(\frac{\eta - \mu}{\sigma}\right), \tag{3.44}$$

where erf(x) denotes the *error function*, given by the integral

$$\text{erf}(x) = \frac{1}{\sqrt{2\pi}} \int_0^x e^{-u^2/2} \, du. \tag{3.45}$$

The expected value and variance are given by

$$\mu_N = \mu \quad \text{and} \quad \sigma_N^2 = \sigma^2, \tag{3.46}$$

respectively. The integral in (3.45) cannot be evaluated in closed form, but it is usually tabulated in mathematics handbooks and can be numerically evaluated using most mathematics, statistics, and engineering software packages.

EXAMPLE 3.6

Distribution functions can be found by integrating their corresponding density function using (3.37).

Question   Can (3.44) be proved by direct integration?

Answer The distribution function of a Gaussian random variable with mean μ and variance σ^2 is

$$P_N(\eta) = \int_{-\infty}^{\eta} p_N(\tau)\, d\tau$$

$$= \int_{-\infty}^{\eta} \frac{1}{\sqrt{2\pi\sigma^2}} e^{-(\tau-\mu)^2/2\sigma^2}\, d\tau$$

$$= \frac{1}{\sqrt{2\pi\sigma^2}} \int_{-\infty}^{\frac{\eta-\mu}{\sigma}} e^{-t^2/2}\sigma\, dt, \quad \text{let } t = \frac{\tau-\mu}{\sigma}$$

$$= \frac{1}{\sqrt{2\pi}} \int_{-\infty}^{0} e^{-t^2/2}\, dt + \mathrm{erf}\left(\frac{\eta-\mu}{\sigma}\right)$$

$$= \frac{1}{2} + \mathrm{erf}\left(\frac{\eta-\mu}{\sigma}\right)$$

To get the last equality, we use the fact that $\frac{1}{\sqrt{2\pi}}e^{-t^2/2}$ is the pdf of the standard Gaussian random variable and is symmetric around $t = 0$. ∎

In general, the mean and variance do not uniquely specify a random variable. This means that, given μ_N and σ_N^2, there might be more than one pdf that produces the same mean and variance. However, in the case of a Gaussian random variable, the pdf is uniquely specified by its mean and variance.

Usually, noise in medical imaging systems is the result of a summation of a large number of independent noise sources. According to the *central limit theorem* of probability, a random variable that is the sum of a large number of independent causes tends to be Gaussian. Therefore, it is often natural to model noise in medical imaging system by means of a Gaussian random variable.

3.4.3 Discrete Random Variables

When the random variable N takes only values $\eta_1, \eta_2, \ldots, \eta_k$, it is said to be a *discrete random variable*. This random variable is uniquely specified by the *probability mass function* (PMF) $\Pr[N = \eta_i]$, for $i = 1, 2, \ldots, k$, where $\Pr[N = \eta_i]$ is the probability that random variable N will take on the particular value η_i. The PMF satisfies the following three properties:

$$0 \le \Pr[N = \eta_i] \le 1, \quad \text{for } i = 1, 2, \ldots, k, \tag{3.47}$$

$$\sum_{i=1}^{k} \Pr[N = \eta_i] = 1, \tag{3.48}$$

$$P_N(\eta) = \Pr[N \le \eta] = \sum_{\text{all } \eta_i \le \eta} \Pr[N = \eta_i]. \tag{3.49}$$

It is permissible for $k \to \infty$, meaning that there will be an infinite (but still countable) number of possible outcomes.

In the case of a discrete random variable, the mean value and variance are given by

$$\mu_N = E[N] = \sum_{i=1}^{k} \eta_i \Pr[N = \eta_i], \tag{3.50}$$

and

$$\sigma_N^2 = \text{Var}[N] = E[(N - \mu_N)^2] = \sum_{i=1}^{k} (\eta_i - \mu_N)^2 \Pr[N = \eta_i], \tag{3.51}$$

respectively. Notice that the integrals in (3.38) and (3.39) have been replaced by sums in (3.50) and (3.51).

Poisson Random Variable Let N be a discrete random variable that takes values $0, 1, \ldots$, and has the PMF

$$\Pr[N = k] = \frac{a^k}{k!} e^{-a}, \qquad \text{for } k = 0, 1, \ldots, \tag{3.52}$$

where $a > 0$ is a real-valued parameter. N is said to be a *Poisson random variable*, and it turns out that its mean equals its variance and

$$\mu_N = a, \tag{3.53}$$

$$\sigma_N^2 = a. \tag{3.54}$$

Poisson random variables play an important role in medical imaging systems, and most particularly in radiographic and nuclear medicine imaging. For example, they are used to statistically characterize the distribution of photons counted per unit area by an x-ray image intensifier, or to characterize the photon counts produced by a radiotracer in nuclear medicine.

EXAMPLE 3.7

In x-ray imaging, the Poisson random variable is used to model the number of photons that arrive at a detector in time t, which is a random variable referred to as a *Poisson process* and given the notation $N(t)$. The PMF of $N(t)$ is given by

$$\Pr[N(t) = k] = \frac{(\lambda t)^k}{k!} e^{-\lambda t}$$

where λ is called the *average arrival rate* of the x-ray photons.

Question What is the probability that there is no photon detected in time t?

Answer The probability that there is no photon detected in time t is

$$\Pr[N(t) = 0] = \frac{(\lambda t)^0}{0!} e^{-\lambda t} = e^{-\lambda t}.$$

∎

EXAMPLE 3.8

For the Poisson process of Example 3.7, the time that the first photon arrives is a random variable, say T.

Question What is the pdf $p_T(\tau)$ of random variable T?

Answer Assume that the first photon arrives in the interval $t < T < t + \Delta t$. For very small Δt, we have

$$\text{Prob}[t < T < t + \Delta t] \approx p_T(t)\Delta t$$

For this to happen, it must be true that no photons arrived in the interval $[0, t]$ and exactly one photon arrived in the interval $[t, t + \Delta t]$. We learned from Example 3.7 that

$$\text{Prob[no photon detected in time } t] = e^{-\lambda t}.$$

The probability that exactly one photon will be detected in the interval $[t, t + \Delta t]$ is

$$\text{Prob[One photon detected in interval } [t, t + \Delta t]] = \frac{(\lambda \Delta t)^1}{1!} e^{-\lambda \Delta t} = \lambda \Delta t \, e^{-\lambda \Delta t}.$$

Using the Taylor series expansion for the exponential function

$$e^x = 1 + x + \frac{x^2}{2!} + \frac{x^3}{3!} + \cdots$$

and the fact that Δt is very small (and will go to zero in the limit) permits the approximation

$$\text{Prob[One photon detected in interval } [t, t + \Delta t]] \approx \lambda \Delta t$$

where second-order terms in Δt have been dropped. Putting all this together yields

$$p_T(t)\Delta t \approx \text{Prob[no photon detected in time } t]$$

$$\cdot \, \text{Prob[One photon detected in interval } [t, t + \Delta t]]$$

$$\approx e^{-\lambda t} \cdot \lambda \Delta t.$$

As $\Delta t \to 0$, all approximations get tighter, and so we recognize by dividing both sides of the above equation by Δt that

$$p_T(t) = \lambda e^{-\lambda t}, \qquad t \in [0, \infty).$$

The random variable T corresponding to this pdf is called the *exponential random variable*. ∎

3.4.4 Independent Random Variables

It is usual in imaging experiments to consider more than one random variable at a time. The theory that is required in order to characterize a collection of random variables follows from the theory of a single random variable, with parallel definitions of distribution, density, and mass functions. In this book, however, we require only a simplified discussion related to *sums of independent random variables*.

Loosely speaking, a collection of random variables are *independent* if knowledge of some of the random variables (i.e., making a partial observation) tells you nothing, statistically speaking, about the remaining random variables.

Consider the collection of random variables $N_1, N_2, \ldots N_m$, having the pdf's $p_1(\eta), p_2(\eta), \ldots, p_m(\eta)$, respectively. The sum of these random variables S is another random variable having another pdf, $p_S(\eta)$. It is always the case that the mean of S is precisely the sum of the means of $N_1, N_2, \ldots, N_m$. That is,

$$\mu_S = \mu_1 + \mu_2 + \cdots + \mu_m, \tag{3.55}$$

where $\mu_1, \ldots, \mu_m$ are the means of the pdf's given above. Independence is not required for this to be true.

When the random variables are independent, we can go a step further. In this case, the variance of S is the sum of the individual variances,

$$\sigma_S^2 = \sigma_1^2 + \sigma_2^2 + \cdots + \sigma_m^2. \tag{3.56}$$

We emphasize that the variances, not the standard deviations, are added. Also, in the case of independence, it is possible to determine the pdf of S by

$$p_S(\eta) = p_1(\eta) * p_2(\eta) * \cdots * p_m(\eta), \tag{3.57}$$

where $*$ is convolution. The facts that the variances add and that the pdf of the sum can be determined easily are powerful results following from the independence of random variables. We will use these facts to derive approximate noise and SNR expressions for CT imaging in Chapter 6.

EXAMPLE 3.9
Consider the sum S of two independent Gaussian random variables N_1 and N_2, each having a mean of zero and variance of σ^2.

Question   What are the mean, variance, and pdf of the resulting random variable?

Answer   The mean of S is zero,

$$\mu_S = \mu_1 + \mu_2 = 0 + 0 = 0,$$

and the variance of S is $2\sigma^2$,

$$\sigma_S^2 = \sigma_1^2 + \sigma_2^2 = \sigma^2 + \sigma^2 = 2\sigma^2.$$

Using the fact that the convolution of two Gaussian waveforms yields a Gaussian waveform, we conclude that the pdf of S has a Gaussian form. Furthermore, since the Gaussian pdf is characterized completely by its mean and variance, we conclude that

$$p_S(\eta) = \frac{1}{\sqrt{2\pi\sigma_S^2}} \exp\left\{\frac{-\eta^2}{2\sigma_S^2}\right\} = \frac{1}{\sqrt{4\pi\sigma^2}} \exp\left\{\frac{-\eta^2}{4\sigma^2}\right\}$$

∎

3.5 Signal-to-Noise Ratio

In this book, we often assume that the output of a medical imaging system is a random variable G (or a collection of random variables), composed of two components, f and N. Component f, which is usually referred to as *signal*, is the (deterministic or nonrandom) "true" value of G, whereas N is a random fluctuation or error component due to noise. The identification of an abnormal condition within the human body most often depends on how "close" an observed value g of G, characteristic to that condition, is to its true value f.

A useful way to quantify this is by means of the *signal-to-noise ratio* (SNR). The SNR describes the relative "strength" of signal f with respect to that of noise N. Higher SNR values indicate that g is a more accurate representation of f, whereas lower SNR values indicate that g is less accurate. Therefore, higher image quality requires that the output of a medical imaging system be characterized by high SNR.

One way of thinking about the "signal" is that it is the modulation or contrast in the image, as discussed in Section 3.2, whereas "noise" is the unwanted, random fluctuations discussed in Section 3.4. As discussed above, blurring reduces contrast and thus SNR; noise also reduces SNR. Figure 3.11 depicts the effects of both blur and noise on SNR; as one moves from the upper left to the lower right, SNR decreases.

3.5.1 Amplitude SNR

Most frequently, the SNR is expressed as the ratio of signal amplitude to noise amplitude:

$$SNR_a = \frac{\text{Amplitude}(f)}{\text{Amplitude}(N)}.$$

(3.58)

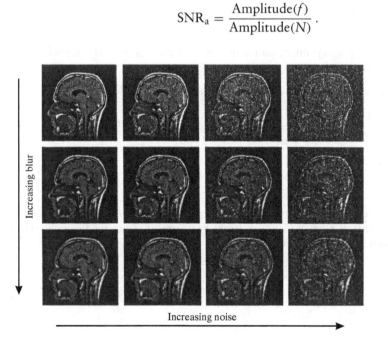

Increasing blur

Increasing noise

Figure 3.11
The effect of increasing blur and noise on SNR.

We refer to SNR_a as the *amplitude SNR*. Notice that there are many ways to specify what we mean by "signal" and "noise." Moreover, exact definition of the amplitude SNR depends on specifying what we mean by "signal amplitude" and "noise amplitude." Thus, the amplitude SNR is case-dependent, and its definition must specifically be adapted to the particular situation at hand.

EXAMPLE 3.10

In projection radiography, the number of photons G counted per unit area by an x-ray image intensifier follows a Poisson distribution, as in (3.52). In this case, we may consider signal f to be the average photon count per unit area (i.e., the mean of G) and noise N to be the random variation of this count around the mean, whose amplitude is quantified by the standard deviation of G.

Question What is the amplitude SNR of such a system?

Answer From (3.52)–(3.54), it follows that the amplitude SNR is given by

$$SNR_a = \frac{\mu_G}{\sigma_G} = \frac{\mu}{\sqrt{\mu}} = \sqrt{\mu}. \tag{3.59}$$

This quantity is known as the *intrinsic SNR* of x-rays, and we will have much more to say about this in Part II of the book. This tells us something very important and practical: The greater the average number μ of photons, the larger the amplitude SNR, and the smaller the *relative* amplitude of random fluctuations in G. Therefore, a higher x-ray exposure generally improves the quality of radiographic images. Keep in mind, however, that greater exposure to ionizing radiation means more risk of radiogenic cancer. ∎

3.5.2 Power SNR

Another way to express the SNR is as the ratio of signal power to noise power:

$$SNR_p = \frac{\text{power}(f)}{\text{power}(N)}. \tag{3.60}$$

We refer to SNR_p as the *power SNR*. Notice that exact definition of the power SNR depends on specifying what we mean by "signal power" and "noise power." Thus, as with the amplitude SNR, the power SNR is case-dependent, and its definition must specifically be adapted to the particular situation at hand.

EXAMPLE 3.11

If $f(x, y)$ is the input to a noisy medical imaging system with PSF $h(x, y)$, then the output at (x, y) may be thought of as a random variable $G(x, y)$, composed of signal $h(x, y) * f(x, y)$ and noise $N(x, y)$, with mean $\mu_N(x, y)$ and variance $\sigma_N^2(x, y)$.

Question What is the power SNR of such a system?

Answer In this framework, the power SNR at the output of this system will be given by

$$SNR_p = \frac{\int_{-\infty}^{\infty} \int_{-\infty}^{\infty} |h(x, y) * f(x, y)|^2 \, dx \, dy}{\sigma_N^2}, \tag{3.61}$$

where the noise power is quantified by means of the variance σ_N^2. Usually, it is assumed that there is no correlation between noise values and that for every (x, y)

$$\mu_N(x, y) = 0 \qquad \text{and} \qquad \sigma_N(x, y) = \sigma_N. \tag{3.62}$$

This type of noise is known as *white noise*. ∎

White noise is a crude approximation of reality, but it is convenient to use since it leads to mathematically simple models. Keep in mind that, in most cases, it is more accurate to consider correlated noise and assume some mathematical expression for such correlation. In this case, if we assume that the noise mean and variance do not depend on (x, y) (a common assumption that characterizes so-called *wide-sense stationary noise*), then it can be shown that

$$\text{SNR}_p = \frac{\displaystyle\int_{-\infty}^{\infty}\int_{-\infty}^{\infty} |h(x, y) * f(x, y)|^2 \, dxdy}{\displaystyle\int_{-\infty}^{\infty}\int_{-\infty}^{\infty} \text{NPS}(u, v) \, dudv}, \tag{3.63}$$

where

$$\text{NPS}(u, v) = \lim_{x_0, y_0 \to \infty} \frac{1}{4x_0 y_0} \text{E}\left[\left|\int_{-x_0}^{x_0}\int_{-y_0}^{y_0} [N(x, y) - \mu_N]\right.\right.$$
$$\left.\left. \exp\left(-j2\pi(ux + vy)\right) \, dxdy \right|^2\right], \tag{3.64}$$

is known as the *noise power spectrum* (NPS). From (3.63) and Parseval's theorem, given by (2.97), we have

$$\text{SNR}_p = \frac{\displaystyle\int_{-\infty}^{\infty}\int_{-\infty}^{\infty} |H(u, v)|^2 \, |F(u, v)|^2 \, dudv}{\displaystyle\int_{-\infty}^{\infty}\int_{-\infty}^{\infty} \text{NPS}(u, v) \, dudv}, \tag{3.65}$$

$$= \frac{\displaystyle\int_{-\infty}^{\infty}\int_{-\infty}^{\infty} \text{SNR}_p(u, v)\text{NPS}(u, v) \, dudv}{\displaystyle\int_{-\infty}^{\infty}\int_{-\infty}^{\infty} \text{NPS}(u, v) \, dudv}, \tag{3.66}$$

where [see also (3.9)]

$$\text{SNR}_p(u, v) = \frac{|H(u, v)|^2 |F(u, v)|^2}{\text{NPS}(u, v)} = \frac{\text{MTF}^2(u, v)}{\text{NPS}(u, v)} |F(u, v)|^2 H^2(0, 0) \tag{3.67}$$

is called the *frequency-dependent power SNR*. The frequency-dependent power SNR quantifies, at a given frequency, the relative "strength" of signal to that of noise at the output of the LSI system under consideration. From (3.66) and (3.67), $\text{SNR}_p(u, v)$ provides a relationship between contrast, resolution, noise, and image quality. For a given output noise level (i.e., a fixed NPS) and a given input $f(x, y)$, better contrast and resolution properties (i.e., a larger MTF) result in better image quality (i.e., a higher output power SNR).

3.5.3 Differential SNR

Consider an object (or target) of interest placed on a background. Let f_t and f_b be the average image intensities within the target and background, respectively. A useful choice for SNR is obtained by taking the "signal" to be the difference in average image intensity values between the target and the background integrated over the area A of the target, and by taking the "noise" to be the random fluctuation of image intensity from its mean over an area A of the background. This leads to the *differential signal-to-noise ratio* (SNR_{diff}), given by

$$\text{SNR}_{\text{diff}} = \frac{A(f_t - f_b)}{\sigma_b(A)}, \tag{3.68}$$

where $\sigma_b(A)$ is the standard deviation of image intensity values from their mean over an area A of the background. From (3.12), we have

$$\text{SNR}_{\text{diff}} = \frac{CAf_b}{\sigma_b(A)}, \tag{3.69}$$

which relates the differential SNR to contrast.

EXAMPLE 3.12

Consider the case of projection radiography. We may take f_b to be the average photon count per unit area in the background region around a target, in which case, $f_b = \lambda_b$, where λ_b is the mean of the underlying Poisson distribution governing the number of background photons counted per unit area. Notice that, in this case, $\sigma_b(A) = \sqrt{\lambda_b A}$.

Question   What is the average number of background photons counted per unit area, if we want to achieve a desirable differential SNR?

Answer   From (3.69),

$$\text{SNR}_{\text{diff}} = \frac{CA\lambda_b}{\sqrt{A\lambda_b}} = C\sqrt{A\lambda_b}. \tag{3.70}$$

From (3.70), the differential signal-to-noise ratio is approximately proportional to contrast as well as to the square root of the object area multiplied by radiation exposure (characterized by the average photon count per unit area λ_b). To achieve a desirable differential SNR, it is required that

$$\lambda_b = \frac{\text{SNR}_{\text{diff}}^2}{C^2 A}. \tag{3.71}$$

This relationship was first suggested by Albert Rose and is known as the *Rose model*. Clearly, to maintain good image quality (i.e., to obtain images with high SNR), high radiation dose is required when viewing small, low-contrast objects.   ∎

Decibels   The SNR is sometimes given in *decibels* (dB). When the SNR is the ratio of amplitudes, such as with the amplitude SNR or the differential SNR, then

$$\text{SNR (in dB)} = 20 \times \log_{10} \text{SNR (ratio of amplitudes)} . \tag{3.72}$$

When the SNR is the ratio of powers, such as with the power SNR, then

$$\text{SNR (in dB)} = 10 \times \log_{10} \text{SNR (ratio of powers)} . \tag{3.73}$$

3.6   Nonrandom Effects

3.6.1   Artifacts

A problem that frequently affects image quality is the creation of image features known as *artifacts* that do not represent valid anatomical or functional objects. Artifacts can obscure important targets, and they can be falsely interpreted as valid image features. Moreover, they can impair correct detection and characterization of features of interest by adding "clutter" to images.

Artifacts are caused by a variety of reasons and can appear at any step of the imaging process. For example, in projection radiography, artifacts can be generated by the x-ray source, by restricting the x-ray beam in order to avoid exposing parts of a patient that need not be imaged, and by nonuniformities in the x-ray image intensifier over the imaging area.

In computed tomography, artifacts may arise from patient motion, which produces streak artifacts throughout the image, known as *motion artifacts*; Figure 3.12(a) illustrates an example. Another typical artifact in computed tomography is known as a *star artifact*. This artifact is generated by the presence of metallic materials in the patient, which results in incomplete projections. An example of this artifact is depicted in Figure 3.12(b). Another artifact is the so-called *beam hardening artifact*. This artifact shows up as broad dark bands or streaks in the image, and it is due to significant beam attenuation caused by certain materials. An example of this artifact is depicted in Figure 3.12(c). Finally, a common artifact in computed tomography is the so-called *ring artifact*, illustrated in Figure 3.12(d). This artifact is caused by detectors that go out of calibration and do not properly record incoming data.

There are many reasons for medical images to be corrupted by artifacts. Evaluation and possibly removal of artifacts should be part of any high quality medical imaging system. Good design, proper calibration, and maintenance of medical imaging systems may control and even eliminate artifacts. Artifacts appearing in each particular imaging modality will be discussed in later chapters.

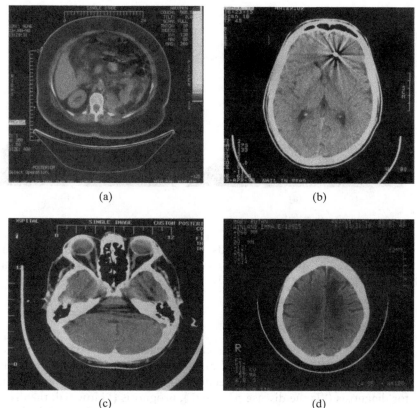

(a)

(b)

(c)

(d)

Figure 3.12
Examples of artifacts in computed tomography: (a) motion artifact, (b) star artifact, (c) beam hardening artifact, and (d) ring artifact. ((b) From A. B. Wolbarst, *Physics of Radiology* (1993).)

3.6.2 Distortion

Medical imaging systems often introduce *distortion*, another factor affecting image quality. Distortion is geometrical in nature and refers to the inability of a medical imaging system to give an accurate impression of the shape, size, and/or position of objects of interest.

In projection radiography, for example, *size distortion* can be present due to *magnification* caused by the distance of the x-ray source from the object being imaged. This is illustrated in Figure 3.13(a). Notice that, although the sizes of the two dark objects are different, their projections are the same.

On the other hand, *shape distortion* can be generated as a result of unequal magnification of the object being imaged. One common cause of shape distortion is the fact that anatomical structures lie at different levels within the human body. In projection radiography, shape distortion is also caused by the divergence of the x-ray beam. This is illustrated in Figure 3.13(b). Notice that, although the two dark objects are the same, the shapes of their projections are different.

Unfortunately, distortion can be very difficult to determine and correct. In order to evaluate distortion, knowledge of the actual shape and size of the object being imaged is required. Moreover, a good understanding of the imaging geometry

Figure 3.13
(a) Size distortion in a radiographic imaging system due to magnification: although the sizes of the two dark objects are different, their projections are the same. (b) Shape distortion in a radiographic imaging system due to x-ray beam divergence: although the two dark objects are the same, the shapes of their projections are different.

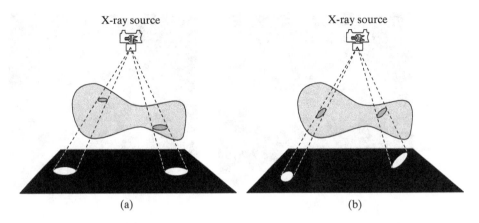

(a)   (b)

is necessary. Developing methods for correcting distortion is very important for increasing image quality and improving diagnosis. Distortions introduced by each particular imaging modality will be discussed in later chapters.

3.7 Accuracy

The preceding discussions in this chapter have focused on quantitative metrics of image quality. Even so, we said at the outset that image quality ultimately must be judged in the context of a specific clinical application. Medical images are used for diagnosis ("Is the disease present?"), prognosis ("How will the disease progress, and what is the expected outcome?"), treatment planning ("Which treatment will work best?"), and treatment monitoring ("Is the treatment reversing the disease, and to what extent?"). Fundamentally, we are interested in the *accuracy* of medical images in the context of these clinical applications. Here, "accuracy" means both conformity to truth (i.e., freedom from error) and clinical utility. In practice, we are usually interested in *quantitative accuracy* and *diagnostic accuracy*.

3.7.1 Quantitative Accuracy

Sometimes, we are interested in the numerical value of a given anatomic or functional feature within an image. For example, we may wish to measure tumor dimensions from a radiograph, or estimate glucose metabolic rate from a nuclear medicine image. In such situations, we need to know the error in our measurement. This error, or difference from the true value, arises from two sources: *bias*, which represents a systematic, reproducible difference from the truth, and *imprecision*, which represents a random, measurement-to-measurement variation.

It is helpful to separate the component of error due to bias from that due to imprecision. If we can establish that the measurement is precise (i.e., reproducible), then we can correct for systematic errors through the use of a calibration standard that converts the measured value to the true value. In practice, error usually arises from both components, and our measurements are never error-free.

3.7.2 Diagnostic Accuracy

We can conceptualize a medical image as representing some parameter (or set of parameters) of interest in a patient. For the sake of simplicity, we assume that the diagnostic process involves a test that extracts this parameter (or set of parameters), which is then used to classify a given patient as either normal or diseased. We further conceptualize two Gaussian distributions of the parameter (or set of parameters), one for normal patients and another for diseased patients, with (usually) some degree of overlap. This implies that some normal patients may be classified as diseased and vice versa.

In a clinical setting, we are interested in two parameters:

- *Sensitivity*, also known as the *true-positive fraction*; this is the fraction of patients with disease who the test (i.e., the medical image) calls abnormal.
- *Specificity*, also known as the *true-negative fraction*; this is the fraction of patients without disease who the test (i.e., the medical image) calls normal.

In practice, sensitivity and specificity are established in a group of patients through the use of a 2 × 2 contingency table, as shown in Figure 3.14, where a and b are, respectively, the number of diseased and normal patients who the test calls abnormal, whereas c and d are, respectively, the number of diseased and normal patients who the test calls normal. In this case, the sensitivity and specificity are given by

$$\text{sensitivity} = \frac{a}{a+c} \quad \text{and} \quad \text{specificity} = \frac{d}{b+d}. \quad (3.74)$$

Typically, the ultimate diagnosis that confirms the presence or absence of disease is through a more-invasive test than medical imaging, a "gold standard" that is used to verify the accuracy of (noninvasive) medical imaging procedures. The *diagnostic accuracy* (DA) is the fraction of patients that are diagnosed correctly, and is given by

$$\text{DA} = \frac{a+d}{a+b+c+d}. \quad (3.75)$$

In order to maximize diagnostic accuracy, both sensitivity and specificity must be maximized. In practice, because of overlap in the distribution of parameter values between normal and diseased patients, a threshold must be established to call a study "abnormal" (see Figure P3.3). A lower threshold implies that more studies

	Disease	
	+	−
Test +	a	b
Test −	c	d

Figure 3.14
A contingency table.

will be called "abnormal," thus increasing sensitivity but decreasing specificity. A higher threshold implies that fewer studies will be called "abnormal," thus increasing specificity but decreasing sensitivity. Therefore, the threshold must be chosen as a balance between sensitivity and specificity. The choice of threshold for a specific test depends on the relative cost-of-errors in calling a normal patient "abnormal" or a diseased patient "normal."

The threshold also depends on the prevalence or proportion of all patients who have disease, because in practice we are interested in two other parameters:

- *Positive predictive value* (PPV), which is the fraction of patients called abnormal who actually have the disease.
- *Negative predictive value* (NPV), which is the fraction of persons called normal who do not have the disease.

Notice that

$$\text{PPV} = \frac{a}{a+b} \qquad \text{and} \qquad \text{NPV} = \frac{d}{c+d}, \tag{3.76}$$

and that both depend on *prevalence* (PR), which is given by

$$\text{PR} = \frac{a+c}{a+b+c+d}. \tag{3.77}$$

EXAMPLE 3.13

Diagnostic accuracy alone cannot tell how good a diagnostic method is. This example shows that a bad diagnostic method can still achieve a high DA.

Question   Consider a group of 100 patients, among which 10 are diseased and 90 are normal. We simply label all patients as normal. Construct the contingency table for this test and determine the sensitivity, specificity, and diagnostic accuracy of the test.

Answer   The contingency table for the test is shown below.

		Disease	
		+	−
Test	+	0	0
	−	10	90

From the table, we have $a = b = 0$, $c = 10$, and $d = 90$. So the sensitivity, specificity, and diagnostic accuracy of the test are

$$\text{Sensitivity} = \frac{a}{a+c} = 0$$

$$\text{Specificity} = \frac{d}{b+d} = 1.0$$

$$\text{DA} = \frac{a+d}{a+b+c+d} = 0.9.$$

It can be seen that although all patients with disease are diagnosed wrong, we still have a relatively high diagnostic accuracy of 0.9. This is simply because the patients without disease comprise the majority of patients in the group being studied. ∎

3.8   Summary and Key Concepts

Image quality characterizes the performance of a medical imaging system and directly affects clinical utility. It is assessed through a combination of specific performance parameters, as measured in images. In this chapter, we presented the following key concepts that you should now understand:

1. *Image quality* refers to the degree to which an image allows a radiologist to accomplish the clinical goals of the imaging study.
2. The six most important factors influencing image quality are *contrast, resolution, noise, artifacts, distortion,* and *accuracy.*
3. *Contrast* refers to the difference in image intensity of an object or target and surrounding objects or background.
4. *Resolution* is the ability of an imaging system to distinguish and depict two signals that differ in space, time, or energy as distinct.
5. *Noise* is any random fluctuation in an image; noise generally interferes with the ability to detect a signal in an image.
6. *Artifacts* are false signals in an image that do not represent any valid structural or functional signal in the patient.
7. *Distortion* is any geometric inaccuracy in size or shape.
8. *Quantitative accuracy* refers to the accuracy, compared with the truth, of numerical values obtained from an image; *diagnostic accuracy* refers to the accuracy of interpretations and conclusions about the presence or absence of disease drawn from image patterns.

Bibliography

Barrett, H. H., and Swindell, W. *Radiological Imaging: The Theory of Image Formation, Detection, and Processing.* New York: Academic Press, 1981.

Carlton, R. R., and Adler, A. M. *Principles of Radiographic Imaging: An Art and a Science*, 3rd ed. Albany, NY: Delmar, 2001.

Cunningham, I. A. "Applied Linear–Systems Theory." pp. 79–159 in *Handbook of Medical Imaging. Vol. 1, Physics and Psychophysics.* Bellingham, WA: SPIE Press, 2000.

Dobbins, J. T. "Image Quality Metrics for Digital Systems." pp. 161–222 in *Handbook of Medical Imaging. Vol. 1, Physics and Psychophysics.* Bellingham, WA: SPIE Press, 2000.

Evans, A. L. *The Evaluation of Medical Images.* Bristol, England: Adam Hilger, 1981.

Krestel, E. *Imaging Systems for Medical Diagnostics.* Berlin, Germany: Siemens Aktiengesellschaft, 1990.

Papoulis, A., and Pillai, S. U. *Probability, Random Variables and Stochastic Processes*, 4th ed. New York: McGraw-Hill, 2002.

Sorenson, J. A., and Phelps, M. E. *Physics in Nuclear Medicine*, 2nd ed. Philadelphia, PA: W. B. Saunders, 1987.

Sprawls, P. *Physical Principles of Medical Imaging*, 2nd ed. Madison, WI: Medical Physics Publishing, 1995.

Wolbarst, A. B., *Physics of Radiology*. Norwalk, CT: Appleton and Lange, 1993.

Problems

Contrast

3.1 Prove Equation (3.5).

3.2 Consider an LSI medical imaging system with PSF given by

$$h(x, y) = \frac{1}{2\pi} e^{-(x^2 + y^2)/2}$$

(a) Calculate the MTF associated with this system.

(b) Plot the MTF as a function of frequency.

(c) If a sinusoidal object $f(x, y) = 2 + \sin(\pi x)$ is imaged through the system, what is the percentage change in modulation caused by this system?

3.3 Let H_1 and H_2 be two 1-D LTI systems whose PSFs are $h_1(x)$ and $h_2(x)$:

$$h_1(x) = e^{-x^2/5}, \qquad h_2(x) = e^{-x^2/10}$$

(a) Find the MTF of system H_1.

(b) Find the MTF of the cascade system of H_1 and H_2.

3.4 Show that the MTF of a nonisotropic medical imaging systems is given by Equation (3.9).

3.5 Consider an organ with a tumor is imaged. In the resulting image, the organ has intensity I_o, and the tumor has intensity $I_t > I_o$. Which of the following methods can improve the local contrast if the organ is treated as background:

(a) Multiplying the image by a constant α.

(b) Subtracting a constant $0 < I_s < I_o$ from the image.

Resolution

3.6 A new imaging system with which you are experimenting has anisotropic properties. You measure the impulse response function as $h(x, y) = e^{-\pi(x^2 + y^2/4)}$. What is the FWHM of the system as a function of polar angle θ?

3.7 A medical imaging system has the following line spread function, where $\alpha = 2$ radians/cm:

$$l(x) = \begin{cases} \cos(\alpha x), & |\alpha x| \leq \pi/2 \\ 0, & \text{otherwise} \end{cases} .$$

(a) Find the FWHM.

(b) Determine the resolution of this imaging system in lines per cm.

3.8 Consider a one-dimensional linear imaging system whose point spread function is given by

$$h(x; \xi) = e^{\frac{-(x-\xi)^2}{2}},$$

which represents the response to the shifted impulse $\delta(x - \xi)$.

(a) What is the PSF of a system having the same basic form of PSF as that above but whose FWHM is one half as large?

(b) Does the resolution of the system improve when you make the above change in FWHM? Why?

(c) What characteristics should be expected from the MTF in an imaging system that has high contrast (low blurring)?

3.9 Consider again the LSI medical imaging system of Problem 3.2.

(a) Calculate its *line spread function* (LSF).

(b) What is the FWHM associated with this system?

3.10 Consider a one-dimensional medical imaging system, which is composed of two subsystems with PSFs given by

$$h_1(x) = e^{-x^2/2} \qquad \text{and} \qquad h_2(x) = e^{-x^2/200}.$$

(a) What is the FWHM associated with each subsystem?

(b) What is the FWHM associated with the overall system?

(c) Which subsystem mostly affects the FWHM of the overall system?

3.11 A bar phantom is imaged by an LSI system, which is modeled as an ideal moving average system. The point spread function of the system is

$$h(x, y) = \text{rect}\left(\frac{x}{\Delta}, \frac{y}{\Delta}\right). \tag{P3.1}$$

(a) If bar separation of the bar phantom is Δ, what is the output of the image system?

(b) If the bar separation is 0.5Δ, what is the output of the image system?

(c) Derive a relation between the contrast of the output image and the bar separation and draw a conclusion about the resolution of the imaging system.

Random Variables and Noise

3.12 Show that, if N is a random variable with mean μ_N and standard deviation σ_N, then

$$M = \frac{N - \mu_N}{\sigma_N}$$

is a random variable with mean $\mu_M = 0$ and standard deviation $\sigma_M = 1$.

3.13 Suppose N random variables X_i, $i = 1, \cdots, N$ are independent with mean μ_i, and variance σ_i^2, $i = 1, \cdots, N$. Show that the mean and the variance of the random variable $X = \sum\limits_{i=1}^{N} X_i$ are given by

$$\mu = \sum_{i=1}^{N} \mu_i \qquad \text{and} \qquad \sigma^2 = \sum_{i=1}^{N} \sigma_i^2.$$

3.14 In the above problem, if X_i, $i = 1, \cdots, N$ are not independent, will the equalities for the mean and variance of the sum of the random variables still hold?

3.15 Show that the expected value and variance of a uniform random variable X over the interval (a, b) are given by

$$\mu_X = \frac{a + b}{2} \qquad \text{and} \qquad \sigma_X^2 = \frac{(b - a)^2}{12}.$$

3.16 Consider two medical imaging systems with PSFs $h_1(x, y)$, $h_2(x, y)$ and MTFs $\text{MTF}_1(u, v)$, $\text{MTF}_2(u, v)$, respectively, such that

$$\text{MTF}_1(u, v) \leq \text{MTF}_2(u, v).$$

Show that the system with larger MTF, and thus with better contrast and resolution properties, is characterized by a larger output power SNR, given by (3.63), and is thus better in terms of image quality.

3.17 Consider the system shown in Figure P3.1, in which an image $f(x, y)$ is corrupted with zero mean white noise $n(x, y)$ with variance σ_n^2. The corrupted image is input to a system with PSF $h(x, y)$ to get an output image $g(x, y)$.

(a) What are the mean and variance of the noise in the output $g(x, y)$?

(b) What are the power SNR for the input and output images of the system?

(c) If the system does not change $f(x, y)$ in any way, under what condition(s) will the system improve the SNR?

Artifacts, Distortion, and Accuracy

3.18 Compare the artifacts and noise, and state the common properties and differences between them.

3.19 In projection radiography, size distortion is an artifact that cannot be ignored. Suppose the x-ray source is a perfect point source located at the origin and the detector plane is the $x = d$ plane. A ball centered on the x-axis between

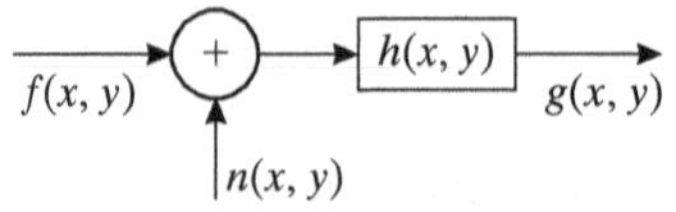

Figure P3.1
A system with additive noise.

the source and the detector plane is imaged by the system. Assuming that the radius of the ball is $r < d/2$, derive the relation between the radius R of the image of the ball on the detector plane and the location of the ball. If the source-detector distance is fixed, what measure(s) can be taken to reduce the size distortion? What is the smallest ratio R/r that can be achieved?

3.20 A medical imaging system has a geometric distortion, which is well modeled as

$$x = \xi + \frac{1}{50}\xi\eta^2$$

$$y = \eta,$$

where x and y are the coordinates in the image plane and ξ and η are coordinates in the physical domain.

(a) If we take measurements on rectangular grids in the image plane, find a way to correct the geometric distortion.

(b) If we want to take measurements on rectangular grids in the physical domain, on what points should we take measurements in the image domain?

3.21 Express the magnification ratio $m = s_1/s_2$ of the projections of the two objects O_1 and O_2 depicted in Figure P3.2, in terms of s, ϕ, d, d_1, and d_2. If $s = 5$ cm, $\phi = 45°$, $d = 120$ cm, $d_1 = 40$ cm, and $d_2 = 80$ cm, what is the value of m?

3.22 Suppose the probability law of a test result for patients with and without a disease are modeled as Gaussian with different mean values and variances, as shown in Figure P3.3. We design a diagnostic test by selecting a threshold t_0. For a patient whose test value is below t_0, we call it normal. If the test value is above t_0, we call it diseased. By selecting different thresholds, we obtain different *diagnostic tests*. Denote the means and variances of the test result of the normal and diseased subjects as μ_0, σ_0^2, and μ_1, σ_1^2.

(a) Write down the expressions for the probability density functions of the test value for normal and diseased subjects.

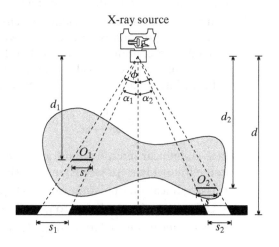

Figure P3.2
An illustration of depth-dependent magnification.

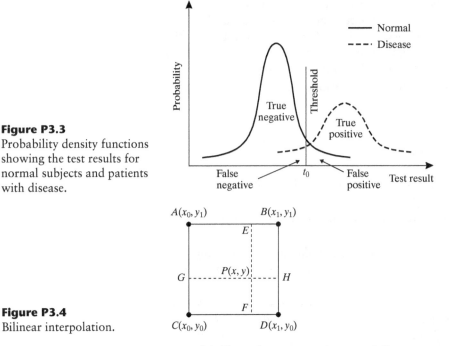

Figure P3.3
Probability density functions showing the test results for normal subjects and patients with disease.

Figure P3.4
Bilinear interpolation.

(b) If we choose $t_0 = (\mu_0 + \mu_1)/2$, compute the sensitivity and the specificity.
(c) Derive an expression for the sensitivity as a function of threshold value.
(d) Derive an expression for the diagnostic accuracy as a function of threshold value.

Applications, Extensions, and Advanced Topics

3.23 Show that a medical imaging system composed of several subsystems will be inferior to each individual subsystem in terms of image quality.

3.24 In most modern medical imaging systems, images are measured at discrete grid points. Interpolation is needed when we want to treat the images as functions defined on a continuous domain. For 1-D signals, linear interpolation is one of the simplest. It can easily be extended to bilinear interpolation in 2-D and trilinear interpolation in 3-D. Linear interpolation is defined as

$$ f(x) = \frac{x_1 - x}{x_1 - x_0} f(x_0) + \frac{x - x_0}{x_1 - x_0} f(x_1), \qquad x_0 \le x \le x_1 $$

In 2-D, bilinear interpolation for $f(p)$ from $f(A)$, $f(B)$, $f(C)$, and $f(D)$ is done in the following steps (see Figure P3.4):
• Using linear interpolation to get $f(E)$ from $f(A)$ and $f(B)$.
• Using linear interpolation to get $f(F)$ from $f(C)$ and $f(D)$.
• Using linear interpolation to get $f(P)$ from $f(E)$ and $f(F)$.

(a) Derive an explicit expression for $f(P)$ using $f(A), f(B), f(C)$, and $f(D)$.

(b) Prove that the results are the same whether we get $f(P)$ from $f(E)$ and $f(F)$, or from $f(G)$ and $f(H)$;

(c) In Problem 3.20, if we take measurements on rectangular grids in the image plane at $(m\Delta x, n\Delta y)$, with $\Delta x = \Delta y = 1$ and m, n are integers, how should we interpolate the value for $\xi = 3$ and $\eta = 3.5$ in the physical domain?

3.25 As we mentioned in Example 3.13, the diagnostic accuracy alone is not enough to determine whether a test is good. In this problem, we will introduce the concept of *receiver operating characteristic* (ROC) curves with an idealized example. Consider the situation described in Problem 3.22; suppose the mean and variance of the test value for the normal subjects are $\mu_0 = 2$ and $\sigma_0^2 = 1$, with arbitrary unit. The mean and variance of the test value for the diseased subjects are $\mu_1 = 8$ and $\sigma_1^2 = 4$, with arbitrary unit.

(a) For threshold value t_0 in the interval $\mu_0 \le t_0 \le \mu_1$, plot the curve of sensitivity against 1-specificity. This is the ROC curve.

(b) What is the ROC curve for a perfect diagnostic test?

(c) For a given ROC curve, if we define the optimal test to be the one that is closest to the perfect test in the sensitivity-(1-specificity) plane, determine the optimal test (threshold) for the above case.

Radiographic Imaging

Overview

Ionizing radiation—radiation capable of ejecting electrons from atoms—forms the basis of a number of important imaging modalities. In some cases, we are interested in the transmission of ionizing radiation through the body. In other cases, we are interested in the emission of ionizing radiation from the body.

In this part of the book, we consider the two main imaging modalities that make use of the transmission of ionizing radiation through the body: projection radiography and computed tomography. Projection radiography and computed tomography rely on the *transmission* of ionizing radiation through the body. Various tissues and organs within the body *attenuate* or decrease the intensity of the beam of ionizing radiation as it passes through the body. Thus, even if the beam entering the body is of uniform intensity, the beam exiting the body contains "shadows" of tissues and organs as a "latent image" of varying beam intensity. Since it is the physical characteristics of the tissues or organs (e.g., effective atomic number and density) that determine that tissue's attenuation abilities, the resulting images depict *structures* within the body, and projection radiography and computed tomography are considered *anatomical* imaging modalities, since they portray anatomy.

In Figure I.1(a), we showed the most classic projection radiograph—a "chest x-ray." In Figure II.1, we show several other transmission images. Figure II.1(a) is also a projection radiograph—this time, a close-up of the spine. Figure II.1(b) is a CT image through the same portion of the spine. This image is a transaxial slice [see Figure I.3(a) and Figure 2.2], and has much higher contrast because of the lack of superposition of out-of-plane tissues. Figure II.1(c) is a projection radiograph of the hand. Figure II.1(d) is also a projection radiograph of the hand, but the arteries are much more visible because *contrast material*—which will be covered in this part of the book—has been injected into the arteries prior to image acquisition.

Projection radiography and computed tomography both rely on an *x-ray tube* to produce the beam of ionizing radiation, and radiation detectors on the "other

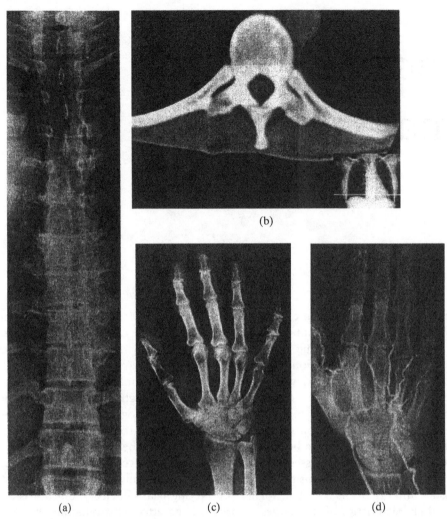

Figure II.1
Representative transmission images: (a) projection radiograph of the thoracic spine, (b) transaxial computed tomograph of the thoracic spine, with the position of the slice indicated on the inset chest x-ray, (c) projection radiograph of the hand, and (d) projection radiograph of the hand after injection of contrast material into the arteries.

side" of the patient to detect the radiation that is passed through without being fully absorbed. In the case of projection radiography, the "detector" is often a piece of film, which then also serves as the "display device." A "chest x-ray" is the most common example of projection radiography. In the case of computed tomography, the patient is usually surrounded by a ring of detectors, and a rotating x-ray tube circles the patient, producing x-ray images at a number of angles around the patient. These images, stored in a computer, form the *projection data* used to *reconstruct* cross-sectional or transaxial images through the patient.

Projection radiography is perhaps the most commonly performed medical imaging approach. It is used any time a physician needs an image of structures within the body, especially to obtain an overall picture of the torso, or to look at bones. Computed tomography produces higher contrast cross-sectional images of structures;

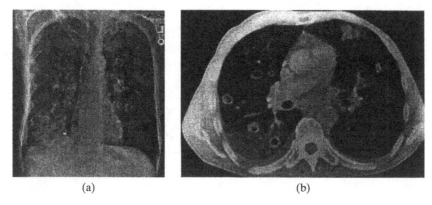

(a)   (b)

Figure II.2
(a) Chest x-ray of a patient with multiple lung metastasizes from a primary cancer in the larynx. (b) Corresponding CT scan.

the use of tomography means that overlying and underlying structures are not superimposed. Figure II.2 depicts a chest x-ray and CT slice of a patient with lung metastases from a primary larynx cancer. The chest x-ray in Figure II.2(a) shows a pronounced "cloudiness" in the lung fields when compared with the normal chest x-ray in Figure I.1(a). The CT slice in Figure II.2(b) shows numerous circular objects in the lung fields, with high contrast compared with the chest x-ray.

4

Physics of Radiography

4.1 Introduction

This chapter provides an introduction to the physics behind x-ray imaging modalities. Broadly speaking, these modalities can be divided into two types, projection radiography and computed tomography, which we cover in the next two chapters. These modalities do not involve radioactivity (a common misconception), so we delay the introduction of radioactivity until Chapter 7, where it is needed for the presentation of the nuclear medicine modalities in Chapters 8 and 9.

As we pointed out in Chapter 1, x-rays were discovered in 1895 by Roentgen while working with a Crooke's tube (a precursor to the modern x-ray tube). He noticed that "rays of mysterious origin," which he named *x-rays*, exposed nearby photographic film. Upon investigation, he determined that these rays were in fact produced by the tube. The first radiograph of a human being—the hand of Roentgen's wife—was made by Roentgen within a month of the discovery. Roentgen's discovery of x-rays and their immediate application to imaging the human body truly mark the "birth" of medical imaging.

Today, we know that x-rays are electromagnetic waves whose frequencies are much higher than light. They are only one form of *ionizing radiation*—i.e., radiation capable of ejecting electrons from atoms—used in medical imaging, however. Other forms of ionizing radiation used in medical imaging include *particulate radiation* and *gamma rays*. Both particulate radiation and gamma rays can be products of radioactive decay, and we will have more to say about this in Chapter 7. It also turns out that an *electron beam*, which is a form of particulate radiation, is used in the production of x-rays. So, we need to study certain aspects of particulate radiation in this chapter as well. Furthermore, gamma rays, like x-rays, are high-frequency electromagnetic waves. So, in studying the propagation properties of x-rays, we are also implicitly studying the same for gamma rays.

This chapter provides a necessary background for all imaging modalities that use ionizing radiation, including projection radiography (Chapter 5), computed tomography (Chapter 6), planar scintigraphy (Chapter 8), and emission computed

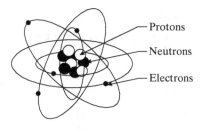

Figure 4.1
The planetary atom: a common (and useful) visualization of atomic structure.

tomography (Chapter 9). We start by describing the basic physical properties of the atom and the concepts of excitation and ionization. We then describe ionizing radiation in a more general context, concluding with its measurement and interaction with biological tissues.

4.2   Ionization

Ionization is the ejection of an electron from an atom, creating a *free electron* and an *ion*. Radiation that carries enough energy to ionize an atom is called *ionizing radiation*. Here, the term *radiation* covers a broad range of physical phenomena. For example, light, x-rays, gamma rays, and electron beams are all examples of radiation. High-energy electromagnetic waves such as x-rays and gamma rays are ionizing radiations; light is not. Particulate radiations such as an electron beam may be ionizing depending on the energy each particle possesses. We now discuss the structure of atoms and the process of ionization in more detail.

4.2.1   Atomic Structure

Today's concept of atomic structure is based on a quantum mechanical picture of the atom (from the 1920s), which in turn arose from the Bohr model of the hydrogen atom (in 1913). An atom consists of a nucleus of protons and neutrons, which together are called *nucleons*, surrounded by orbiting electrons. A common visualization of atomic structure, sometimes called the *planetary atom*, is shown in Figure 4.1. The *atomic number Z* is equal to the number of protons in the nucleus and defines the element. Each proton has a positive charge equal in magnitude and opposite in charge to that of each electron. Since an atom as a whole is electrically neutral, there are an equal number of electrons and protons in an atom; hence, Z also represents the number of electrons in the atom. A short table of the elements is given in Table 4.1.

The *mass number A* of an atom is equal to the number of nucleons in the nucleus. The term *nuclide* refers to any unique combination of protons and neutrons which forms a nucleus. Nuclides are typically denoted by either $_{Z}^{A}X$ or X-A, where X is the element symbol. For example, $_{6}^{12}C$ and C-12 are both symbols for the most abundant carbon atom. Notice that the element symbol and atomic number are redundant.

As indicated in Table 4.1, the number of neutrons in a given nucleus is approximately equal to the number of protons; and certain combinations are stable, while

TABLE 4.1

Abbreviated Table of the Elements				
Element	Symbol	Atomic Number (Z)	Mass Numbers of Stable Isotopes (A)	Mass Numbers of Unstable Isotopes (A)
Hydrogen	H	1	1, 2	3
Helium	He	2	3, 4	5, 6, 8
Lithium	Li	3	6, 7	5, 8, 9, 11
Beryllium	Be	4	9	6, 7, 8, 10, 11, 12
Boron	B	5	10, 11	8, 9, 12, 13
Carbon	C	6	12, 13	9, 10, 11, 14, 15, 16
Nitrogen	N	7	14, 15	12, 13, 16, 17, 18
Oxygen	O	8	16, 17, 18	13, 14, 15, 19, 20

From Johns and Cunningham, 1983.

others are unstable. Unstable nuclides are called *radionuclides*, and their atoms are *radioactive*. Radionuclides are statistically likely to undergo radioactive decay, which causes a rearrangement of the nucleus, which in turn gives off energy and results in a more stable nucleus. For example, $^{14}_{6}$C or C-14 denotes a radioactive carbon atom; it is statistically likely to decay into $^{14}_{7}$N, a stable nitrogen atom, during which a beta particle will be emitted from the nucleus. We will have a lot more to say about radionuclides and radioactivity in Chapters 7, 8, and 9.

The electrons orbiting the nucleus are organized into so-called orbits or *shells*. The "K" shell is closest to the nucleus, the "L" shell is next, then the "M" shell, and so on. Electrons are restricted to specific quantum states within each shell. Only one electron is permitted to be in each state, and this leads to a maximum number of electrons per shell, given by $2n^2$, where n is the shell number ($K = 1$, $L = 2$, etc.). For example, only two electrons are permitted in the K shell, 8 in the L shell, 18 in the M shell, and 32 in the N shell. Each atom has a so-called *ground state* configuration for its electrons, which corresponds to the lowest energy configuration of the atom—nature's preferred arrangement of electrons in a given atom. Generally speaking, in the ground state, the electrons will be in the lowest orbital shells and within the lowest energy quantum states within each shell. A diagram of the arrangement of electrons in shells for the carbon atom is shown in Figure 4.2.

4.2.2 Electron Binding Energy

It is energetically more favorable for an electron to be *bound* in an atom rather than to be *free*. In other words, the total energy of the atom is less than the total energy of the atom (minus the electron) and the (free) electron. The difference between these two energies is called the *electron binding energy*. Binding energy is usually specified in units of *electron volts* (eV). Recall that one electron volt is equal to the kinetic energy gained by an electron when accelerated across one volt potential. For comparison with more conventional units of energy, $1 \text{ eV} = 1.6 \times 10^{-12}$ ergs $= 1.6 \times 10^{-19}$ J.

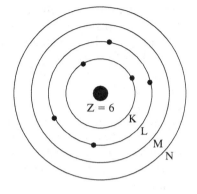

Figure 4.2
The arrangement of electrons into shells for the carbon atom.

A given electron's binding energy depends on the element to which the electron is bound and the shell within which the given electron resides. For any given element, electron binding energy decreases with increasing shell number. The binding energy of the sole electron in a hydrogen atom is 13.6 eV, the smallest binding energy among all "lighter" atoms (i.e., those having smaller atomic numbers). There are even smaller electron binding energies in the outer orbits of the heavier elements. For example, some electrons in the O-shell of mercury atoms have binding energies equal to 7.8 eV. For our purposes, however, it is sufficient to consider an "average" binding energy of the electrons in a given atom or even in a given molecule. For example, the average binding energy for air is about 34 eV. Metals have significantly larger average electron binding energies. For example, the average binding energy of lead is about 1 keV, and for tungsten it is about 4 keV.

4.2.3 Ionization and Excitation

If radiation (particulate or electromagnetic) transfers energy to an orbiting electron (in an atom of the material the radiation is passing through) which is equal to or greater than that electron's binding energy, then the electron is ejected from the atom. This process, illustrated in Figure 4.3, is called *ionization*. It yields an *ion* (an atom with a +1 charge in this case) and the electron that came from it, which together

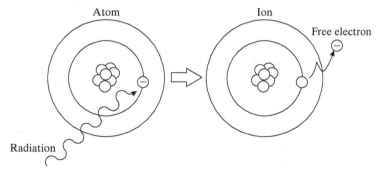

Bound energy < Unbound energy + Electron energy

Figure 4.3
After an ionization event, the ejected electron plus the ion have more energy than the original atom.

are called an *ion pair*. By convention, radiation with energy greater than or equal to 13.6 eV is considered *ionizing*; all other radiations are considered *nonionizing*.

It is possible for a single ionizing radiation emission to ionize many atoms along its path before its energy is exhausted. The ionizing radiation we use in medical imaging has energies ranging from about 25 keV to 500 keV. Broadly speaking, the "easiest" atom to ionize is a hydrogen atom, since it has the smallest average electron binding energy. Tungsten may be about the "most difficult" to ionize in the substances normally encountered in medical imaging. Using the range of ionizing radiations energies given above and the binding energies for these two elements, we can conclude that a single ionizing particle or ray used in medical imaging is capable of ionizing between 10 and 40,000 atoms before its energy is exhausted.

If an ionizing particle or ray transfers some energy to a bound electron but less than the electron's binding energy, then the electron is raised to a higher energy state—e.g., a more outer orbit—but is not ejected. This process is called *excitation*. In both ionization and excitation, an electron shell is left with a "hole" that must be filled in order to return the atom to a lower energy state. The filling of these open "holes" comprises an important source of secondary radiation called *characteristic radiation*, which we will explore below.

EXAMPLE 4.1

Suppose an electron is accelerated within a vacuum from a heated cathode held at ground potential to an anode held at 120 kV (DC).

Question   If the anode is made of tungsten, what is the maximum number of tungsten atoms that can be ionized on average?

Answer   The electron will have kinetic energy equal to 120 keV upon reaching the anode (recall the definition of eV). The average binding energy of tungsten is 4 keV. Therefore, the maximum number of tungsten atoms that can be ionized on average is 120 keV/ 4 keV = 30.   ∎

4.3   Forms of Ionizing Radiation

Ionizing radiation can be divided into two broad categories: *particulate* and *electromagnetic*. We now describe these types of radiation in some detail.

4.3.1   Particulate Radiation

Any subatomic particle (e.g., a proton, neutron, or electron) can be considered to be ionizing radiation if it possesses enough kinetic energy to ionize an atom. In calculating the kinetic energy of these particles, relativistic effects cannot be ignored since the particles often travel at a significant fraction of the speed of light. As a particle approaches the speed of light, its mass increases, which significantly increases its kinetic energy above that which would be calculated using the usual low-speed approximation.

From Einstein's theory of relativity, we know that the (relativistic) mass of a particle is given by

$$m = \frac{m_0}{\sqrt{1 - v^2/c^2}}, \tag{4.1}$$

where m_0 is the rest mass of the particle, v is the speed of the particle, and c is the speed of light. Einstein's theory also tells us that there is an equivalence between the energy E of a particle and its mass, which is given by

$$E = mc^2. \tag{4.2}$$

The kinetic energy of a particle is the difference in energy between the moving particle and the stationary particle:

$$KE = E - E_0$$
$$= mc^2 - m_0 c^2. \tag{4.3}$$

It is straightforward to show (see Problem 4.3) that when v is small relative to c, the kinetic energy reduces to

$$KE = \frac{1}{2} mv^2, \qquad v \ll c, \tag{4.4}$$

the usual (nonrelativistic) equation for kinetic energy.

In nuclear medicine, we consider modes of radioactive decay that give rise to different kinds of particulate ionizing radiations (see Chapter 7). In projection radiography and computed tomography, however, the only particulate radiation we need to consider is electrons that are not bound in an atom and have gained kinetic energy. As we shall see, these electrons arise from a heated filament and a potential difference between a cathode and anode within an x-ray tube. In this case, we know the energies of these particles by the manner in which they were created. For example, electrons accelerated across a 80 kV potential will have energies equal to 80 keV. The speed and mass of these particles is interesting to know but largely irrelevant from an imaging point of view.

EXAMPLE 4.2

Consider an electron that has been accelerated between a cathode and anode held at a 120 kV potential difference. Assume the situation is *nonrelativistic*.

Question What is the speed of the electron when it "slams" into the anode?

Answer The rest mass of an electron is 9.11×10^{-31} kg. Because it was accelerated across a 120 kV potential, the electron must have kinetic energy given by KE = 120 keV. Therefore, assuming nonrelativistic speeds, we must have

$$KE = \frac{1}{2} \times 9.11 \times 10^{-31} \text{ kg} \times v^2 = 120 \text{ keV}$$

Noting that $1\ eV = 1.602\ 177\ 33 \times 10^{-19}$ J and recalling that $1\ J = 1 kg\ m^2/s^2$, we find that

$$v = \sqrt{\frac{2 \times 120 \times 10^3 \times 1.602\ 177\ 33 \times 10^{-19}\ J}{9.11 \times 10^{-31}\ kg}}$$

$$= 2.054 \times 10^8\ m/s\,.$$

This speed is a significant fraction of the speed of light, so our calculations are not accurate; they must be redone (see Problem 4.2). Looking ahead, this example tells us that the electrons in a typical x-ray tube are traveling at relativistic speeds by the time they strike the anode. ∎

4.3.2 Electromagnetic Radiation

Electromagnetic radiation comprises an electric wave and a magnetic wave traveling together at right angles to each other. Radio waves, microwaves, infrared light, visible light, ultraviolet light, x-rays, and gamma rays are all examples of electromagnetic radiation. Electromagnetic radiation has no rest mass and no charge, and can act like either a particle or a wave. When treated as a "particle," electromagnetic radiation is conceptualized as "packets" of energy termed *photons*. The energy of a photon is given by

$$E = \hbar v\,, \tag{4.5}$$

where $\hbar = 6.626 \times 10^{-34}$ joule-sec is Planck's constant and v is the *frequency* of the radiation (in Hz). Viewed as a wave, electromagnetic radiation has a wavelength given by

$$\lambda = c/v\,, \tag{4.6}$$

where $c = 3.0 \times 10^8$ meters/sec is the speed of light.

Table 4.2 summarizes the frequencies, wavelengths, and photon energies of the electromagnetic spectrum relevant to medical imaging. Radio waves are very low frequency, very long wavelength electromagnetic radiations; their photon energy

TABLE 4.2

The Electromagnetic Spectrum			
Frequency Range	Wavelengths	Photon Energies	Description
1.0×10^5–3.0×10^{10} Hz	3 km–0.01 m	413 peV–124 μeV	Radio waves
3.0×10^{12}–3.0×10^{14} Hz	100–1 μm	12.4 meV–1.24 eV	Infrared radiation
4.3×10^{14}–7.5×10^{14} Hz	700–400 nm	1.77–3.1 eV	Visible light
7.5×10^{14}–3.0×10^{16} Hz	400–10 nm	3.1–124 eV	Ultraviolet light
3.0×10^{16}–3.0×10^{18} Hz	10 nm–100 pm	124 eV–12.4 keV	Soft x-rays
3.0×10^{18}–3.0×10^{19} Hz	100–10 pm	12.4–124 keV	Diagnostic x-rays
3.0×10^{19}–3.0×10^{20} Hz	10–1 pm	124 keV–1.24 MeV	Gamma rays

Source: Adapted from Johns and Cunningham, 1983.

ranges from 10^{-10} to 10^{-2} eV. Since the photon energies of radio waves are below 13.6 eV, radio waves are not considered to be ionizing radiation. Visible light is moderate frequency, long wavelength electromagnetic radiation. The photon energy of light is about 2 eV, which is also not large enough to be considered ionizing. (See Problem 4.6 for consideration of ultraviolet light.) X-rays and gamma rays, on the other hand, are higher frequency, shorter wavelength electromagnetic radiation, having energies in the keVs to MeV range. Both x-rays and gamma rays are clearly ionizing radiations.

Although it is suggested by Table 4.2, x-rays and gamma rays are *not* distinguished by their frequency or photon energies. Instead, they are distinguished by their point of origin. In particular, x-rays are created in the electron cloud of atoms while gamma rays are created in the nuclei of atoms, which in turn are undergoing reorganization due to radioactive decay. Thus, gamma rays are associated with radioactivity and x-rays are not. Although gamma rays tend to have higher frequencies (energies) than x-rays, there is a large overlap in the frequencies (energies) of x-rays and gamma rays used in medical imaging. Furthermore, once produced, x-rays and gamma rays behave the same in terms of their propagation properties and interaction with matter. Therefore, our understanding of the propagation and detection of electromagnetic waves in the x-ray–gamma ray photon energy range is essential background material for all the medical imaging modalities using ionizing radiation.

It turns out that both the production and detection of electromagnetic radiation over a large range of frequencies play major roles in medical imaging. Radiofrequency waves are used to stimulate nuclei in magnetic resonance imaging and are then detected as the same nuclei generate electromagnetic radiation. Visible light is used in radiography to improve the efficiency of photographic film in the detection of x-rays. The attenuation of x-rays is the primary mechanism used to create images in projection radiography and computed tomography. Finally, gamma rays are detected in order to locate radiotracers in nuclear medicine.

4.4 Nature and Properties of Ionizing Radiation

Particulate and electromagnetic ionizing radiations interact with the materials through which they are traveling, imparting energy to the material, losing energy from and redirecting their own radiation, and generating new types of particles and radiation. Generally speaking, the effects that we care about fall into two broad categories: (1) those that are used in imaging or that affect the imaging process and (2) those that are not used in imaging but contribute to *dose*—i.e., have biological consequences. Specific concepts related to these two broad categories tend to apply separately to particulate radiation or electromagnetic radiation. Table 4.3 lists specific concepts related to ionizing radiation, positioned according to the categories for which they largely apply. We will now begin to discuss each of these concepts, starting with those important in forming and analyzing medical images (column 1) and ending with those important in understanding the biological effects of radiation

TABLE 4.3

Radiation Concepts		
	Imaging	Dose
Particulate	Bremsstrahlung Characteristic radiation *Positron annihilation** *Range*	Linear energy transfer Specific ionization
Electromagnetic	Attenuation Photoelectric effect Compton scatter Characteristic radiation Polyenergetic	Air kerma Dose Dose equivalent Effective dose f-factor

*Italicized entries are discussed in Chapter 7.

(column 2). Concepts that are relevant only in nuclear medicine are listed in italics and will be discussed in Chapter 7.

4.4.1 Primary Energetic Electron Interactions

The only particles of direct consequence to the *formation* of medical images are electrons and positrons.[1] Since positrons are used solely in nuclear medicine, we postpone discussion of the interaction of positrons with matter until Chapter 7. Here we consider only the interaction of *energetic electrons* with matter.

Energetic electrons interact and transfer energy to an absorbing medium by two modes: *collisional transfer* and *radiative transfer*. In collisional transfer, by far the most common type of interaction at the electron energies used in medical imaging, a (typically small) fraction of the electron's kinetic energy is transferred to another electron in the target medium with which it collides. As the affected atom returns to its original state, infrared radiation is generated, producing heat in the target medium. As shown in Figure 4.4(a), the incident electron's path may be redirected as a result of the collision, and many more interactions, both collisional and radiative, may subsequently take place, until the incident electron's kinetic energy is exhausted. Occasionally, a large amount of energy may be transferred to a struck electron, creating a new energetic electron, which forms a new path of ionization, called a *delta ray*, and a new set of collisional and radiative energy transfers.

In radiative transfer, the energetic electron's interaction with an atom produces x-rays. This can happen in two ways: *characteristic radiation* or *bremsstrahlung radiation*. In the generation of characteristic radiation, as shown in Figure 4.4(b), the incident electron collides with a K-shell electron, exciting or ionizing the atom, and temporarily leaving a "hole" in that shell. (This can also happen to higher-orbit electrons but is not significant in imaging.) The K-shell "hole" is then filled by an electron from the L-shell, M-shell, or N-shell. Since the electron binding energy of

---

[1] Other particles, most notably alpha particles, are relevant in the *production* of radiotracers for use in nuclear medicine (see Chapter 7).

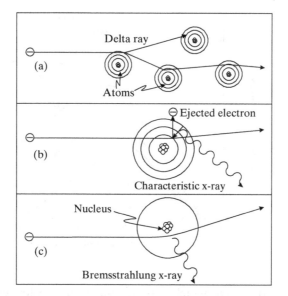

Figure 4.4
Energetic electrons can (a) collide with other electrons until they lose their energy; (b) eject a K-shell electron, generating a characteristic x-ray; or (c) be "braked" by a nucleus, generating bremsstrahlung radiation.

the K-shell is higher than that of the L-, M-, or N-shell, there is a *loss* of energy as the electron fills the K-shell hole or vacancy. The loss of energy experienced by the outer-shell electron creates an electromagnetic photon. The x-ray is called a *characteristic x-ray*. The energy of the characteristic x-ray photon is exactly equal to the difference in electron binding energies between the two shells. In fact, the particular electron "subshells," which have slightly different energies according to quantum physics, are distinguished. Since these electron binding energies are determined by the particular atom and those transitions allowed by quantum physics, these x-ray energies are "characteristic" of particular atoms, which is the reason for the name *characteristic x-rays*. A given atom can be identified by the nature (energies and proportions) of the characteristic x-rays it generates.

Bremsstrahlung radiation is caused by the interaction of an energetic electron with the nucleus of an atom, as shown in Figure 4.4(c). As the electron approaches the nucleus, the positive charge of the nucleus attracts the electron, causing it to bend, as if to go into orbit around the nucleus. As the electron decelerates around the nucleus, it loses energy in the form of an electromagnetic photon, which comprises *bremsstrahlung radiation* (German for "braking radiation"). Unless the electron actually strikes the nucleus, the electron departs the atom with a kinetic energy reduced by the energy of the emitted photon. In the rare event of a collision with the nucleus, the electron is annihilated, and the photon emitted has energy equal to the kinetic energy of the incident electron. Otherwise, the emitted photon must have energy lower than that of the incident electron. The intensity of bremsstrahlung radiation increases with the energy of the incident electron and the atomic number of the atom with which it interacts.

As will be seen in Chapter 5, bremsstrahlung radiation is the primary source of x-rays from an x-ray tube. This is because an x-ray tube works by accelerating electrons across a voltage potential in a vacuum. When the electrons strike the *target*

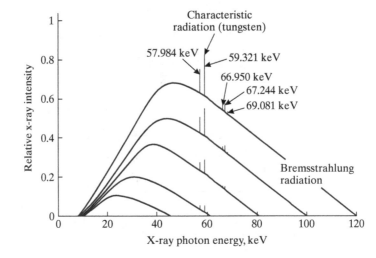

Figure 4.5
When energetic electrons bombard a target, two kinds of x-rays are produced: characteristic x-rays and bremsstrahlung x-rays. Low-energy x-rays of both types are absorbed by the medium.

or *anode*, which is usually tungsten, they lose energy by both collisional and radiative transfer; so heat, characteristic x-rays, and bremsstrahlung x-rays are produced. The x-ray energies produced by the characteristic and bremsstrahlung interactions differ in character, as shown in Figure 4.5. The bremsstrahlung spectrum, sometimes called the *continuous spectrum*, has its highest energy equal to the anode-to-cathode potential. In this figure, different potentials are considered: 45 kV, 61 kV, 80 kV, 100 kV, and 120 kV. For a given fixed potential, x-rays at the highest energies are produced only by the rare direct collisions between energetic electrons and nuclei. It is increasingly more likely for lower energy bremsstrahlung interactions to take place; it is approximately a linear increase with frequency decreasing from the maximum energy. At the lower energies, bremsstrahlung radiation is still produced, but is absorbed by the anode itself, so the spectrum goes to (effectively) zero.

Characteristic x-rays can be generated only if the incident electrons have sufficient energy to eject K-shell electrons. Thus, for a tungsten anode, if the kinetic energy of the incident electrons is below 57.984 keV—i.e., the potential difference between the anode and cathode is below 57.984 kV—there will not be any characteristic x-rays. When present, the characteristic x-rays add spectral lines to the underlying bremsstrahlung spectrum. The particular spectral lines shown in Figure 4.5 correspond to tungsten, the usual target (anode) in x-ray tubes. Their heights are in proportion to the likelihood that such transitions take place during reconfiguration of the electron cloud after a K-shell electron is ejected.

4.4.2 Primary Electromagnetic Radiation Interactions

Electromagnetic (EM) ionizing radiation interacts with matter through significantly different mechanisms than particulate ionizing radiation. The three main mechanisms by which EM ionizing radiation interacts with materials are the (1) photoelectric effect, (2) Compton scatter, and (3) pair production. Since a photon must have at least 1.02 MeV of energy for pair production to occur, and since the range of photon energies in medical imaging is about 25–500 keV, we will consider only the photoelectric effect and Compton scattering here.

In both the photoelectric effect and Compton scattering, an incident x-ray photon interacts with the electron cloud of an atom. The primary difference between the two interactions is that in the photoelectric effect, the photon is completely absorbed by the atom, while in Compton scattering, the photon is not absorbed but instead loses energy and changes its direction. Both effects are important in imaging human tissues. The photoelectric effect is the primary mechanism that provides contrast between different types of tissues; Compton scattering is a primary mechanism for limiting the resolution of x-ray images, both in projection radiography and computed tomography. We now study both interactions in some detail.

Photoelectric Effect   In the *photoelectric effect*, a photon with energy $\hbar\nu$ interacts with the coulomb field of the nucleus of an atom, causing the ejection of an electron, usually a K-shell electron, from the atom. This process is illustrated in Figure 4.6 (a) and (b). The incident photon is completely absorbed by the atom, and the ejected electron, called a *photoelectron*, propagates away with energy

$$E_{e^-} = \hbar\nu - E_B,\qquad(4.7)$$

where E_B is the binding energy of the ejected electron. The remaining atom is now an ion, having a "hole," typically in the K-shell, which must be filled. The hole is filled by electron transitions from higher-orbits, which produces characteristic radiation, as discussed above and shown in Figure 4.6(a). Sometimes, the characteristic x-ray transfers its energy to an outer-orbit electron, called an *Auger electron*, which is

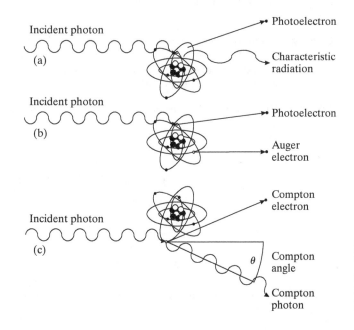

Figure 4.6
The photoelectric effect, shown in (a) and (b), and Compton scattering, shown in (c).

ejected from the atom, accompanied by a readjustment of the remaining electron orbits, as shown in Figure 4.6(b). Photoelectrons and Auger electrons are energetic electrons that are free to interact with matter in the ways we described in the previous section. We will see below that these energetic electrons contribute to the detrimental biological effects of ionizing electromagnetic radiation.

Compton Scattering   In *Compton scattering*, as shown in Figure 4.6(c), a photon with energy $\hbar v$ ejects a valence (outer-shell) electron, yielding a new energetic electron called a *Compton electron*. As a result of this interaction, the incident photon loses energy (to the Compton electron) and changes its direction. The energy of the scattered photon, the so-called *Compton photon*, is given by

$$\hbar v' = \frac{\hbar v}{1 + (1 - \cos\theta)\hbar v/(m_0 c^2)}, \tag{4.8}$$

where $m_0 c^2 = 511$ keV is the energy equivalent to the rest mass m_0 of an electron, and θ is the angle through which the photon is scattered [see Figure 4.6(c)]. We see that the amount of energy remaining with the Compton photon depends on the scatter angle. The more the photon is deflected, the more energy it loses. This implies that the maximum energy loss occurs when the photon is deflected 180° back towards the source, i.e., a 180° *backscatter*. Accordingly, the kinetic energy of the Compton electron is given by

$$E_{e^-} = \hbar v - \hbar v'. \tag{4.9}$$

EXAMPLE 4.3

Compton scattering is usually undesirable in medical imaging. In planar scintigraphy, the energy of a photon is used to determine whether it has been scattered prior to arrival at the detector.

Question   Suppose a photon with energy $\hbar v = 100$ keV is incident to some material and exits with energy $\hbar v'$. A detector decides that the photon has not been scattered if $\hbar v' > 98$ keV. What is the maximum angle by which the photon is scattered but is still being treated as a photon traveling along a straight path?

Answer   The energy of the photon after Compton scattering is

$$\hbar v' = \frac{\hbar v}{1 + (1 - \cos\theta)\hbar v/(m_0 c^2)}$$

$$= \frac{100 \text{ keV}}{1 + (1 - \cos\theta)100 \text{ keV}/(511 \text{ keV})}$$

$$= 98 \text{ keV}$$

$$\theta = \cos^{-1}\left(1 - \frac{511}{100 * 49}\right)$$

$$= 26.4°$$

So a photon scattered by an angle up to 26.4° is still considered as traveling along a straight path by the detector.   ∎

Probability of EM Interactions We now consider the factors that make photoelectric and Compton events more likely to occur. There are many reasons why this subject is important. For example, the formation of a useful projection radiograph depends on the *differential attenuation* of an x-ray beam passing through a patient, by the photoelectric effect and Compton scattering, as we shall see in the next chapter. As another example, we observed above that the photoelectric effect causes complete absorption of an incident photon. Therefore, if we find a way to increase the likelihood of photoelectric events, we can *block* or *shield* objects (or patients, staff, and physicians) from the source of ionizing EM radiation. We should therefore be able to understand what makes lead a good x-ray shield while plastic is not; we should also be able to predict whether lead will shield high-energy photons better than low-energy photons. As another example, the reason why Compton scattering is detrimental to imaging is that it causes photons to deviate from straight-line paths. We would like to know the conditions under which Compton scattering is large so that instrumentation can be designed to compensate for its effects. We might choose to use x-ray energies that minimize the Compton scattering effect to image patients.

Consider a thin "slab" of material having a fixed thickness. Suppose that an incident photon can either pass through the material or experience a photoelectric or Compton event. What are the factors that make a photoelectric effect more likely? Since a photoelectric event occurs as a result of an interaction with the coulomb field of the nucleus of an atom and if there are more (positively charged) protons in the nucleus, the likelihood should increase. The probability of a photoelectric event is related to the atom's atomic number Z. In a compound comprising a variety of different atoms, we define an *effective atomic number* Z_{eff}, which characterizes the average atomic number in a certain sense. The probability of a photoelectric event is (approximately) proportional to Z_{eff}^4.[2]

The other factor affecting the probability of a photoelectric event is the incident photon energy. Intuitively, we might expect that high-energy photons are more penetrating—i.e., they pass farther into an object before being absorbed. This intuition is correct; in fact, the probability of a photoelectric event (given all else held constant) is proportional to $1/(\hbar\nu)^3$. Putting all this together, we find

$$\text{Prob[photoelectric event]} \propto \frac{Z_{eff}^4}{(\hbar\nu)^3}. \qquad (4.10)$$

Finally, we note that when the photon energy rises above the binding energy of L-shell or K-shell electrons, experiments reveal that the probability of a photoelectric interaction increases abruptly, then begins to diminish as the energy is further increased. This sudden increase occurs because a large number of electrons has suddenly become available to be ejected from their host atoms. This property will be important in the use of *contrast agents* in radiography and computed tomography.

Now we ask another question: What are the factors that make a Compton scattering event more likely? Since Compton events occur with very loosely bound

---

[2]For high Z materials (as opposed to human tissues), it is more nearly proportional to Z_{eff}^3.

TABLE 4.4

Physical Properties of Some Materials			
Material	Density (kg/m^3)	Z_{eff}	Electron Density (electrons/kg)
Hydrogen	0.0899	1	5.97×10^{26}
Carbon	2250	6	3.01×10^{26}
Air	1.293	7.8	3.01×10^{26}
Water	1000	7.5	3.34×10^{26}
Muscle	1040	7.6	3.31×10^{26}
Fat	916	6.5	3.34×10^{26}
Bone	1650	12.3	3.19×10^{26}

From A. B. Wolbarst, p. 119, 1993.

(or "free") electrons in the outer shells, what matters is the number of electrons per kilogram of material—the *electron density* (ED). The electron density is given by

$$ED = \frac{N_A Z}{W_m}, \qquad (4.11)$$

where N_A is Avogadro's number (atoms/mole), Z is the atomic number (electrons/atom), and W_m is the molecular weight of the atom (grams/mole). As shown in Table 4.4, except for hydrogen, the electron density for various biological materials is nearly the same, 3×10^{26} electrons/kg (see Problem 4.7). Therefore, the probability of Compton scattering is nearly independent of the atomic number (actual or effective).

The other factor that might influence the probability of Compton scatter is the photon energy. There is a fairly complicated relationship described by the *Klein-Nishina formula*, which predicts that the probability of a Compton event generally decreases with increasing energy. However, this decrease is a very gradual one, taking place over the very highest energies of interest in diagnostic imaging. Over the range of energies of importance in x-ray imaging, the probability is reasonably constant. In summary, we conclude that

$$\text{Prob[Compton event]} \propto ED. \qquad (4.12)$$

Since photoelectric and Compton events affect diagnostic images differently, it is also important to understand the *relative* frequency of occurrence of these two events in body tissues. There are two ways to look at this: by a relative frequency of occurrence or by a ratio of energy deposited. Table 4.5 summarizes these two characterizations over the diagnostic range of energies. We see that the relative frequency of Compton interactions increases as diagnostic x-ray energy increases, becoming the dominant type of interaction above 30 keV. However, even at 60 keV, when Compton events occur over 90 percent of the time (whenever either event occurs), these events account for only *55 percent* of the energy deposited in the

TABLE 4.5

Photoelectric Versus Compton Interactions in Water		
Photon Energy (keV)	Percentage of Compton Interactions	Percentage of Compton Energy
10.0	3.2	0.1
15.0	11.8	0.4
20.0	26.4	1.3
30.0	58.3	6.8
40.0	77.9	19.3
50.0	88.0	37.2
60.0	93.0	55.0
80.0	97.0	78.8
100.0	98.4	89.6
150.0	99.5	97.4

Adapted from Johns and Cunningham, p. 163, 1983.

tissue. This apparent discrepancy occurs because photoelectric events deposit *all* of their incident photon energy, while Compton events deposit only a fraction of their incident photon energy.

4.5 Attenuation of Electromagnetic Radiation

Attenuation is the process describing the loss of strength of a beam of electromagnetic radiation. Tissue-dependent attenuation is the primary mechanism by which contrast is created in radiography modalities. Although this is actually a statistical process, we study attenuation here as though it were deterministic. In Chapter 5, we will include the statistical aspects in a very natural way, as we study noise in projection radiographs.

4.5.1 Measures of X-Ray Beam Strength

In radiography, we are generally concerned with a brief burst of x-rays, also called an x-ray *beam*. For many reasons, we are also concerned with the "strength" of the x-ray burst. For example, in the design of x-ray detection systems used to form images, it is critical to characterize the strength of the x-ray beam in order to characterize the inherent noise in the system and to adjust the dynamic range of the detection system. Measures of x-ray beam strength are also used in estimating the (adverse) biological effects of ionizing radiation. This is critical because ionizing radiation is potentially harmful. At high doses, it can cause burns or cataract formation, for example. Even at low doses, it can increase the risk of cancer. It turns out that there are many ways to define and measure the strength of x-rays. In this section, we consider several of the most common and important measures of x-ray beam strength.

The conceptually simplest measure of the "strength" of an x-ray burst is simply the number of photons N in the burst. It is also important to consider the *area* over which these photons are spread. Accordingly, we define the *photon fluence* Φ as the number of photons N per unit area A,

$$\Phi = \frac{N}{A}, \tag{4.13}$$

where the area is oriented at a right angle to the direction of the radiation beam propagation. It is sometimes important to account for *time*, since measurements may take place over a fixed interval Δt. Accordingly, *photon fluence rate* ϕ is defined as

$$\phi = \frac{N}{A\Delta t}. \tag{4.14}$$

It may be of interest to know how much energy the burst is carrying and could deposit into a material if totally absorbed. If each photon in the burst has the same energy $\hbar v$—i.e., the beam is monoenergetic—then the total energy of the burst is simply $N\hbar v$. We then define the *energy fluence* Ψ and *energy fluence rate* ψ as

$$\Psi = \frac{N\hbar v}{A}, \tag{4.15}$$

$$\psi = \frac{N\hbar v}{A\Delta t}. \tag{4.16}$$

The energy fluence rate is also known as the *intensity* of an x-ray beam, and it is often given the symbol I. From (4.14) and (4.16), we see that

$$I = E\phi, \tag{4.17}$$

where $E = \hbar v$. Intensity has units of energy per unit area per unit time.

In x-ray imaging, a photon burst is always polyenergetic, since it arises (mainly) from bremsstrahlung. In principle, if only the photon count were important, photon fluence and photon fluence rate could still be used as a valid measure of radiation strength. However, these measures are not very important in practice, since photons are not counted in x-ray imaging modalities, but instead their total energy is measured. It is more useful, therefore, to capture the idea of intensity for polyenergetic sources.

Since each x-ray photon carries its own discrete energy, a plot of N as a function of E for polyenergetic sources is a *line spectrum*. Because of the random nature of bremsstrahlung, the fine details of a line spectrum change with each photon burst; however, the line density—e.g., number of photons per unit energy as a function of E—remains constant for a given source. The *x-ray spectrum* $S(E)$ is this line density per unit area per unit time. Examples of x-ray spectra were shown in Figure 4.5. The energies at which $S(E)$ is large have larger numbers of x-ray photons per unit energy.

Since the integral of a line density yields the number of lines, from the definition of $S(E)$ and (4.14), the integral of the spectrum yields the photon fluence rate

$$\phi = \int_0^\infty S(E')\,dE', \tag{4.18}$$

a quantity that does not depend on energy. Furthermore, by analogy to (4.17), it is natural to define the intensity of a polyenergetic source as

$$I = \int_0^\infty E' S(E') \, dE' . \tag{4.19}$$

EXAMPLE 4.4

It is often desirable to model a polyenergetic x-ray beam as a monoenergetic source.[3]

Question What energy would a (hypothetical) monoenergetic source have to be in order to produce the same intensity as the (true) polyenergetic source using the same number of photons?

Answer For a given area A and time interval Δt, the number of photons in the polyenergetic source is

$$N_p = A \Delta t \int_0^\infty S(E') \, dE' .$$

The intensity of the polyenergetic source is

$$I_p = \int_0^\infty E' S(E') \, dE' .$$

The equivalent monoenergetic beam has intensity

$$I_p = E \frac{N_p}{A \Delta t} .$$

Therefore, the energy of this equivalent beam must be

$$E = \frac{\int_0^\infty E' S(E') \, dE'}{\int_0^\infty S(E') \, dE'} .$$

This expression should look vaguely familiar; it is the expression for center of mass of a sheetlike object having mass density $S(E)$. Alternatively, the numerator is the expression for the mean of a random variable if $S(E)$ were its pdf. The denominator normalizes $S(E)$ so that the ratio effectively does compute the mean. The energy computed this way is sometimes called the *effective energy* of a polyenergetic source. We will see an alternate definition in Chapter 6. ∎

4.5.2 Narrow Beam, Monoenergetic Photons

Consider a beam of N monoenergetic photons incident upon a thin slab of homogeneous material, as shown in Figure 4.7(a). The depicted geometry is called *narrow beam geometry* because the photon beam is no wider than the detector used to detect or *count* the photons. In this case, if the slab were not present (and ignoring statistical effects and problems of detector efficiency), the detector would record N photons. With the slab present, however, some photons will be absorbed within the slab by

---

[3]In a monoenergetic beam, all photons have the same energy, $h\nu$.

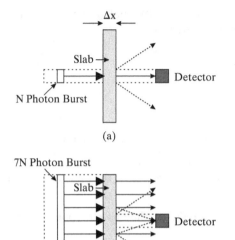

Figure 4.7
Two geometries by which to study attenuation: (a) narrow beam geometry and (b) broad beam geometry.

the photoelectric effect, and other photons will be deflected away from the detector by Compton events. Overall, the detector count N' will be smaller than N, which is the primary characteristic of *attenuation*, which we now explore in more detail.

Suppose n photons are "lost" in the above experiment. Provided that Δx is small, we would expect that doubling Δx should double n, and doubling N should also double n. In other words, n is proportional to both N and Δx, or

$$n = \mu N \Delta x, \qquad (4.20)$$

where μ is a constant of proportionality called the *linear attenuation coefficient*. Rearranging (4.20) yields a definition of the linear attenuation coefficient,

$$\mu = \frac{n/N}{\Delta x}, \qquad (4.21)$$

which can be interpreted as the fraction of photons that are lost per unit length.

The change in the number of photons upon interaction with the slab is

$$\Delta N = N' - N$$

$$= -n$$

$$= -\mu N \Delta x. \qquad (4.22)$$

Letting the slab become incrementally thin and treating N as a continuous quantity leads to the differential equation

$$\frac{dN}{N} = -\mu \, dx, \qquad (4.23)$$

which can be integrated (see Problem 4.9) to yield

$$N = N_0 e^{-\mu \Delta x}, \tag{4.24}$$

where N_0 is the number of photons at $x = 0$. This equation can be called the *fundamental photon attenuation law*; it is of central importance throughout all of x-ray radiography. It should be noted that in the monoenergetic case, this law can also be written in terms of intensity as

$$I = I_0 e^{-\mu \Delta x}, \tag{4.25}$$

where I_0 is the intensity of the incident beam.

Using (4.24), we can compute the fraction of (monoenergetic) photons that will be stopped or transmitted by a noninfinitesimal layer of homogeneous material. One question can be asked: What thickness of a given material will attenuate half of the incident photons? This thickness, called the *half value layer* (HVL), satisfies

$$\frac{N}{N_0} = \frac{1}{2} = e^{-\mu \text{HVL}}. \tag{4.26}$$

A simple series of manipulations of (4.26) yields

$$\text{HVL} = \frac{\ln 2}{\mu} = \frac{0.693}{\mu}. \tag{4.27}$$

EXAMPLE 4.5

We will see in Chapter 7 that 140 keV gamma rays are generated by the radioactive decay of technetium-99m, and that sodium iodide crystals are used to detect such gamma rays. Assume that the HVL of sodium iodide at 140 keV is 0.3 cm.

Question   What percentage of gamma rays will pass right through a 1.2 cm sodium iodide crystal?

Answer   We need to determine

$$\text{Percent transmitted} = 100\% \times \frac{N}{N_0}$$

$$= 100\% \times e^{-\mu 1.2 \text{ cm}}.$$

Since $\mu = \ln 2 / \text{HVL}$, we have

$$\text{Percent transmitted} = 100\% \times e^{-\ln 2 (1.2 \text{ cm})/(0.3 \text{ cm})}$$

$$= 100\% \times e^{-4 \ln 2}$$

$$= 6.25\%. \qquad \blacksquare$$

Suppose the slab is not homogeneous. In this case, the linear attenuation coefficient depends on the position x, and we are faced with solving

$$\frac{dN}{N} = -\mu(x)\,dx\,. \tag{4.28}$$

Straightforward integration yields the number of photons at position x,

$$N(x) = N_0 \exp\left\{-\int_0^x \mu(x')\,dx'\right\}\,, \tag{4.29}$$

where x' is a dummy variable for integration. The analogous relationship holds for intensity as well,

$$I(x) = I_0 \exp\left\{-\int_0^x \mu(x')\,dx'\right\}\,. \tag{4.30}$$

This form can be thought of as the *integral form* of the fundamental x-ray attenuation law. This equation is the most important physical model for projection radiography and computed tomography. It is also a valid model for positron emission tomography.

4.5.3 Narrow Beam, Polyenergetic Photons

In the narrow beam case, photons (of any energy) are either completely absorbed or scattered away from the detector. Therefore, we can consider the attenuation principles in the narrow beam, polyenergetic case to be the same as in the monoenergetic case, except that they take place at each energy independently.

In general, the linear attenuation coefficient is different for different materials, and it also varies as a function of energy for the same material. A plot showing these relationships for bone, muscle, and fat is given in Figure 4.8. This plot shows that bone is more attenuating than muscle, which is in turn more attenuating than fat. This plot shows as well that x-rays experience less attenuation at higher energies—i.e., they are more penetrating at higher energies. For a polyenergetic x-ray beam, we must treat the linear attenuation coefficient of a given material as a function of E—i.e., $\mu(E)$.

Consider a homogeneous slab of material of thickness Δx with an incident x-ray beam having spectrum $S_0(E)$. The spectrum leaving the slab is attenuated in a fashion identical with that of the monoenergetic case [see (4.25)], except that the linear attenuation coefficient depends on E,

$$S(E) = S_0(E)e^{-\mu(E)\Delta x}\,. \tag{4.31}$$

For a heterogenous slab, the linear attenuation coefficient is dependent on both position along the line and energy. In this case, the integral form of the attenuation law obeys

$$S(x; E) = S_0(E) \exp\left\{-\int_0^x \mu(x'; E)\,dx'\right\}\,. \tag{4.32}$$

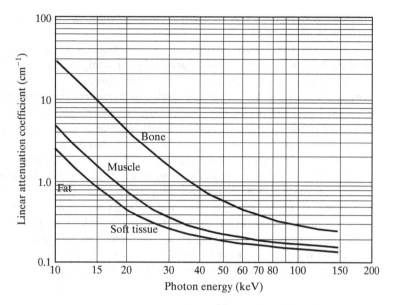

Figure 4.8
Linear attenuation coefficient for bone, muscle, and fat as a function of incident x-ray photon energy.

If the overall intensity of the beam is of interest, then we can use (4.19) together with (4.31) and (4.32) to yield

$$I = \int_0^\infty S_0(E')E' \exp\left\{-\mu(E')\Delta x\right\} dE' \tag{4.33}$$

and

$$I(x) = \int_0^\infty S_0(E')E' \exp\left\{-\int_0^x \mu(x'; E') dx'\right\} dE'. \tag{4.34}$$

Although (4.34) is a valid model for heterogenous materials, including the body, we will not find it useful in developing imaging equations for the x-ray modalities. The energy dependence of μ, while useful in helping to understanding the basic imaging properties, is intractable from a mathematical standpoint. The concept of effective energy (see Example 4.4) is more useful, as we shall see.

4.5.4 Broad Beam Case

Now consider the *broad beam geometry* in Figure 4.7(b), which should be compared with the narrow beam geometry in Figure 4.7(a). In (b), the x-ray beam is seven times wider and has $7N$ as many photons (the number 7 being used here for illustration purposes). When the slab is removed, the detector, having the same width as in (a), will detect only N photons, since the remaining photons miss the detector. When the slab is in place, photons previously directed at the detector might be absorbed by photoelectric interactions or scattered by Compton events—just as they are in the narrow beam case. However, there is an additional possibility here: photons from outside the detector's line-of-sight geometry might get scattered toward the detector by Compton interactions.

Thus, in the broad beam case, more photons are generally detected than is predicted by a monoenergetic, narrow beam analysis. The general attenuation law [(4.24) and (4.25)] does not hold in this case, and the assumption of straight propagation of x-rays is also violated. Furthermore, the photons that comprise the detected burst are no longer monoenergetic, since the Compton scattering process reduces photon energy. In such a situation, the average or effective energy of the x-ray beam is reduced, a process called *beam softening*.

Fortunately, most x-ray imaging modalities use *detector collimation*, which reduces the number of x-rays from nonnormal directions that can hit the detectors. Therefore, from an imaging standpoint, the narrow beam geometry assumption can be viewed as fairly accurate. We make the assumption of narrow beam geometry throughout this book. From the point of view of calculating the dose to the patient or determining how much lead is needed to shield a radiologist or technician, however, broad beam geometry must be considered. After all, there is no detector collimation on the "human detectors" involved in the imaging process.

4.6 Radiation Dosimetry

There are many ways to report the presence and amount of radiation. We have already described the ideas related to photon and energy flow via *fluence* and *intensity* concepts. We did so with a focus on the production of medical images. However, radiation also produces adverse biological effects. Here, we are now more concerned with what EM radiation *does* than what it *is*. That is the essence of the concept of *dose*.

4.6.1 Exposure

By definition, ionizing radiation is capable of ionizing the hydrogen atom. Therefore, it is capable of producing ions in air. It is straightforward to build and calibrate an *ionization chamber* that measures the current produced between two plates held at a fixed potential due to radiation producing ions in the air between the two plates.

The term *exposure*, which is given the symbol X, refers to the number of ion pairs produced in a specific volume of air by electromagnetic radiation. The International System unit for exposure is coulombs per kilogram of air (C/kg); however, a more useful unit in medical imaging is the roentgen (R), which equals 2.58×10^{-4} C/kg. The conversion between the two is 1 C/kg = 3876 R.

EXAMPLE 4.6
For a point source of radiation, the exposure at a distance d from the source follows an inverse square law.

Question If the exposure at $d = 30$ cm from a point source is 1 R, what is the exposure at $d = 5$ cm from the source?

Answer The exposure at $d = 5$ cm is 36 $((30/5)^2 = 36)$ times that at $d = 30$ cm. So the exposure at $d = 5$ cm is $X_5 = 36 X_{30} = 36$ R. ∎

4.6.2 Dose and Kerma

As ionizing EM radiation passes through a material, it deposits energy into the material by both the photoelectric effect and Compton scattering. This is the concept of *dose*. The unit of *absorbed dose* is the *rad*, which is defined as the absorption of 100 ergs per gram of material. It is given the symbol D. Notice that this unit refers to an energy-deposition concentration rather than a total amount of energy. The SI unit for absorbed dose is the *gray* (Gy); 1 Gy = 1 J/kg = 100 rads. For the degree of accuracy required in biomedical dosimetry, one (1) roentgen of exposure yields one (1) rad of absorbed dose in soft tissue.

There is a another quantity tightly related to dose, called *kerma*, which is given the symbol K. Kerma is defined as the amount of energy per unit mass imparted directly to the electrons in a given material. It is also measured in units of gray (Gy). As a practical matter, at diagnostic x-ray energies, kerma and dose are essentially equivalent. When kerma is being used in air for calibration purposes, it is referred to as *air kerma* and is given the symbol K_{air}.

4.6.3 Linear Energy Transfer

The *linear energy transfer* (LET) is a measure of the energy transferred by radiation to the material through which it is passing per unit length; higher LET radiation tends to produce greater adverse biological consequences. *Specific ionization* (SI) is the number of ion pairs formed per unit length. LET and SI are related to each other by the average amount of energy required to form one ion pair (often referred to as W), which is a characteristic of the material.

4.6.4 The *f*-Factor

It is often useful to be able to measure exposure but to express the results as dose to an individual in that radiation field. Because of the way in which the roentgen and rad are defined, a relationship between exposure in air and dose in air may be easily derived: 1 R = 0.87 rad. To compute the dose to a material other than air, the *f*-factor is used as a conversion,

$$D = fX. \tag{4.35}$$

The *f*-factor is defined as

$$f = 0.87 \frac{(\mu/\rho)_{\text{material}}}{(\mu/\rho)_{\text{air}}}, \tag{4.36}$$

where μ is the linear attenuation coefficient and ρ is the mass density of the material (numerator) and air (denominator). The quantity (μ/ρ) is called the *mass attenuation coefficient*.

4.6.5 Dose Equivalent

We are all exposed to ionizing radiation, from cosmic rays, radioactivity in building materials and soil, and even from radioactivity in our bodies. Although dose itself is a well-defined quantity, different types of radiation, when delivering the same dose, can actually have different effects on the body. In order to account for this, the concept of *dose equivalent*, given the symbol H, is used. Dose equivalent is defined as

$$H = D\,Q, \tag{4.37}$$

where Q is the so-called *quality factor*, a property of the type of radiation used. For example, the quality factor of x-rays, gamma rays, electrons, and beta particles is $Q \approx 1$, whereas that for neutrons and protons is $Q \approx 10$, and for alpha particles $Q \approx 20$.

Since $Q \approx 1$ for radiation used in medical imaging, dose equivalent is equal to dose. When D is measured in rads, H is considered to have the units *rems*. In SI units, the dose is measured in grays. For a dose of 1 gray, and $Q = 1$, the dose equivalent H is 1 sievert (Sv).

EXAMPLE 4.7

Consider a chest x-ray at an energy of 20 keV. For simplicity ignore the tissues other than lung.

Question   If we want to keep the absorbed dose below 10 mrads (i.e., 0.01 rads), what should be the limit on the exposure?

Answer   From Table 4.6, we find that at 20 keV,

$$(\mu/\rho)_{\text{air}} = 0.78 \text{ cm}^2/\text{g},$$

and

$$(\mu/\rho)_{\text{lung}} = 0.83 \text{ cm}^2/\text{g}.$$

TABLE 4.6

X-Ray Mass Attenuation Coefficients for Some Materials		
Material	20 keV (cm²/g)	100 keV (cm²/g)
Air	0.7779	0.1541
Water	0.8096	0.1707
Bone	0.4001	0.1855
Muscle	0.8205	0.1693
Lung	0.8316	0.1695
Brain	0.8281	0.1701

Source: Hubbell and Seltzer, NIST online tables.

So the f-factor is

$$f = 0.87 \frac{(\mu/\rho)_{\text{lung}}}{(\mu/\rho)_{\text{air}}} = 0.93.$$

The absorbed dose is related to the exposure by (4.35). If we want to limit the dose to be under 10 mrads, the exposure must be under $10/0.93 = 10.8$ mR. ∎

4.6.6 Effective Dose

For the purpose of relating dose of ionizing radiation to risk, an extension of the dose equivalent is used to express dose as that which would have been received if the whole body had been irradiated uniformly. The *effective dose* is obtained as the sum of dose equivalents to different organs or body tissues weighted in such a fashion as to provide a value proportional to radiation-induced somatic and genetic risk even when the body is not uniformly irradiated. This can be expressed as

$$D_{\text{effective}} = \sum_{\text{organs}} H_j w_j \qquad (4.38)$$

where $D_{\text{effective}}$ is effective dose, H_j is the dose equivalent for organ j, and w_j is the weighting factor for organ j. The sum is over all organs; note that

$$\sum_{\text{organs}} w_j = 1. \qquad (4.39)$$

In this fashion, risks may be compared for different radiations and different target tissues.

On the average, our annual effective dose is about 300 mrems. Medical imaging procedures produce doses over a wide range, depending on the specific procedure. At the low end, a typical chest x-ray might result in a dose of 10 mrem; at the high end, a fluoroscopic study might yield several rem.

The main risk from ionizing radiation at the doses involved in medical imaging is cancer production (*radiogenic carcinogenesis*). In practice, we assume that any dose of radiation increases the risk of getting cancer, with larger doses proportionately increasing risk. Of importance, the physician and patient together usually make the decision that the medical benefits of the imaging procedure outweigh any potential risks.

4.7 Summary and Key Concepts

Ionizing radiation is used in several imaging modalities, including projection radiography and CT. The interaction of particulate radiation creates x-rays used in these modalities; the interaction of the x-rays in the patient creates the radiograph. In this chapter, we presented the following key concepts that you should now understand:

1. *Ionization* is the ejection of an orbiting electron from an atom; *ionizing radiation* has sufficient energy to produce ionization.

2. Ionizing radiation may be *particulate* or *electromagnetic*; the main ionizing radiations of interest in medical imaging are *x-rays, gamma rays, energetic electrons*, and *positrons*.

3. Particulate ionizing radiation transfers energy via *collisional transfer* and *radiative transfer* (which results in *bremsstrahlung x-rays*).

4. The *probability of radiative transfer* increases with increasing effective atomic number of the material through which the particulate radiation passes.

5. Electromagnetic ionizing radiation transfers energy in medical imaging applications via the *photoelectric effect* or *Compton scattering*.

6. The *probability of electromagnetic interactions* depends on the photon's energy, and the effective atomic number, density, and molecular weight of the material through which the electromagnetic radiation passes.

7. *Attenuation* of the intensity of a beam of electromagnetic radiation by some material is described by a monoexponential relation, and is a function of the thickness of the material and the *linear attenuation coefficient*, which itself depends on characteristics of the material and photon energy.

8. *Radiation dose* to tissues is ultimately characterized by *effective dose*, which itself includes considerations of deposited energy, the biological effectiveness of a given ionizing radiation, and the relative radiosensitivity of different tissues.

Bibliography

Attix, F. H. *Introduction to Radiological Physics and Radiation Dosimetry*. New York: Wiley, 1986.

Bushberg, J. T., Seibert, J. A., Leidholdt Jr., E. M., and Boone, J. M. *The Essential Physics of Medical Imaging*, 2nd ed., Philidelphia, PA: Lippincott Williams & Wilkins, 2002.

Carlton, R. R., and Adler, A. M. *Principles of Radiographic Imaging: An Art and a Science*, 3rd ed. Albany, NY: Delmar, 2001.

Hubbell, J. H., and Seltzer, S. M. *Tables of X-Ray Mass Attenuation Coefficients and Mass Energy-Absorption Coefficients*, version 1.4. Online at http://physics.nist.gov/xaamdi, Gaithersburg, MD:

National Institute of Standards and Technology, 2004.

Johns, E. J., and Cunningham, J. R. *The Physics of Radiology*. 4th ed. Springfield, IL: Charles C Thomas Publisher, 1983.

Sprawls Jr., P. *Physical Principles of Medical Imaging*, Madison, WI: Medical Physics Publishing, 1995.

Wang, C. H., Willis, D. L., and Loveland, W. D. *Radiotracer Methodology in the Biological, Environmental, and Physical Sciences*. Englewood Cliffs, NJ: Prentice Hall, 1975.

Wolbarst, A. B. *Physics of Radiology*. Norwalk, CT: Appleton & Lange, 1993.

Problems

Physics of Atoms

4.1 The rest masses of a proton, neutron, and electron are $1.6726216 \times 10^{-27}$kg, $1.6749272 \times 10^{-27}$ kg, and 9.10938×10^{-31}kg, respectively.

(a) The *atomic mass unit* (amu) is defined as one-twelfth of the mass of a carbon-12 atom. The mass of a carbon-12 atom is less than the sum of the masses of its constituents; the difference is the mass defect. Calculate the mass defect (in kg or amu).

(b) The binding energy is the energy equivalent to the mass defect. What is the binding energy of carbon-12?

4.2 (a) Calculate the mass-equivalent energy of an electron at rest.

(b) Determine the voltage potential difference between a cathode and anode in which electrons accelerated between the two will obtain speeds equal to 1/10 the speed of light.

(c) Using relativistic equations, determine the speed of an electron that is accelerated across a 120 kV potential.

4.3 Show that when a particle's speed v is much smaller than the speed of light, its kinetic energy can be computed using the formula $\mathrm{KE} = mv^2/2$.

Ionizing Radiation

4.4 Compare characteristic radiation and bremsstrahlung radiation.

4.5 (a) Explain why radiation with energy smaller than 13.6 eV is not considered ionizing.

(b) What are the differences between ionization and excitation?

4.6 Ultraviolet light is defined as electromagnetic waves having wavelengths in the range of 4–400 nanometers.

(a) Determine the frequency range of ultraviolet light.

(b) Determine the photon energy range of ultraviolet light.

(c) Determine whether ultraviolet light is ionizing radiation or not.

4.7 (a) Explain why the electron density of hydrogen is very nearly 6×10^{26} electrons/kg.

(b) Explain why the electron density of all other low atomic number materials is nearly 3×10^{26} electrons/kg.

(c) Speculate why the electron density of materials might differ slightly from 3×10^{26} electrons/kg.

Attenuation of Electromagnetic Radiation

4.8 (a) Calculate the thickness of shielding material needed to block out 99.5 percent of the incident radiation for a material with linear attenuation coefficient μ.

(b) Range is defined as the reciprocal of μ, the linear attenuation coefficient. It represents the distance at which the intensity of a beam has been reduced to $1/e$ of its original value. What approximate order of magnitude would you want the range of an ionizing beam in tissue to be? Microns? Millimeters? Centimeters? Meters? Kilometers? Why? What approximation order of magnitude would you want the half-life of a radioactive tracer used in nuclear medicine to be? Milliseconds? Seconds? Minutes? Hours? Days? Weeks? Years? Why?

4.9 Prove Equation (4.24).

4.10 A bar phantom of thickness 0.4 cm is uniformly irradiated by monoenergetic x-ray photons, and a screen is placed behind the phantom to detect the x-ray photons. The bars of the phantom are made from a material that has a HVL of 0.1 cm. In the space between the bars, x-ray photons pass without attenuation. Assuming the intensity of the image is proportional to the x-ray photons hit the screen in a unit area, what is the contrast of the resulting image?

4.11 Compton scattering of photons causes image blurring since the secondary x-ray photon diverges from the primary beam. You are designing an x-ray detector system and wish to eliminate all photons that have scattered more than 25 degrees in an attempt to improve the resulting image quality. You are using a monoenergetic x-ray source that emits photons with wavelength 8.9×10^{-2} angstroms. Your detector is capable of discriminating the energy of the incoming photon by examining the pulse height. What range of photon energies will your system accept?

Radiation Dosimetry

4.12 Consider a chest x-ray given by a point source operating at an energy of 20 keV. The exposure at $d = 1$ cm is 10R. If we want to keep the dose equivalent to be under 10 mrems, how far should the patient be away from the x-ray source? (Ignore the tissues other than the lung, and ignore the size of the lung.)

4.13 Consider the case of hand x-ray imaging. For simplicity, assume that the hand consists of only bone and muscle, and the weighting factors for the bone and the muscle in the hand are $w_{bone} = w_{muscle} = 0.002$. What is the effective dose for an exposure of $X = 1$ roentgen for x-rays at 20 keV?

Projection Radiography

5.1   Introduction

Projection radiography, sometimes called *conventional radiography*, refers to the most commonly used method of medical imaging utilizing x-rays. A conventional *radiograph* represents a projection of the three-dimensional volume of the body onto a two-dimensional imaging surface. The term *projection* will be defined more rigorously shortly, but conceptually, the projection radiograph represents the transmission of the x-ray beam through the patient, weighted by the integrated loss of beam energy due to scattering and absorption in the body. The primary radiographic image is a two-dimensional display of these transmitted intensities. It is useful to think of a projection radiograph as a *shadow* cast by a semitransparent body illuminated by x-rays.

There are many types of projection radiography systems in common use. While their overall system designs have common elements, there are different clinical requirements, which make the details differ. Three examples, representative of the breadth of possibilities required by different clinical requirements, are shown in Figure 5.1.

Overall, the advantages of projection radiographic systems include short exposure time (0.1 second), the production of a large area image (e.g., 14×17 in.), low cost, low radiation exposure (30 mR for a chest radiograph, equivalent to about one-tenth of the annual background dose), and excellent contrast and spatial resolution. Projection radiography is used to screen for pneumonia, heart disease, lung disease, bone fractures, cancer, and vascular disease. Typically, the *chest x-ray* is the most common imaging examination performed in a hospital. Its major limitation is the lack of depth resolution—superimpositions of shadows from overlying and underlying tissues sometimes "hide" important lesions, which limits contrast.

In this chapter, we describe the components of standard projection radiography systems, develop mathematical models to characterize image formation, describe the factors affecting image quality, and look at several applications and special techniques for projection radiography systems (see again Figure I.2).

Figure 5.1
(a) A general projection
radiography system; (b) a
fluoroscopy system; and (c) a
mammography system.
(Used with permission of GE
Healthcare.)

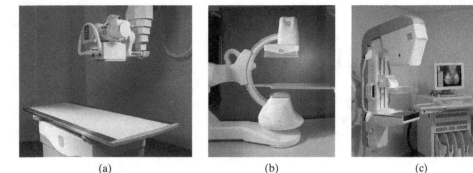

(a)   (b)   (c)

Figure 5.2
A conventional projection
radiographic system.

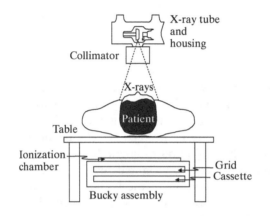

5.2 Instrumentation

A diagram of a conventional projection radiographic system is shown in Figure 5.2. The x-ray tube generates a short pulse of x-rays as a *beam* that travels through the patient. X-ray photons that are not absorbed within the patient or scattered outside the region of the detector impinge upon the large area detector, ultimately creating an image on a sheet of film (or alternatively within an electronic detector). We now examine each element of this imaging system in detail.

5.2.1 X-Ray Tubes

X-rays are used in projection radiography and computed tomography because they penetrate the body well and their wavelengths are small enough for high-resolution imaging. X-rays are generated using an *x-ray tube*, as shown in Figure 5.3. The operation of an x-ray tube is shown schematically in Figure 5.4. A current, typically 3 to 5 amperes at 6 to 12 volts, is passed through a thin thoriated tungsten wire, called the *filament*, contained within the *cathode assembly*. Electrical resistance causes the filament to heat up and discharge electrons in a cloud around the filament through a process called *thermionic emission*. These electrons are now available

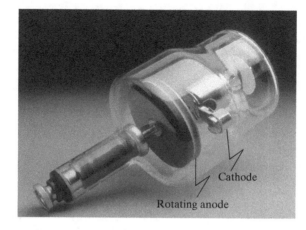

Figure 5.3
An x-ray tube.

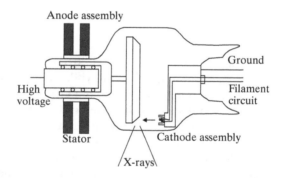

Figure 5.4
Schematic diagram of an
x-ray tube.

to flow (i.e., be accelerated) toward the anode when the anode voltage is applied, producing the *tube current*, which is referred to as the *mA*. The filament current directly controls the tube current, because the current controls filament heat, which in turn determines the number of discharged electrons. The x-ray control console is calibrated according to tube current, which typically ranges between 50 and 1,200 mA.

Once the filament current is applied, the x-ray tube is primed to produce x-rays. This is accomplished by applying a high voltage, the *tube voltage* or *kVp*, between the anode and cathode for a brief period of time. The tube voltage is typically generated by transforming the alternating current (AC) line voltage to a higher voltage and then rectifying this voltage. The abbreviation kVp refers to the *peak* kilo-voltage applied to the anode; the voltage ripple (temporal variation) below the peak value depends on the specific type of high-voltage generator in use. Typical values for the tube voltage lie in the range 30–150 kVp.

While the tube voltage is being applied, electrons within or near the cathode are accelerated toward the anode. The *focusing cup*, a small depression in the cathode containing the filament, is shaped to help focus the electron beam toward a particular spot on the anode. This target, or focal point, of the electron beam is a bevelled edge of the anode disk, which is coated with a rhenium-alloyed tungsten.

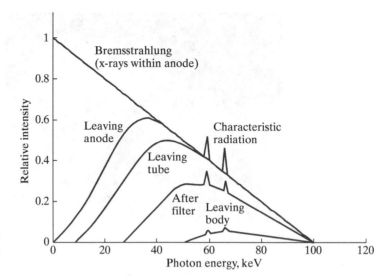

Figure 5.5
Relative intensity of x-ray
photons. (Adapted from
Webster, 1998. This material
is used by permission of John
Wiley & Sons, Inc.)

The anode disk itself is made from molybdenum, and in the case of x-ray tubes used in mammography, molybdenum is used in the target area as well. These energetic electrons bombard the target transferring energy by both collisional and radiative transfer. As discussed in Chapter 4, this results in both characteristic and bremsstrahlung x-rays. As can be seen in Figure 5.5, the vast majority of the x-rays produced by an x-ray tube are from bremsstrahlung.

Besides x-rays, heat is also produced at the anode by the bombardment of electrons. In fact, only about 1 percent of the energy deposited by the electron beam is turned into x-rays; the remaining 99 percent turns into heat. So, at the same time the filament current is applied, the anode is set into rotation by applying alternating current to the stator electromagnets located outside the tube's glass housing. The rotors of this induction motor are connected to the anode inside the glass housing, and arranged around bearings so that the anode can rotate freely. Anode rotation is necessary for nearly all diagnostic radiography equipment in order to avoid melting the anode target area because of the high-energy transfer from the accelerated electrons to the target. In most x-ray tubes, the anode rotates at 3,200 to 3,600 revolutions per minute (rpm), but higher rpm tubes are also available for even better heat dissipation.

In order to maintain a constant tube current (mA) while the tube voltage (kVp) is being applied, the filament circuit is very carefully controlled to keep a consistent electron cloud around the filament (i.e., a consistent amount of thermionic emission). The overall exposure is then determined by the duration of the applied kVp, which is controlled by either a fixed timer circuit or an automatic exposure control (AEC) timer. A fixed timer is generally a silicon-controlled rectifier (SCR) switch timed by a microprocessor. Timing accuracy for these circuits is approximately 0.001 seconds. AEC timers have 5 mm thick parallel-plate ionization chambers placed between the patient table and the film-screen cassette (see Figure 5.2). The voltage achieved across the plates is used to trigger the SCR, which shuts off the tube voltage and terminates the exposure.

The product of the tube current and the exposure time yields the milliampere-second (mAs) value for the exposure. When a fixed timer is used, the radiologist controls both the mA and the exposure time directly, and thereby determines the mAs for the exposure. In AEC timers, the mAs is set by the radiologist—typically using film density indicators—and the exposure time is determined automatically by the AEC circuitry. Maximum times are set to prevent accidental overdose in the event the AEC circuit malfunctions or the ionization chamber is missing or incorrectly positioned.

5.2.2 Filtration and Restriction

The bremsstrahlung x-rays that are generated within the anode do not all enter the patient, and not all that enter the patient end up leaving the patient. In this section, we discuss modifications to the x-ray beam that take place before the x-rays enter the body. *Filtration* is the process of absorbing low-energy x-ray photons before they enter the patient. *Restriction* is the process of absorbing the x-rays outside a certain field of view. We now discuss these processes in detail.

Filtration   The maximum energy of the emitted x-ray photons is determined by the tube voltage. For example, if the tube voltage is 100 kVp, then the maximum photon energy is 100 keV (recall the definition of eV). However, there is a *spectrum* of lower energy x-ray photons that represent bremsstrahlung, as shown in Figure 5.5. The top line depicts the theoretical spectral distribution of bremsstrahlung radiation (normalized to 0.5 at 50 keV). The intensity "spikes" also shown in Figure 5.5 are *characteristic radiations*, and they are due to electrons making transitions between specific electron shells in the anode atoms (as discussed in Chapter 4). Because the x-ray photons emitted from an x-ray tube have a distribution of energies, the x-ray sources used in medical imaging systems are *polyenergetic*.

In practice, it is very undesirable for low-energy photons to enter the body, as these photons are almost entirely absorbed within the body, thus contributing to patient radiation dose but not to the image. There are three important filtering processes in a projection x-ray machine, which reduce the number of low energy x-ray photons that enter the body (see Figure 5.5). First, the tungsten anode itself absorbs a large fraction of low-energy x-ray photons before they even leave the anode. Second, the glass housing of the x-ray tube and the dielectric oil that surrounds it filter out more low-energy photons. This effect might be accentuated over time, since aging x-ray tubes tend to accumulate a tungsten film on the inside of the housing due to vaporization of the filament during repeated heating. These first two processes are referred to as *inherent filtering*, since they are caused by the x-ray tube itself.

The third x-ray filtering process is called *added filtering*, since it arises from metal placed in the x-ray beam outside of the tube. The most common material used to filter x-rays is aluminum (1–2 mm thick), which is considered the standard x-ray filter material. Other filter materials are often rated in terms of the equivalent aluminum (Al/Eq) that would have to be used to achieve the same attenuation. For higher-energy systems, copper might be used because it yields more attenuation for an equivalent actual thickness of aluminum; but copper must be followed by aluminum in order to attenuate the 8 keV characteristic x-ray photons created within the copper. Another filtering effect comes from the silvered mirror placed within the collimator (as discussed in the next section). This might provide an additional 1.0 mm Al/Eq of filtering.

Figure 5.5 shows a progressive shift in the position of the spectrum "to the right" (i.e., to higher average energies) as the beam passes through successive materials. This increase in the beam's "effective energy" (see Example 4.4) is called *beam hardening*. Beam hardening is caused by the preferential absorption of lower energy photons, for which the attenuation is higher in most materials (see Figure 4.8). These materials include the filters discussed here, as well as the tissues of the body, which further "harden" the beam.

EXAMPLE 5.1

For radiography systems operating above 70 kVp, the National Council on Radiation Protection and Measurements (NCRP) recommends a minimum total filtration of 2.5 mm Al/Eq at the exit port of the x-ray tube. Although such filtration reduces high-energy as well as low-energy x-rays, thus requiring longer exposure times to properly expose the x-ray film, the overall dose to the patient is reduced because of the reduction in low-energy x-rays that are absorbed almost entirely by the patient.

Question   At 80 kVp, what thickness of copper would provide 2.5 mm Al/Eq of filtration?

Answer   The mass attenuation coefficient of aluminum at 80 kVp is $\mu/\rho = 0.02015$ m^2/kg. The density of aluminum is $\rho = 2699$ kg/m^3. Therefore,

$$\mu(Al) = 0.02015 \text{ m}^2/\text{kg} \times 2699 \text{ kg/m}^3$$

$$= 54.38 \text{ m}^{-1}$$

For copper at 80 kVp: $\mu/\rho = 0.07519$ m^2/kg, $\rho = 8960$ kg/m^3. So,

$$\mu(Cu) = 0.07519 \text{ m}^2/\text{kg} \times 8960 \text{ kg/m}^3$$

$$= 673.7 \text{ m}^{-1}$$

Since attenuation is determined by the exponential $e^{-\mu\Delta x}$, the x-ray attenuation is equal if the exponents are equal. Hence, the following relation must be satisfied:

$$\mu(Al)x(Al) = \mu(Cu)x(Cu).$$

The copper thickness equivalent to 2.5 mm of aluminum at 80 kVp is therefore given by

$$x(Cu) = \frac{54.38 \text{ m}^{-1} \times 2.5 \text{ mm}}{673.7 \text{ m}^{-1}}$$

$$= 0.2 \text{ mm}$$

(Remember that, as stated above, copper must be followed by a thin layer of aluminum (typically 1 mm thick) to stop the 8 keV characteristic photons generated by the copper.)   ∎

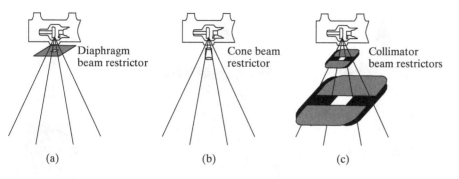

Figure 5.6
Different beam restrictors.

Restriction Beam   X-ray tubes generate x-rays in all directions. Many photons are absorbed by the anode, others by the x-ray housing itself. By design, the x-rays that exit through the tube window form a cone that is ordinarily much larger than the desired body region to be imaged. The exiting beam must be further restricted both to avoid exposing parts of the patient that need not be imaged and to help reduce the deleterious effects of Compton scatter (see below). *Beam restriction* is used for this purpose.

There are three basic kinds of beam restrictors: diaphragms, cones or cylinders, and collimators. *Diaphragms* are flat pieces of lead with holes cut into them, centered on the x-ray beam, and usually placed close to the tube window, as shown in Figure 5.6(a). They are simple and inexpensive, but result in a fixed geometry that can usually be used only in dedicated systems that have only one purpose (such as chest radiography). *Cones* or *cylinders*, as shown in Figure 5.6(b), are fixed in their geometry but can have somewhat better performance, as will be discussed below. *Collimators* are more expensive, but are so much more flexible and better performing that they are used nearly all the time in projection x-ray systems. As shown in Figure 5.6(c), collimators have variable diaphragms comprised of movable pieces of lead. Most often, there are two collimators, one near the tube and one farther away from the tube. Typically, there is a scored mirror in between these two collimators so that a light coming from the side will shine through the second collimator, illuminating the field of view with an alignment grid.

5.2.3 Compensation Filters and Contrast Agents

Attenuation is the process by which x-rays are absorbed or redirected (scattered) within the body or other objects in the field of view. Body tissues attenuate x-rays in different amounts depending on their linear attenuation coefficients and the x-ray energies. It is this differential attenuation that gives rise to contrast—and therefore the ability to differentiate tissues—in an x-ray image. In certain circumstances, we may want to artificially change the natural attenuation of the body prior to detecting the x-rays. We now discuss compensation filters and contrast agents, both used for this purpose.

Compensation Filters   Thick body parts or dense materials (such as metal or bone) stop more x-rays than thinner body parts and normal soft tissues. For example, the

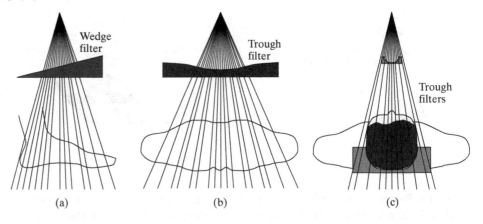

Figure 5.7
Various compensation filters.
(Adapted from Carlton and
Adler, 2001, and Wolbarst,
1993.)

torso attenuates more x-rays in the middle than at its edges, simply because it is
thicker in the middle. Imaging both locations—i.e., a location having large numbers
of transmitted photons and a location having low numbers of photons—within the
same exposure is difficult because of the limited dynamic range of x-ray detectors.
One region would be overexposed while the other is underexposed, and neither
would yield a usable diagnostic image.

In these cases, a *compensation filter*, comprising a specially shaped aluminum
or leaded-plastic object, can be placed between the x-ray source and the patient, or
in some cases between the patient and the detector. Several types of compensation
filters are shown in Figure 5.7. Notice that the compensation filter is thicker where
the body part is thinner and vice versa, so that the x-ray detector requires a smaller
dynamic range.

Contrast Agents Differential attenuation in the body gives rise to contrast in the
x-ray image. Put another way, in order to see the difference between anatomical
structures on an x-ray film, these structures must have different x-ray attenuations.
Often, however, different soft tissue structures are difficult to visualize due to
insufficient intrinsic contrast. This situation can be improved by using *contrast
agents*—chemical compounds that are introduced into the body in order to increase
x-ray absorption within the anatomical regions into which they are introduced,
thereby enhancing x-ray contrast (compared with neighboring regions without
such agents).

The two most common contrast agents used in x-ray diagnosis are iodine
($Z = 53$) and barium ($Z = 56$). In addition to their relatively high atomic numbers,
iodine and barium also have K-shell electrons whose binding energy falls within
the diagnostic x-ray energy range; $E_k = 33.2$ keV for iodine and $E_k = 37.4$ keV for
barium. As we have already discussed, when the energy of x-ray photons slightly
exceeds the K-shell binding energy of a material, the probability of photoelectric
interaction significantly rises. Such a photoelectric interaction will cause electrons
from the K-shell to be ejected, and the x-ray photons will be completely absorbed.
This effect, called *K-edge absorption*, significantly increases the attenuation coefficient of the material in x-ray energies slightly higher than the K-shell energy. The

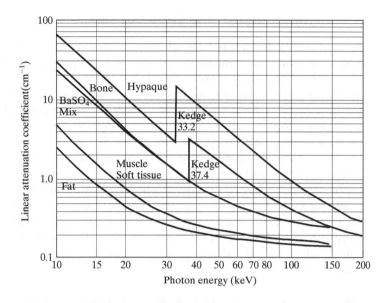

Figure 5.8
Linear attenuation
coefficients of bone, muscle,
fat, and two contrast agents.
(From Johns and
Cunningham, 1983.)

linear attenuation coefficients of bone, muscle, fat, and two contrast agents, one made using iodine and one using barium, are shown in Figure 5.8. This stepwise increase in attenuation apparent in these plots is called the *K absorption edge*; the location of these edges on the energy scale is directly related to the energies of the characteristic radiations of the contrast agents themselves. There is a dramatically higher absorption of x-rays that are slightly above the K-edge, leading to very high differential absorption between the contrast agent and the surrounding tissues. These physical properties make iodine and barium radiologically ideal compounds for enhancing contrast.

Because iodine is naturally abundant in the thyroid gland, the thyroid itself is readily imaged in x-ray studies. Fortunately, iodine can also be synthesized into soluble compounds that are safely introduced into the body through intravascular injection or ingestion. Iodinated contrast compounds are used for enhancing the appearance of blood vessels, heart chambers, tumors, infections, and other abnormalities on x-ray images (projection radiography as well as CT scans). Because these iodine compounds are concentrated by the kidney and excreted into the urine, they are also used for diagnosing problems of the kidneys and bladder using x-ray imaging.

Barium is used exclusively to enhance contrast within the gastrointestinal tract. It is typically administered as a chalky "milkshake," which passes through the gastrointestinal tract without being absorbed or altering the basic function of the stomach and bowel. Barium is the standard agent for enhancing the contrast of the stomach, small and large bowel in conjunction with projection radiography studies.

Air itself can also be used as a contrast agent. Because it does not readily absorb x-rays, it is used as an "opposite" type of contrast to that of iodine and barium. For example, by inflating the lungs, air provides contrast for lung tissue. Air is also used with barium to provide a double contrast between regions of the bowel containing barium and those containing air.

5.2.4 Grids, Airgaps, and Scanning Slits

X-ray photons that are not absorbed or scattered arrive at the detector from along a line segment originating at the x-ray source (anode of the x-ray tube). If an x-ray photon is scattered, however, and still manages to hit the detector, its point of origin will not intersect the x-ray source (unless a highly unlikely multiple scattering sequence occurs). Since scattering is a random phenomenon, this process will cause a random "fog" throughout the image if left uncorrected, thereby reducing the contrast of the direct image (see Section 5.4.3).

In nuclear medicine, where monoenergetic radiation is used, energy-sensitive detectors can be used to discriminate between the primary radiation and the scattered radiation, which has lower-energy photons. This concept cannot be employed in conventional radiography, however, because the radiation is polyenergetic and because an integrated measurement of beam intensity is made. Instead, there are three other methods used in conventional radiography to reduce the effects of scatter: (1) grids, (2) airgaps, and (3) scanning slits. We now describe these in more detail.

Grids Scatter-reducing *grids* use thin strips of lead alternating with highly transmissive interspace material, typically aluminum and sometimes plastic. A typical x-ray grid is shown in Figure 5.9. This so-called *linear, focused grid* has lead strips that are arranged in lines and angled toward the x-ray tube. By careful placement of the grid, photons that travel from the source to the detector pass through the interspace material, and are not severely attenuated by the grid. Photons that arise from Compton scattering within the patient, however, travel off-axis, intercept the lead strips, and are absorbed in the grid.

Other combinations of grid geometries are sometimes used as well. For example, "crosshatch" grids have lead strips arranged in an actual grid pattern (as in the original design of Gustav Bucky) and "parallel" grids have lead strips that are parallel to each other (so that the focal radius is infinite). Linear grids, however, have the advantage that the x-ray beam can be angulated along the grid line direction, allowing adjustments of the viewing angle. Furthermore, focused grids have the advantage that very few x-rays are actually blocked by the geometry of the grid.

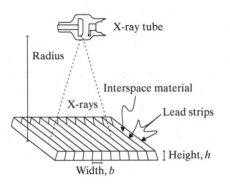

Figure 5.9
A typical x-ray grid.

The effectiveness of the grid for reducing scatter is a function of the *grid ratio*, defined as

$$\text{grid ratio} = \frac{h}{b}, \tag{5.1}$$

where h is the height of the lead strips and b is the spacing between the lead strips (see Figure 5.9). This fraction is generally reported as a normalized ratio in which the denominator is unity. Grid ratios range from 6:1 to 16:1, are used in conventional radiographic systems, and can be as low as 2:1 in mammography systems. Grid spacing is generally reported using its reciprocal, which is known as *grid frequency*, which ranges from 60 cm^{-1} for conventional radiographic systems to as much as 80 cm^{-1} for mammography systems.

Grids that have tall lead strips or fine lead strip spacing—i.e., higher grid ratios—are more capable of stopping off-axis radiation. However, there is a price to be paid for using grids. As the lead strips become wider or are packed more closely in an effort to stop more scattered photons from reaching the detector, they will also stop some photons in the primary beam. To maintain a high-quality image, it then becomes necessary to increase the amount of radiation arriving at the grid, which increases the radiation dose to the patient. The *grid conversion factor* (GCF) characterizes the amount of additional exposure required for a particular grid. GCF is defined as

$$\text{GCF} = \frac{\text{mAs with the grid}}{\text{mAs without the grid}}. \tag{5.2}$$

Typical GCF values range from 3 to 8.

Having GCFs in the range 3–8 implies that some grids can absorb a large fraction of the incident radiation, so the impact of a grid should not be ignored. Since Compton scattering events are more isotropic at lower energies, it may not be necessary to use a grid at lower energies. A rule of thumb in projection radiography is to use a grid when the tube voltage is above 60 kVp. Compton scattering also increases with the thickness of the body part. Another rule of thumb in projection radiography is to use a grid when imaging a body part thicker than 10 cm.

Grids can be mounted as *stationary* grids or used with a *Potter-Bucky diaphragm*, which moves the grid during exposure. The reason that Potter-Bucky grids are usually desirable is that stationary grids—especially those with low grid frequency—can introduce visible artifacts—lines or grid lines—into the image. The Potter-Bucky diaphragm causes the grid to move 2 to 3 cm during exposure in a linear or circular path. Since the lead strips within the grid move during exposure, their x-ray shadow also moves across the image plane, and its image is blurred out.

Grids are used for nearly all x-ray film/screen examinations. The type of grid used varies with the examination. For example, thin body parts (e.g., extremities) generate less scatter than thick body parts (e.g., abdomen or chest), so a lower grid ratio might be employed for extremity exams than for abdominal exams.

Airgaps Physical separation of the detector and the object—i.e., leaving an *airgap* between the patient and the detector —is an effective means of scatter rejection. Because the scattered photons diverge radially from their point of origin inside the object, while the primary beam photons diverge from the source, separating the source and detector reduces the percentage of scattered photons that reach the plane of the detector. This can be demonstrated by simple geometry. The price that is paid for this airgap method of scatter reduction is increased geometric magnification and increased blurring or unsharpness due to x-ray focal spot size effects (see Section 5.4).

Scanning Slits The use of mechanical lead slits that are placed in front of and/or in back of the patient has been evaluated, and appear in some systems. These slits are moved together during the x-ray exposure, effectively providing a linear "scan" of the patient. Because the x-ray beam is collimated to a thin line, the amount of scatter generated is small, and because the detector is also collimated to a thin line, the amount of scatter accepted is small. Hence, these systems can provide greater than 95% scatter reduction, at the expense of a more complex and costly system, and longer exposure times.

5.2.5 Film-Screen Detectors

In 1895, Roentgen discovered x-rays and made the first radiograph by allowing the x-rays to directly expose a photographic plate. X-rays can directly expose today's modern photographic film, but this is a very inefficient way to create a radiograph. In fact, only about 1 to 2 percent of the x-rays are stopped by the film, so creating radiographs by direct film exposure would require an unnecessarily large x-ray dose to the patient.

   To greatly improve their efficiency, modern diagnostic x-ray units always have *intensifying screens* on both sides of the radiographic film. The intensifying screens stop most of the x-rays and convert them to light, which then exposes the film. This is a very efficient process, and the screens cause only a small amount of additional image blurring.

   In this section, we first describe the process of luminescence, and the composition and dimensions of modern intensifying screens. We then describe the optical properties of radiographic film. Finally, we conclude with a discussion of the radiographic cassette, which holds the intensifying screens and the film.

Intensifying Screens A cross-section of a typical radiographic intensifying screen is shown in Figure 5.10. Since screens are used on both the front and back of the radiographic film, all parts of the screen except the phosphor must be uniformly radiolucent. The phosphor is the active part of an intensifying screen; its purpose is to transform x-ray photons into light photons. The light photons then travel into the film, causing it to be exposed and to form a latent image. The *latent image* is the "virtual" image resident in the film, but it is not yet viewable by an observer.

   The base of the screen is provided for mechanical stability, but it must be somewhat flexible so that it can be pushed tightly against the film. It is typically

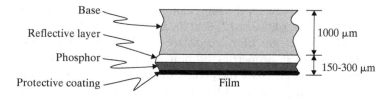

Figure 5.10
Geometry of an intensifying screen.

made of polyester plastic. The reflective layer is provided so that light from the phosphor is reflected back into the film rather than getting lost in the base. It is typically about 25 μm thick and made of magnesium oxide or titanium dioxide. A plastic protective coating is applied to the film side of the screen to protect it from repeated film loading and unloading.

Phosphors are examples of materials that are *luminescent*; that is, they convert one form of energy, in this case x-rays, into light. Two types of luminescence are distinguished: *fluorescence*, in which the emission of light takes place entirely within 1×10^{-8} second of the excitation, and *phosphorescence*, in which light emission can be delayed and extended over a longer period of time. For screens, it is desirable to use a luminescent material that is much more fluorescent than phosphorescent. This way, there is little chance of an "afterglow" that might spoil the exposure by either motion after exposure or by light from a previous exposure.

Among the many luminescent materials, good intensifying screen phosphors should also be highly x-ray attenuating and should emit many light photons for every x-ray photon that is stopped. Thus, the best phosphors have high atomic numbers (so the linear attenuation coefficient is large) and high conversion efficiencies. The *conversion efficiency* is a measure of the number of light photons emitted per incident x-ray photon. Typical conversion efficiencies are between 5 and 20 percent depending on the type of phosphor used and its thickness.

A typical conversion efficiency for modern screens yields 1×10^3 light photons per incident 50 keV x-ray photon. This number depends on the energy of the incident x-ray photon as well, since the conversion of a higher energy x-ray photon will produce more light. Total light output depends on a combination of the attenuation of the material at a given x-ray energy and the conversion efficiency at that energy. The so-called *speed* of a screen is really just a measure of its conversion efficiency. If the conversion efficiency is higher, then the screen is "faster," because the larger numbers of light photons emitted by the phosphor will expose the film faster.

Thomas Edison explored many phosphors in the early 1900s for their possible use in radiography. He discovered that calcium tungstate ($CaWO_4$) is an excellent fluorescent phosphor for radiological applications. Until relatively recently, nearly all intensifying screens were made from calcium tungstate. In the late 1970s, rare earth phosphors were introduced and have revolutionized the diagnostic x-ray industry.

Radiographic Film After Roentgen discovered x-rays by observation of a mysteriously glowing phosphor, he quickly learned that x-rays could also expose a photographic emulsion. Today, film is still the most commonly used image receptor in radiology. Because of the nearly exclusive use of scintillation screens, radiographic film is really just optical film, designed to capture the optical image created within the screens that sandwich the film. Common sizes in the United States include 14 × 17, 14 × 14, 11 × 14, 10 × 12, 8 × 10, and 7 × 17 inches.

Many different film speeds are available, depending on the application. The detailed optical properties of film greatly influence the resultant quality of the developed image. Also, the precise details of chemical development play a large role in the final appearance of the film as well. We will only consider these issues from the highest level, in enough detail so that we can understand the broad issues relating x-ray exposure to film appearance (see Section 5.3.4).

Radiographic Cassette   A radiographic cassette, as shown in Figure 5.11, is really just a holder for two intensifying screens and the film "sandwiched" between. One side of the cassette is radiolucent, while the other usually includes a sheet of lead foil. (Therefore, the cassette can be loaded only one way into the x-ray machine.) At least one side of the interior of the cassette contains a spongy, foam material that applies a uniform pressure against the screen so that a uniform contact is created between the film and all parts of each of the screens. Optical mirrors are located outside the screens, pointed inward so that nearly all the light produced in the screens ends up exposing the film.

Film must be loaded in a darkroom in order to avoid premature exposure. The cassette is opened like a suitcase or briefcase only as much as is required to slip in the radiographic film. This minimizes the amount of dust and other contaminants that might end up inside the cassette. When the cassette is closed, pressure is applied so that the foam inside compresses, creating a uniform contact between the screens and the film. Cassettes and screens must be cleaned regularly in order to avoid contaminants that might appear in the images.

5.2.6 X-Ray Image Intensifiers

X-ray image intensifiers (XRIIs) are used in fluoroscopy, where low-dose, real-time projection radiography is required. A diagram depicting an XRII is given in Figure 5.12.

X-rays pass through an *input window* made of aluminum or titanium, between 0.25 and 0.5 mm thick. This type of material creates minimal loss of x-ray photons, yet it is capable of holding a vacuum. The x-rays then strike a 0.5 mm thick *input phosphor* [typically CsI(Na)], which is circular with a diameter ranging between 15

Figure 5.11
A radiographic cassette.
(Courtesy of American
X-Ray & Medical Supply.)

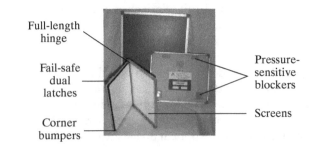

Full-length
hinge

Fail-safe
dual
latches

Corner
bumpers

Pressure-
sensitive
blockers

Screens

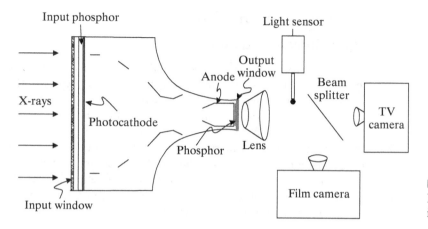

Figure 5.12
Block diagram of an x-ray image intensifier (XRII).

and 40 cm. The x-ray photons that are absorbed in the input phosphor generate flashes of light that are channelled toward the photocathode. Photons that travel backward toward the input window are reflected by a 0.5 mm thick aluminum sheet that supports the input phosphor.

The remainder of the XRII resembles a standard image intensifier, such as that found in night vision devices. The light photons generated in the input phosphor strike a photocathode, which in turn generates free electrons within the vacuum tube. The electrons are accelerated through a series of electrodes (called *dynodes*) toward the anode, which is kept at a voltage of 25 to 35 kV relative to the cathode. The free electrons do not strike the dynodes as in photomultiplier tubes (PMTs; see Chapter 8), but they are instead shaped by them into an (inverted) electron intensity image that represents the intensity of x-rays entering the XRII. The voltage profile of the dynodes can be altered—continuously in some systems and in discrete steps in other systems—to provide variable image magnification. For example, a 40 cm XRII could be used to image a 20 cm field of view.

The electrons accelerate through the XRII until they are absorbed by either the anode or the *output phosphor screen*. The output phosphor is a P20-type phosphor deposited on the *output window*, typically a 15 mm piece of glass 25 to 35 mm in diameter. A thin aluminum film placed on the inner side of the phosphor acts as the anode and a reflector, keeping light from passing back into the XRII, which might otherwise cause secondary excitation of the photocathode. The light passing through the output window typically encounters a *lens*, which magnifies the image.

The magnified light image leaving the XRII can be used in several ways, three of which are depicted in Figure 5.12. First, a light sensor can be put into the field to provide an automatic gain control—i.e., feedback to the x-ray tube current—to provide a relatively constant image brightness throughout an exam. Second, the main function of a fluoroscope is achieved by feeding the light image into a TV camera, which provides real-time viewing capability by sending the camera's signal into a standard TV monitor. Finally, using a splitter, the image can also be fed into a film camera, which can capture selected still images onto film.

5.3   Image Formation

5.3.1 Basic Imaging Equation

The x-ray tube emits a burst of x-rays that, after filtration and restriction, are incident upon the patient. These x-rays are then attenuated as they pass through the body in a spatial pattern that depends on the linear attenuation coefficient distribution in the body. Consider a particular line segment through the object starting at the x-ray origin and ending on the detector plane at point (x, y), as shown in Figure 5.13. The linear attenuation is a function of x, y, and z, in general; on the line, it can be considered to be a function of its distance s from the origin. Suppose that the length of the line segment is $r = r(x, y)$, where it is made explicit that the length of the line depends on the position (x, y). Then, referring to (4.34), the intensity of x-rays incident on the detector at (x, y) is given by

$$I(x, y) = \int_0^{E_{\max}} S_0(E')E' \exp\left\{-\int_0^{r(x,y)} \mu(s; E', x, y)ds\right\} dE', \qquad (5.3)$$

where $S_0(E)$ is the spectrum of the incident x-rays. Given a different line, the linear attenuation function is usually different, since it passes through different materials; that is why μ is made explicitly dependent on x and y here. Therefore, if we were to actually evaluate this double integral, we would have to first determine the function $\mu(s; E, x, y)$ for the particular line of interest, and then proceed with the integration. Any photon "surviving" its passage through the body then hits the radiographic cassette, and either it is converted to a burst of light (thousands to ten thousands of photons) in one of the screens, it is absorbed within the film itself through photoelectric absorption, it is absorbed in the lead film on the back of the cassette, or it exits the cassette.

5.3.2 Geometric Effects

X-ray images are created from a diverging beam of x-rays (see again Figure 5.13); this divergence produces a number of undesirable effects that arise from geometric considerations. These effects are multiplicative; we now study them in sequence.

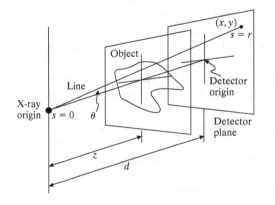

Figure 5.13
The geometry of a conventional projection radiographic system.

Inverse Square Law The *inverse square law* states that the net flux of photons (i.e., photons per unit area) decreases as $1/r^2$, where r is the distance from the x-ray origin. Assume that the beam intensity integrated over a small sphere surrounding the source is given by I_S. Let the source-to-detector distance be d, as shown in Figure 5.13. Then, assuming there is no object causing beam attenuation between the source and detector, the intensity at the origin of the detector—i.e., where $x = 0$, $y = 0$—is

$$I_0 = \frac{I_S}{4\pi d^2} .$$

(5.4)

The intensity at an arbitrary point (x, y) on the detector is smaller than that at the detector origin simply because it is farther away from the x-ray source. Let $r = r(x, y)$ be the distance between the x-ray origin and the detector point (x, y). Then the intensity at (x, y) is

$$I_r = \frac{I_S}{4\pi r^2} ,$$

(5.5)

where, again, we have assumed that there is no object attenuation. These relations, and the fact that $\cos\theta = d/r$ (see Figure 5.13), yield

$$I_r = I_0 \frac{d^2}{r^2} = I_0 \cos^2\theta .$$

(5.6)

Thus, the inverse square law causes a $\cos^2\theta$ drop-off of x-ray intensity away from the detector origin, even without object attenuation. Without compensation, this effect could be falsely interpreted as object attenuation in a circular pattern around the detector origin.

EXAMPLE 5.2

The inverse square law has a very practical use in radiography. Suppose an acceptable chest radiograph was taken using 30 mAs at 80 kVp from 1 m. Suppose that it was now requested that one be taken at 1.5 m at 80 kVp.

Question What mAs setting should be used to yield the same exposure?

Answer The intensity at the detector should remain constant in order to have the same exposure. We determine that

$$I_{old} = \frac{I_S(old)}{4\pi d_{old}^2} = I_{new} = \frac{I_S(new)}{4\pi d_{new}^2}$$

or

$$I_S(new) = I_S(old)\frac{d_{new}^2}{d_{old}^2},$$

where I_S refers to the intensity at the x-ray source. X-ray intensity at the source is directly proportional to mAs, the product of the tube current and the exposure time. Therefore,

$$\text{mAs(new)} = \text{mAs(old)} \frac{d^2_{\text{new}}}{d^2_{\text{old}}} \tag{5.7}$$

$$= 30 \text{ mAs} \times \frac{(1.5 \text{ m})^2}{1 \text{ m}^2}$$

$$= 67.5 \text{ mAs}.$$

Equation (5.7) is called the *density maintenance formula*; it is of significant practical utility in maintaining equivalent x-ray exposures while varying patient distance. ∎

Obliquity   *Obliquity* is a second factor that acts to decrease the beam intensity away from the detector origin. The obliquity effect is caused by the detector surface not being orthogonal to the direction of x-ray propagation (except at the detector origin). As shown in Figure 5.14, this fact implies that x-rays passing through a unit area orthogonal to the direction of x-ray propagation actually pass through a larger area on the detector. Thus, the x-ray flux is lower, which directly results in a lower measured x-ray intensity on the detector surface.

Given an area A orthogonal to the direction of x-ray propagation, the projected area on the detector is $A_d = A/\cos\theta$. Therefore, the measured intensity due to obliquity alone is

$$I_d = I_0 \cos\theta, \tag{5.8}$$

where we have again assumed that there is no object attenuation.

Beam Divergence and Flat Detector   The effects of beam divergence and the flat detector act together to reduce intensity at the detector plane in two ways: (1) reduction in beam intensity due to the inverse square law effect [(5.6)], and (2)

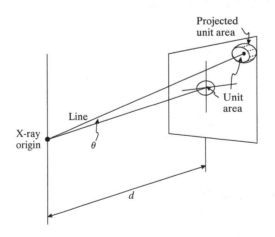

Figure 5.14
The effect of obliquity on spot size.

reduction in beam intensity due to obliquity [(5.8)]. The combination of these two effects is multiplicative. So, the overall beam intensity relative to the intensity I_0 at the detector origin, assuming no object attenuation, is given by

$$I_d(x, y) = I_0 \cos^3 \theta . \tag{5.9}$$

If θ is small, then $\cos^3 \theta \approx 1$, and both effects can be ignored. This is a good approximation for most projection radiography systems, where either the tube is very far away from the detector or the detector is small enough that θ is small even at the edges of the detector.

EXAMPLE 5.3

Suppose a chest x-ray is taken at 2 yards using 14 inch by 17 inch film.

Question What will be the smallest ratio I_d/I_0 across the film (assuming no object attenuation)?

Answer The corner of the film is distance r_d away from the detector origin, where

$$r_d = \sqrt{7^2 + 8.5^2} = 11.0 \text{ in.}$$

The cosine of the largest angle θ made by the beam is therefore

$$\cos \theta = \frac{d}{\sqrt{d^2 + r_d^2}}$$

$$= \frac{72}{\sqrt{72^2 + 11^2}}$$

$$= 0.989 .$$

The smallest intensity ratio is therefore

$$\frac{I_d}{I_0} = \cos^3 \theta = 0.966 ,$$

which represents about a 3 percent variation. Given the other effects we will be exploring—anode heel effect, path length, depth-dependent magnification, noise, and scattering—this effect is a negligible intensity perturbation across the film. ∎

Anode Heel Effect We have heretofore modeled the beam coming from an x-ray tube as having uniform intensity within a cone. The *anode heel effect*, however, predicts a stronger intensity beam in the cathode direction and no variation orthogonal to the cathode-anode direction. The reason for this effect is due to the geometry of the anode, as shown in Figure 5.4. Although the generation of x-rays within the anode is essentially isotropic at the atomic level, those x-rays traveling out of the anode in the forward (cathode-to-anode) direction have more anode material through which to propagate before leaving the anode itself than those leaving the anode in the backward direction. Thus, the beam intensity is higher in the cathode direction; the variation can be as much as 45 percent intensity variation in the cathode to anode direction. Thus, the anode heel effect far outweighs the effects of obliquity and inverse square law in its overall effect on the uniformity of intensity across the detector surface.

The anode heel effect can be compensated by using an x-ray filter that is thicker in the cathode direction than in the anode direction. If it is not compensated by filtration, then it must be recognized when positioning patients relative to the x-ray tube. In particular, the heel effect should be used as a kind of inherent compensation for natural gradients in the body's x-ray attenuation profile. For example, when imaging the foot, the toes should be positioned toward the anode since that part of the foot is thinner; the cathode end will have more x-rays to penetrate the thicker (proximal) part of the foot. This principle can be applied throughout the body. In our development of x-ray imaging equations, we will assume that the anode heel effect is compensated by filtration.

Path Length   Consider imaging a slab of material with a constant linear attenuation coefficient μ and thickness L arranged parallel to the plane of the detector, as shown in Figure 5.15. Assume a monoenergetic x-ray spectrum for the present analysis. The central ray of the x-ray beam—i.e., the ray that goes through the center of the object and hits the detector origin—encounters a net path length L through the object. Thus, the x-ray intensity at $(x, y) = (0, 0)$ will be $I_0 \exp(-\mu L)$, where I_0 is the intensity of the beam that would be present at the detector origin if the slab were not present.

The x-rays propagating toward the detector point (x, y), however, experience a different path length through the slab. The path length through the slab along the line between the x-ray source and (x, y) is given by

$$L' = \frac{L}{\cos \theta} . \tag{5.10}$$

Since $L' > L$, more x-rays will be attenuated along this line than along the central path, and the x-ray intensity at the detector will be smaller. Ignoring the inverse

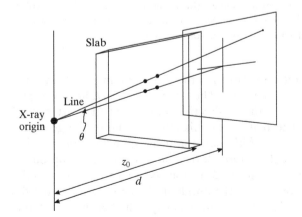

Figure 5.15
Imaging a uniform slab.

square law, obliquity, and the anode heel effect, the intensity is given by

$$I_d(x, y) = I_0 e^{-\mu L / \cos \theta} . \tag{5.11}$$

If the inverse square law and obliquity are included, then the intensity is given by

$$I_d(x, y) = I_0 \cos^3 \theta \, e^{-\mu L / \cos \theta} . \tag{5.12}$$

If the anode heel effect is compensated by filtration, then (5.12) represents a valid imaging equation for a homogeneous slab parallel to the image plane within the field of view.

The interpretation of radiographs, both visually and through automatic image analysis techniques, should be informed by (5.12) and by the anode heel effect (if not compensated). Since materials with larger attenuation are "brighter" when viewed on a light box (or a computer screen if digitized), these effects will cause a *shading artifact* within a homogeneous object. This could be interpreted as representing a different attenuation within the object or different object thickness. It is truly an ambiguous situation. Radiologists are trained to study radiographic intensities locally, and to not compare absolute image values across the field of view. This is not a desirable situation for automatic (computer-based) image analysis methods, in general, which requires such methods to be more complicated than would otherwise be the case.

Depth-dependent Magnification   Another consequence of divergent x-rays is object magnification, which is specifically called *depth-dependent magnification* in radiography. Consider the "stick" object of height w shown in Figure 5.16. It is clear from the geometry that the object will always appear larger on the detector than it is in reality. Furthermore, the height of the object on the detector is different depending on the position of the object in the field-of-view.

Using the concept of similar triangles, it is easy to show that when the object is at position z, its height w_z on the detector will be

$$w_z = w \frac{d}{z} . \tag{5.13}$$

From this, we see that the *magnification* $M(z)$ is given by

$$M(z) = \frac{d}{z} . \tag{5.14}$$

The magnification is a function of z, which is why this magnification is referred to as *depth-dependent*.

There are three important consequences of depth-dependent magnification. First, two objects within the body of the same size may appear to have different sizes on the radiograph. Thus, judgments about relative sizes of anatomical features must be made with caution and knowledge. Second, anatomical features studied longitudinally—i.e., using radiographs of the same patient taken over several months

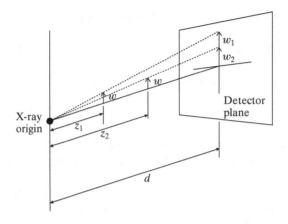

Figure 5.16
Depth-dependent magnification of a "stick" object.

or years—can only be compared in size if the same radiographic conditions and patient positioning is used. A change in the patient's weight, for example, may position a physiological landmark differently relative to the x-ray source causing an apparent change in size which could be falsely associated with progression of disease. Third, the boundaries of a single object can be blurred simply because the "front" of the object is closer to the radiograph than its "back." This effect, which might be called *depth-dependent blurring*, is studied in the following example.

EXAMPLE 5.4

Consider imaging the rectangular prism shown in Figure 5.17, defined by

$$\mu(x, y, z) = \mu_a \operatorname{rect}(y/W) \operatorname{rect}(x/W) \operatorname{rect}([z - z_0]/L), \tag{5.15}$$

where [see (2.16)]

$$\operatorname{rect}(x) = \begin{cases} 1 & |x| \leq 1/2 \\ 0 & \text{otherwise} \end{cases}.$$

Question How is this rectangular prism portrayed in a radiograph?

Answer In the center of the figure, the appearance of the image is governed by the inverse square law, obliquity, and path length variations, so (5.12) applies,

$$I_d(x, y) = I_0 \cos^3 \theta \, e^{-\mu_a L/\cos\theta}.$$

When the x-rays are passing through the edges of the object, however, there is a loss of object path length, and corresponding reduction in attenuation. Suppose a ray passes through the edge of the prism at range z'. Then, using simple trigonometry, we find that the path length through the prism is $[z' - (z_0 - L/2)]/\cos\theta$. (Notice that if $z' = z_0 + L/2$, this yields the path length $L/\cos\theta$, as above.) In this case,

$$I_d(x, y) = I_0 \cos^3 \theta \, e^{-\mu_a(z' - (z_0 - L/2))/\cos\theta}.$$

The third possibility is that the ray misses the prism entirely. In this case,

$$I_d(x, y) = I_0 \cos^3 \theta.$$

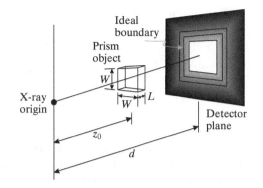

Figure 5.17
Imaging a rectangular prism.

Given (x, y) on the detector, what is required to complete this analysis is to determine (1) which of the three ranges apply, and (2) if at the edge, the value of z'. This analysis is left as an exercise for the reader in Problem 5.9(a). ∎

This example shows that divergent rays lead to edge blurring, an additional distortion to understand when viewing radiographs. This effect can be reduced by keeping the x-ray source as far from the detector plane as possible, and keeping the object (the patient) as close to the detector as possible. For example, for routine chest radiography, the distance between the x-ray source and the detector (film) is typically six feet. However, for other examinations (such as imaging extremities), the source/detector distance may be much shorter.

Imaging Equation with Geometric Effects It is useful to develop an imaging equation that incorporates the geometric effects we have presented in this section. To do this, we utilize an idealized object $t_z(x, y)$ that is infinitesimally thin and located in a single plane given by the coordinate z, and is capable of differentially attenuating x-rays as a function of x and y. This object can be thought of as a sheet of paper impregnated with a distribution of lead; a pictorial representation is given in Figure 5.13.

We should think of $t_z(x, y)$ as a *transmittivity*, rather than an attenuation; it replaces the entire exponential factor, rather than μ itself. Accordingly, if the object is located at the detector face—so that there is no magnification—the recorded intensity would be

$$I_d(x, y) = I_0 \cos^3 \theta \, t_d(x, y), \qquad (5.16)$$

where

$$\cos \theta = \frac{d}{\sqrt{d^2 + x^2 + y^2}}. \qquad (5.17)$$

In general, when the object is located at arbitrary z, where $0 < z \leq d$, the magnification must be taken into account. This effect is just a scaling of the size of the object, and it is

$$I_d(x, y) = I_0 \cos^3 \theta \, t_z(x/M(z), y/M(z)) \,. \tag{5.18}$$

Putting (5.18) together with (5.14) and (5.17) yields

$$I_d(x, y) = I_0 \left(\frac{d}{\sqrt{d^2 + x^2 + y^2}} \right)^3 t_z(xz/d, yz/d) \,. \tag{5.19}$$

This equation is a reasonable approximation for relatively thin objects that have nearly the same magnification and no variation in their attenuation in the z direction. This equation does not take into account the effect of integration along ray paths through thick objects (such as the human body).

5.3.3 Blurring Effects

It was shown above that divergent rays will blur edges of objects that are thick in the range direction. There are two effects, however, that will blur objects even if they do not have a z extent: extended sources and the intensifying screen. These processes can both be modeled as convolutional effects that degrade image resolution. They are not present in (5.18) and (5.19) because in these equations there is only magnification and intensity changes made to the infinitesimally thin object $t_z(x, y)$. We now explore these two effects in detail.

Extended Sources   Another characteristic of projection radiographs is blurriness arising from the nonzero extent of the x-ray source—the so-called *spot size* of the site of x-ray emissions from the anode. As shown in Figure 5.18, this blurring contributes to both a fuzziness at the edge of the field-of-view and a fuzziness of an object boundary. Here, we care about the blurring caused to the object. We will show that the extended source effect is a convolution of the source shape with the object shape (with magnification effects also included). Because of this, extended sources produce a significant loss of resolution of x-ray images. (This problem cannot be solved by shrinking the spot size on the anode because of heat dissipation requirements in the anode material.)

The physical extent of blurring caused by an extended source depends on the size of the source spot and the location of the object. Consider, for example, the image of the point hole shown in Figure 5.19. From the geometry, if the source has diameter D, then the image of the point hole located at range z will have diameter D', given by

$$D' = \frac{d - z}{z} D \,. \tag{5.20}$$

The factor multiplying D in (5.20) is the absolute value of the *source magnification* $m(z)$, which is given by

$$m(z) = -\frac{d - z}{z} \,. \tag{5.21}$$

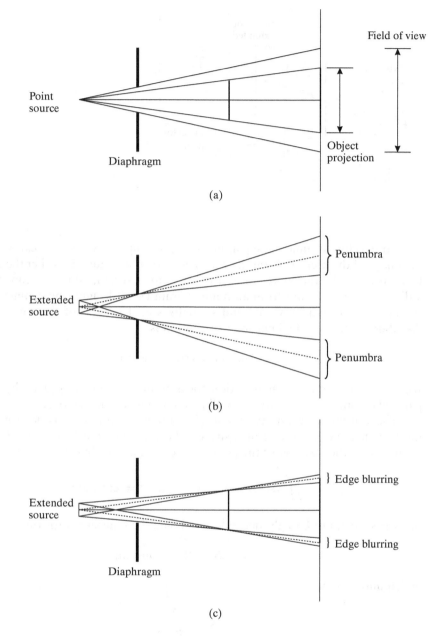

Figure 5.18
Effects of extended source.
(a) Ideal field of view and
object projection (with
magnification). (b) Penumbra
at edges of field of view due
to extended source. (c)
Blurred object edges due to
extended source.

The source magnification is negative because it inverts the image of the source, as can be appreciated by studying Figure 5.19. From (5.14), the depth-dependent (object) magnification and the source magnification are related as

$$m(z) = 1 - M(z). \qquad (5.22)$$

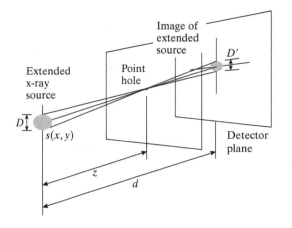

Figure 5.19
Principle of source
magnification.

To develop an imaging equation incorporating source magnification, consider the image of the point hole on the z-axis, as shown in Figure 5.19. Let the x-ray spot be represented by the source intensity distribution $s(x, y)$, which we can think of as a disk emanating x-rays from all points within the disk. Ignoring geometric effects, the image will be an inverted and spatially scaled version of the source intensity distribution. In fact, the intensity distribution is

$$I_d(x, y) = ks(x/m, y/m),\qquad (5.23)$$

where m is the source magnification for a point at range z, as given by (5.21). In particular, since m is negative, I_d will be the symmetric image of s.

The amplitude scaling term k in (5.23) must be found. It is determined using the fact that the integrated intensity on the detector plane must remain constant regardless of the position of the point hole on the z-axis. In other words,

$$\iint ks(x/m(z), y/m(z))\,dx\,dy = \text{constant},\qquad (5.24)$$

regardless of z, but k might depend on z. Using the Fourier transform,

$$km^2(z)S(0, 0) = \text{constant},\qquad (5.25)$$

which implies that

$$k \propto \frac{1}{m^2(z)}.\qquad (5.26)$$

This implies that the amplitude of the blurring caused by the extended source depends on the location of the object relative to the detector plane. In particular, if the object approaches the detector, then $m(z) \to 0$, and

$$\frac{s(x/m, y/m)}{m^2} \to S(0, 0)\delta(x, y).\qquad (5.27)$$

In this case, there is no loss of resolution or amplitude; the image of the extended source is actually perfect.

The above analysis holds for a point hole located anywhere in the same z-plane. We can then use superposition [see (2.34)] to understand the behavior of a spatial distribution of attenuating objects within a given z plane, such as the transmittivity function $t_z(x, y)$ used before in describing imaging under object magnification [see (5.19)]. This analysis leads to an imaging equation that is the convolution of the magnified object by the magnified and scaled source function,

$$I_d(x, y) = \frac{\cos^3 \theta}{4\pi d^2 m^2} s(x/m, y/m) * t_z(x/M, y/M) \,. \tag{5.28}$$

Notice that the term $4\pi d^2$ is included to account for the fact that there is loss of source intensity due to the inverse square law. For objects close to the detector, $M \approx 1$ and $m \approx 0$; the object will have unity magnification and will not be blurred by the source focal spot, regardless of the size of that focal spot. This is just one more motivation for putting the patient directly against the film-screen cassette.

Film-Screen Blurring The detailed geometry of a film-screen detector is shown in Figure 5.20. The film is sandwiched between two phosphors—i.e., the active parts of intensifying screens. For each x-ray photon that is absorbed by a phosphor (by photoelectric absorption), a large number of lower-energy light photons are produced. Unlike x-ray photons, these light photons are efficiently detected by film emulsions (see Section 5.2.5). The light photons travel (approximately) isotropically from the point where the x-ray was absorbed, as shown in Figure 5.20. They can then be absorbed in the film at locations far from the x-ray's path. The union of detected light photons form a "spot" on the film, which is effectively an impulse response to the x-ray "impulse." Using superposition, an imaging equation incorporating film-screen blurring is readily derived. To good approximation, it is simply (5.28) with an additional convolution with the film-screen impulse response function $h(x, y)$,

$$I_d(x, y) = \frac{\cos^3 \theta}{4\pi d^2 m^2} s(x/m, y/m) * t_z(x/M, y/M) * h(x, y) \,. \tag{5.29}$$

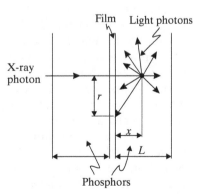

Figure 5.20
A film-screen detector.

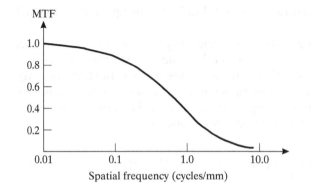

Figure 5.21
A typical MTF for a
film-screen detector.

Typically, the impulse response of a film-screen detector system is circularly symmetric. The MTF of a typical film-screen detector is shown in Figure 5.21.

Use of a thinner phosphor yields less film-screen blurring. However, thin screens do not stop as many x-rays, reducing the efficiency of the detector system. Detector *efficiency* η is defined as the fraction of photons captured by the detector on average. Calcium tungstate is often used for intensifying screens because it has reasonably good efficiency for stopping x-ray photons ($\eta \approx 0.30$). In general, the thicker the screen, the higher the efficiency and the worse the resolution. We will see in Section 5.4.1 that maintaining high efficiency is key to keeping image noise low. Thus, while thinner screens are desirable for good resolution, there is a tradeoff below which the improvements in resolution will be offset by increased noise.

5.3.4 Film Characteristics

The intrinsic spatial resolution of film is generally significantly better than that of the intensifying screen; so, for most situations, we can ignore the spatial frequency characteristics of the film. If necessary or desirable, the spatial resolution of film can be modeled by a point spread function analogous to that of the intensifying screen above; the overall performance would be a cascade of subsystems, as depicted in (2.47), (3.29), and (3.30). Rather than a film's spatial resolution properties, it is its intensity transformation properties that are of primary interest here.

Because film is a very poor direct detector of x-rays, we will ignore the image created directly on the film by the x-rays themselves, and only consider the image created by the light produced in the phosphors adjacent to the film. The parameters of interest are the contrast or *gamma* of the film, and its dynamic range or *latitude*; we will define these terms in a moment. We use many different types of film with varying contrast and latitude depending upon the type of x-ray examination desired. Those of you who are amateur photographers may have a practical appreciation of these film characteristics.

We consider the relationship between the amount of light that is incident upon the film and the final image that is viewed when the film is exposed, developed, and placed on a light source ("light box"). The light photons that are captured by the film produce a *latent image* (i.e., it is not visible). When the film is developed, that latent image produces a blackening of the film. We therefore characterize film as a transformation between exposure to light and the degree of blackening of the film, which is characterized by the so-called *optical density* of the film.

When a developed film is viewed on a light box, the blacker parts of the film absorb more of the light coming from the box, creating the impression of a gray level image when viewed by eye. The *optical transmissivity* is defined as the fraction of light transmitted through the exposed film, or

$$T = \frac{I_t}{I_i},\qquad (5.30)$$

where I_i is the irradiance of the incident light and I_t is the irradiance of the transmitted light, both in units of energy per unit area per second. For example, the darker parts of the film might have a transmissivity of 0.1, while the more transparent parts might have a transmissivity of 0.9. The *optical opacity* is the inverse of the transmissivity, and the *optical density* is defined to be the common logarithm of the optical opacity,

$$D = \log_{10} \frac{I_i}{I_t}.\qquad (5.31)$$

We see that the optical density characterizes how black the film is on a logarithmic scale. Optical densities in the range of 0.25 to 2.25 are usable, but the human eye discriminates shades of gray best when $1.0 < D < 1.5$.

We now appreciate that the appearance of the x-ray film after exposure and development is determined by the spatial distribution of optical density, $D(x, y)$. How does x-ray exposure relate to optical density? We first recall from Chapter 4 that x-ray exposure X is measured in units of roentgens (R); in particular, 3876 R of exposure is equivalent to 1 C of charge per kilogram of air; the charge comes from the ions produced by the x-ray interactions. For direct exposure (no screens) and $D < 2$, D is directly proportional to X. When screens are used and $D > 2$, however, the relationship between X and D is nonlinear, and it is given by the classic *H&D curve* (named after Hurter and Driffield, who first described this function), as shown in Figure 5.22. A typical H&D curve is S-shaped, with a low exposure/low D *toe*, a linear portion, and a high exposure/high D *shoulder*. Notice that D never goes to zero; there is always a so-called *base fog* optical density, even in the absence of exposure.

In the linear region of an H&D curve, the response between common logarithm and optical density is approximately linear,

$$D = \Gamma \log_{10}(X/X_0),\qquad (5.32)$$

where Γ is the slope of the H&D curve in the linear region and X_0 is the exposure at which the linear region would hit the horizontal axis $(D = 0)$ of the H&D curve.

The quantity Γ is called the *film gamma*. It is unique to a particular film and a particular method of developing that film (i.e., a function of the developing chemicals, temperature, time of development, etc.), and it is typically in the range of

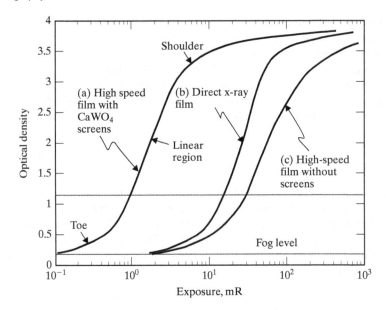

Figure 5.22
An H&D curve.

0.5 to 3. Increasing the gamma of the film increases contrast, but reduces the range of exposures over which the optical density response is linear. Correct exposure of a film occurs only in the linear range of its H&D curve. The *latitude* of a film is the range of exposures over which the H&D curve is linear. The *speed* of a film is the inverse of the exposure at which $D = 1 +$ fog level. Consider the radiographic films shown in Figure 5.22. Clearly, the film/screen combination (a) is the fastest film. The direct x-ray exposure film (b) is next fastest and has the largest gamma and lowest latitude. Finally, standard high-speed optical film (c) without screens is the slowest; it would require the largest x-ray exposure of all three films in order to achieve optical densities in the typical range of 1 to 1.5.

5.4 Noise and Scattering

Until now, we have only considered the effects of the x-ray source and the composition of the object. We have implicitly assumed that the intensity of the x-ray beam at the detector plane is faithfully reproduced by the detector. In reality, of course, the detector does not faithfully reproduce the incident intensity distribution. Furthermore, x-rays arrive in discrete packets of energy (*quanta* or *photons*, as discussed in Section 4.3.2). The discrete nature of their arrival can lead to fluctuations in the image. In the following sections, we will discuss some of the additional factors introduced by x-ray physics and the image recorder.

5.4.1 Signal-to-Noise Ratio

Assuming unity magnification and infinitesimal source size, a rectangular object, like that depicted in Figure 5.17, will cast a rectangular shadow on the detector

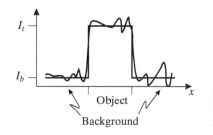

Figure 5.23
Detector intensities from a rectangular object.

with dimensions equal to the object dimensions. A 1-D slice through the intensities on the detector will look like a rect function, as defined by (2.19) and shown in Figure 5.23. Let I_b be the background intensity and I_t be the object (target) intensity on the detector plane, with $I_t > I_b$ in this example. As discussed in Section 3.2, the *local contrast* of this object is given by

$$C = \frac{I_t - I_b}{I_b}. \tag{5.33}$$

Because the x-rays arrive in discrete quanta (photons), there will be random fluctuations in the number of photons arriving in each small area of the detector, leading to *noise*, as discussed in Section 3.4. This effect is called *quantum mottle*, and it is responsible for the imprecision of detector measurements of x-ray intensity.

We can quantify the effect of noise on image formation using the concept of signal-to-noise ratio (SNR), as presented in Section 3.5. The higher the SNR, the less evident will be the granularity in the image resulting from this quantum effect. In this scenario, the "signal" is $I_t - I_b$, the difference between the target intensity and the background intensity. The noise is caused by quantum-related fluctuations in the background, and it can be characterized by a standard deviation σ_b. Therefore, the basic SNR in this scenario is given by

$$\text{SNR} = \frac{I_t - I_b}{\sigma_b} = \frac{CI_b}{\sigma_b}. \tag{5.34}$$

Suppose that all photons arriving at the detector have the same energy, $\hbar v$, which might be called the *effective energy* in the actual case of a polyenergetic x-ray beam (see Example 4.4). In this case [combining (4.14) and (4.17)], the intensity is related to the number of photons by

$$I = \frac{N\hbar v}{A\Delta t}, \tag{5.35}$$

where A is a small area on the detector—i.e., a pixel—and Δt is the duration of the x-ray burst. The number of photons N is a Poisson random variable, and in the background, we can denote the average number of photons per burst per area A as N_b. Then, the average background intensity is given by

$$I_b = \frac{N_b\hbar v}{A\Delta t}. \tag{5.36}$$

Also, the variance of the number of photons per burst per area A in the background is given by

$$\sigma_b^2 = N_b \left(\frac{\hbar\nu}{A\Delta t} \right)^2 . \tag{5.37}$$

From (5.34), (5.35), (5.36), and (5.37), the local SNR is given by

$$\text{SNR} = C\sqrt{N_b} . \tag{5.38}$$

Equation (5.38) reveals a fundamental tradeoff in x-ray imaging. In order to improve the visibility of a particular structure in a radiograph, it is necessary to either increase the contrast of the structure or to increase the number of photons used in the visualization or analysis (or both). The contrast of the structure might be improved by changing the energy of the x-rays or by using a contrast agent or dual-energy techniques. The number of photons might be increased by several means: by increasing the filament current, the duration of the x-ray pulse, or the energy of the x-rays (which would then penetrate the body better), or by using larger area elements (pixels) or a more efficient detector.

Recall that the x-ray tube provides a spectrum of x-ray energies ranging from low energies to the maximum energy determined by the peak kilovoltage applied to the x-ray tube anode. By changing the kVp, we modify the x-ray energy. Contrast, signal-to-noise ratio, and patient radiation exposure (dose) are markedly affected by changes in the kVp. For example, at low kVp and therefore low x-ray energies, image contrast is high because the difference between the attenuation of different body tissues increases as energy decreases. However, the body is less transparent to x-rays at lower energies, so that fewer x-ray photons penetrate the body. Hence, although C is high, N is low, and it turns out that SNR is low. Because the body is highly absorbent to these low energy x-ray photons, the patient radiation dose is high.

At high energies, on the other hand, contrast is low, as the attenuation coefficients for different types of body tissues become similar. While the body is highly transparent at high energies, the number of x-ray photons per Roentgen decreases, and therefore for a given radiation exposure, the SNR will be low at very high energies. Between high and low x-ray energies, the SNR achieves a maximum at a point where there is reasonably good tissue contrast, the body is still relatively transparent to x-rays, and the number of photons per Roentgen is high.

We can express the SNR in more detail by adding several additional concepts. Since we indicated above that SNR is proportional to the number of photons per unit area of detector, if we increase the unit area, we will increase the number of photons. Therefore, the detailed expression for SNR becomes

$$\text{SNR} = C\sqrt{\Phi A R t \eta} , \tag{5.39}$$

where Φ is the number of photons per Roentgen per cm^2, A is the unit area, R is the body's radiation exposure in roentgens, t is the fraction of photons transmitted through the body, and η is the detector efficiency.

EXAMPLE 5.5

Consider the following parameters, which are from a typical chest x-ray:

$$\Phi = 637 \times 10^6 \text{ photons R}^{-1} \text{ cm}^{-2}$$

$$R = 50 \text{ mR}$$

$$t = 0.05$$

$$\eta = 0.25 \text{ (25\% efficiency)}$$

$$A = 1 \text{ mm}^2$$

Question What is the SNR of a lesion having 10% contrast, i.e., $C = 0.1$?

Answer Using (5.39), we compute that SNR = 16 dB. ∎

5.4.2 Quantum Efficiency and Detective Quantum Efficiency

In projection radiography, x-rays that are not stopped by the body should be detected. Ideally, all photons incident upon a detector would be detected and characterized by their location, energy, and time-of-arrival. No detector can achieve these ideals, however. Instead, detector designs must emphasize certain performance measures at the expense of others, and these tradeoffs influence the quality of the measured radiographs. In this section, we develop the concept of *detective quantum efficiency* (DQE), which represents one way to characterize the performance of a detector that has a direct relation on the SNR of the images it produces.

Quantum Efficiency Ideally, all photons incident upon a detector should be detected. What does this really mean? In order to be "detected," an incident photon must interact with the detector—e.g., it must be stopped by a photoelectric interaction—and that interaction must produce some measurable output—e.g., a flash of light, a collection of ionized atoms, or an electrical current. *Quantum efficiency* (QE) is the probability that a single photon incident upon the detector will be detected; it is a basic property of any x-ray detector.

A detector with a higher QE does not necessarily mean that the detector or the image it produces is better, however. Consider two detectors having the same thickness and linear attenuation coefficients. They are capable of stopping the same fraction of incident photons; their basic "stopping power" is the same. Suppose that the output of a detected event is highly predictable in the first detector and highly variable in the second. We intuitively understand that the first detector is better. In order to more accurately characterize the quality of a detector, we must account for these variations in detector output given a single event. That is the goal of the parameter DQE.

Detective Quantum Efficiency To better characterize detector performance, detective quantum efficiency (DQE) considers the transformation of SNR from a detector's input to its output. In this way, DQE moves away from simply counting photons as in QE. The *detective quantum efficiency* is defined as

$$DQE = \left(\frac{SNR_{out}}{SNR_{in}}\right)^2,\tag{5.40}$$

where SNR_{in} is the intrinsic SNR of the incident radiation and SNR_{out} is the SNR of the measured quantity—e.g., flash of light, detector voltage, film density, etc. DQE can be viewed as a measure of the degradation in the SNR due to the detection process. It can also be viewed as the fraction of photons that are detected "correctly." In general $DQE \leq QE \leq 1$.

EXAMPLE 5.6

Consider a hypothetical detector having QE = 0.5 and the ability to perfectly localize every photon that is stopped by the detector.

Question What is the DQE of this detector?

Answer Suppose a photon burst with an average of $\overline{N}$ is incident upon the detector. The SNR of the input is just the intrinsic SNR of the photon burst, which is given by

$$SNR_{in} = \sqrt{\overline{N}}.$$

Since QE = 0.5, we know that $0.5 \times \overline{N}$ photons (on average) are stopped by the detector. These photons are perfectly localized, meaning that the detector produces an accurate count of the number and position of all detected photons no matter what pixel resolution is desired. Therefore, the output SNR is

$$SNR_{out} = \sqrt{0.5\overline{N}},$$

and the DQE is

$$\begin{aligned}DQE &= \left(\frac{SNR_{out}}{SNR_{in}}\right)^2 \\ &= \left(\frac{\sqrt{0.5\overline{N}}}{\sqrt{\overline{N}}}\right)^2 \\ &= 0.5\end{aligned}$$

In this case, $DQE = QE$, which occurs because of the hypothetical perfect localization of all detected photons. ∎

EXAMPLE 5.7

Suppose that an x-ray tube is set up to fire n 10,000-photon bursts at a detector and the detector's output x is recorded as x_i, $i = 1, \ldots, n$. Suppose that the mean and variance of $\{x_i\}$ is found to be $\bar{x} = 8,000$ and $\sigma_x^2 = 40,000$, respectively.

Question What is the DQE of this detector?

Answer   Think of $\bar{x} = 8,000$ as an estimate of the average number of photons in the output of the detector. If the detector were simply counting photons, then the variance should also be 8,000, since photon count is a Poisson random variable. Yet, the measured variance is 40,000, five times the Poisson variance. This is conclusive evidence that there is variation in the detector's response to detected photons. The SNR of the input is simply the intrinsic SNR of the photon burst,

$$\text{SNR}_{\text{in}} = \frac{10,000}{\sqrt{10,000}} = 100 \, .$$

The SNR of the output is

$$\text{SNR}_{\text{out}} = \frac{8,000}{\sqrt{40,000}} = 40 \, .$$

Therefore, the detective quantum efficiency for this detector is

$$\text{DQE} = \left(\frac{40}{100} \right)^2 = 0.16 \, .$$

Only about 16 percent of the incident photons are detected "correctly."   ∎

5.4.3 Compton Scattering

Earlier, we noted that Compton scattering degrades image quality. The reason for this is that Compton photons are deflected from their ideal straight-line path, and some are detected in locations away from the correct, straight-line location. This produces two unwanted results: a decrease in image contrast and a decrease in SNR.

Effect on Image Contrast   Compton scatter has an negative effect on image contrast. Consider a target with local contrast [see (3.12)] given by

$$C = \frac{I_t - I_b}{I_b} \, .$$

Compton scatter adds a constant intensity I_s to both target and background intensity, yielding a new contrast of

$$\begin{aligned} C' &= \frac{(I_t + I_s) - (I_b + I_s)}{I_b + I_s} \\ &= C \frac{I_b}{I_b + I_s} \\ &= \frac{C}{1 + I_s/I_b} \, . \end{aligned} \tag{5.41}$$

Therefore, the effect of scatter is to reduce contrast by the factor $1/(1 + I_s/I_b)$. The ratio I_s/I_b is called *scatter-to-primary ratio*.

Signal-to-Noise Ratio with Scatter The derivation of SNR in the presence of Compton scattering follows the Compton-free derivation very closely:

$$\text{SNR}' = \frac{I_t - I_b}{\sigma_b}$$

$$= C\frac{I_b}{\sigma_b}$$

$$= C\frac{N_b}{\sqrt{N_b + N_s}}$$

$$= \frac{C\sqrt{N_b}}{\sqrt{1 + N_s/N_b}} \,. \tag{5.42}$$

Here, the symbol N_s stands for the number of Compton scattered photons per burst per area A on the detector, and the symbol C is the underlying contrast (not the Compton scatter-reduced contrast).

Equation (5.42) shows the reduction of SNR due to Compton scattering. The SNR with Compton scattering is related to the SNR without Compton scattering by

$$\text{SNR}' = \text{SNR}\frac{1}{\sqrt{1 + I_s/I_b}} \tag{5.43}$$

Thus, in addition to the loss in contrast, there is an additional loss of $1/\sqrt{1 + I_s/I_b}$ in SNR due to the effects of noise.

EXAMPLE 5.8

Suppose 20 percent of the incident x-ray photons have been scattered in a certain material before they arrive at detectors.

Question What is the scatter-to-primary ratio? By what factor is the SNR degraded?

Answer The x-ray photons that contribute to the background intensity are those that hit the detectors without Compton scattering. The number of these photons is $0.8N$, where N is the number of incident x-ray photons. The number of scattered photons is $0.2N$. Because the intensity of the image is proportional to the number of photons detected, we have

$$I_b \propto 0.8N, \quad I_s \propto 0.2N.$$

The scatter-to-primary ratio is

$$\frac{I_s}{I_b} = \frac{0.2N}{0.8N} = \frac{1}{4}$$

From (5.42), we can see that the loss of SNR is $1 - 1/\sqrt{1 + I_s/I_b} = 0.11$. So the Compton scattering introduces 11 percent loss of SNR. ∎

5.5 Summary and Key Concepts

Projection radiography is the oldest and most fundamental medical imaging modality. This modality uses x-ray tubes and film to record images of the "shadow" created by body tissues. In this chapter, we presented the following key concepts that you should now understand:

1. *Projection radiography* produces *radiographs*, which are 2-D projections of a 3-D object.
2. A *projection radiography system* consists of an x-ray tube, devices for beam filtration and restriction, compensation filters, grids, and (usually) a film-screen detector.
3. The *basic imaging equation* describes the energy- and material-dependent attenuation of the x-ray beam produced by the system as it passes through the patient.
4. This equation must be modified by several geometric effects, including the *inverse square law, obliquity, beam divergence, anode heel effect, path length,* and *depth-dependent magnification.*
5. The film-screen detector produces an *optical image on film*; the degree of film blackening—the *optical density*—depends on film exposure in a nonlinear way characterized by the H&D curve.
6. *Noise* arising from the random nature of x-ray production and transmission reduces an image's signal-to-noise ratio, and thus the *detective quantum efficiency* of the system.
7. Acceptance of *Compton scattered photons* reduces image contrast, and thus signal-to-noise ratio as well.

Bibliography

Bushberg, J. T., Seibert, J. A., Leidholdt Jr., E. M., and Boone, J. M. *The Essential Physics of Medical Imaging*, 2nd ed., Philadelphia, PA: Lippincott Williams & Wilkins, 2002.

Carlton, R. R. and Adler, A. M. *Principles of Radiographic Imaging: An Art and a Science*, 3rd ed. Albany, NY: Delmar, 2001.

Johns, E. J., and Cunningham, J. R. *The Physics of Radiology*, 4th ed. Springfield, IL: Charles C Thomas Publisher, 1983.

Macovski, A. *Medical Imaging Systems*, Englewood Cliffs, NJ: Prentice Hall, 1983.

Webster, J. G. *Medical Instrumentation: Application and Design*, 3rd ed. New York: Wiley, 1998.

Wolbarst, A. B. *Physics of Radiology*. Norwalk, CT: Appleton & Lange, 1993.

Problems

Instrumentation

5.1 Examine the projection radiography system in Figure 5.13. Is this system linear? Is it shift invariant?

5.2 (a) What determines the highest energy of x-ray photons emitted from an x-ray tube? What determines the energy spectrum of the x-ray photons?

   (b) In radiographic imaging, why are low-energy photons undesired? What measures can be taken to reduce the number of the low-energy photons entering the human body?

   (c) What is beam hardening? What are the causes of beam hardening?

5.3 We are designing an x-ray tube and want to use a lighter filter. Suppose we have, in addition to copper, a new material with a density of 5000 kg/m^3, and a mass attenuation coefficient of 0.08 m^2/kg at 80 kVp. In order to satisfy the NCRP recommendation of 2.5 mm Al/Eq at 80kVp, what should we use?

5.4 (a) Why are iodine and barium commonly used as contrast agents?

   (b) Explain with a diagram why an airgap can be used to reject scattered photons. What is the problem with airgaps?

5.5 (a) Why is Compton scattering bad for the images produced using projection radiography?

   (b) For a film with a given H&D curve, why is it better to have a range of x-ray exposures that are in the linear region of the curve?

   (c) Why do we have to filter X-rays from the X-ray source?

   (d) Lead strips can be used (in a grid, for example) to reduce scatter in radiography systems. If the height of the grid is eight times its width, what is the maximum scatter angle of the photon after which the photon cannot pass through the grid?

Image Formation

5.6 Suppose a chest x-ray imaging system with a perfect point source can provide images with acceptable quality when the intensity variation is smaller than 5 percent when imaging a slab of material with uniform attenuation coefficient. If the source-detector distance is 2 m, what is the maximal size of the image?

5.7 (a) Derive a simple expression for the magnification of a thin object on a projection radiograph. Assume x-ray source-to-object and source-to-detector distances.

   (b) In practice, what simple strategies can an x-ray technician use to reduce the magnification and distortion effects of the projection radiography system?

5.8 You are designing a digital x-ray detector panel with an array of discrete detectors. You decide that you are able to compensate for the beam divergence effect by weighting the outputs of your detectors appropriately. This way, the effective beam intensity should be uniform throughout the entire image.

   (a) Assume your detectors are very small and very close together and that the detector panel is a distance d away from the x-ray source. Find an appropriate weighting (as a function of r_d) that will compensate for the beam divergence variation.

   (b) Do these adjustments improve the image quality? Explain.

5.9 Consider the radiographic image of the prism, as developed in Example 5.4.

(a) Determine the exact expression for the intensity image $I_d(x, y)$ of the prism.

(b) Plot the intensity profile along the line $y = 0$ on the detector plane.

(c) Write an expression for the image on the developed radiographic film of the prism.

5.10 Derive (5.28) from first principles.

Image Quality

5.11 Assume that each photon incident upon an x-ray image intensifier is detected with probability (quantum efficiency) p, independently from other photons. If the number $N(t)$ of photons that arrive at the detector within the time interval $[0, t)$ follows a Poisson distribution with mean μt, find the PMF of the number $D(t)$ of photons detected in the time interval $[0, t)$.

5.12 Consider two 1-D functions

$$h_1(x) = e^{-x^2/5}, \qquad h_1(x) = e^{-x^2/10}.$$

(a) Suppose $h_1(x)$ is the PSF of the extended source in a projection radiography system for an object at range $z = 3d/4$, where d is the source-to-detector distance. What is the PSF of the extended source as function of z?

(b) In this same projection radiography system, the intensifier screen causes a blurring that can be modeled as $h_2(x)$. Find the MTF of the overall blurring as a function of source magnification, m.

(c) What is the FWHM of the system as a function of m?

5.13 We examine the effect on the SNR of an increase in the recorded scatter fraction from 0.35 to 0.8. Suppose that a 1 mm thick bone presents a scatter-free subject contrast of 0.08, and suppose that 1,000 photons/mm^2 were collected. What would happen if the detector absorption efficiency were halved?

5.14 Show that, for a given detector, its quantum efficiency QE will be at least as large as its DQE ; i.e.,

$$\mathrm{DQE} \leq \mathrm{QE}.$$

5.15 Suppose that an x-ray tube fires, on the average, 10,000 photons per second at an x-ray image intensifier. If the detector provides an average count of 10,000 photons per second, calculate and plot the variance of the detector's output as a function of DQE . What DQE value is required for an output variance of 2,000?

5.16 Calculate the DQE as a function of (u, v) for a nonideal detector with PSF

$$h(x, y) = \frac{1}{2\pi} e^{-(x^2+y^2)/2},$$

by assuming that the input to the detector is white Poisson noise.

5.17 Figure P5.1 shows a projection radiograph of an off-axis, hollow, plastic cylinder. Make a simple sketch of the cylinder's position and orientation relative to the source and detector, and explain the distortion in the image.

Figure P5.1
A projection radiograph of a plastic cylinder. From Macovski (1983).

Applications

5.18 Interesting properties of tissues can sometimes be revealed by imaging them at two x-ray energies. Suppose you are going to use your conventional chest x-ray setup to make two films, one where the highest x-ray energies are at 30 keV and the other where the highest energies are at 100 keV.

(a) What physical parameter will you change, and to what values will you set it, in order to make these two images.

(b) Assuming nothing else changes, which film would you expect to be more exposed than the other? Explain.

(c) At which energy is Compton scattering more of a problem? Describe how it will affect that image.

(d) At which energy is the absorbed dose to the patient higher?

(e) Suppose you took the two exposed films and subtracted their optical densities, creating a third image: $D(x, y) = D(x, y; E_h) - D(x, y; E_l)$. Describe in mathematical terms what $D(x, y)$ measures.

5.19 Suppose you have a 20 cm thick slab of material. You measure the fraction $t(E)$ of x-rays that get through this material (on average) over the energy band 50–150 keV, and discover that it is well-modeled by the relation

$$\log_{10} t(E) = -\frac{(E[\text{keV}] - 150)^2}{5000}.$$

(a) Find an expression for the linear attenuation coefficient of this material as a function of x-ray energy E.

Suppose a 5 cm cube of a second material replaced a 5 cm cube of original material from the center of the slab and suppose this second material had a linear attenuation coefficient of 0.15 cm^{-1} at 75 keV.

(b) What is the intrinsic contrast of this new material relative to the slab at 75 keV? (Intrinsic contrast is defined by the linear attenuation coefficients alone, as if you could image them directly.)

(c) What is the contrast of this new material if imaged through a standard radiographic system at 75 keV? (Ignore photon noise, detector efficiency, beam divergence, scatter, and magnification effects.)

5.20 A hypothetical x-ray tube has been designed. An x-ray burst from this tube yields exactly 10^4 x-ray photons at energy $E_1 = 60$ keV and 10^5 x-ray photons at energy $E_2 = 65$ keV.

(a) Draw the energy spectrum associated with this x-ray pulse. Explain why the units associated with the numbers 10^4 and 10^5 in the spectrum must be photons-keV. In the following parts of this question, assume that the above x-ray tube is used as a source in an imaging system. In addition, make the following simplifying assumptions:

- Assume a parallel beam of x-rays that are uniformly distributed over an extended source (see Figure P5.2).
- Assume that the x-rays passing through the phantom undergo the photoelectric effect only—i.e., they neglect the effect of Compton scattering.

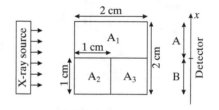

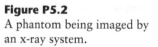

Figure P5.2
A phantom being imaged by an x-ray system.

It consists of three areas A_1, A_2, A_3 as labeled. The linear attenuation coefficient of these three areas—namely, μ_1, μ_2, μ_3 (units of cm-1) for the energy levels present in the X-ray beam—are given in the table below:

	E_1	E_2
μ_1	0.2	0.4
μ_2	0.3	0.1
μ_3	0.5	0.4

(b) Draw the total number of x-ray photons per centimeter that hit the detector as a function of position x.

(c) Assume the x-ray intensity is proportional to the number of photons. Calculate the contrast of the image observed at the detector as a function of position x assuming that A is the target and B is the background.

(d) Make a rough sketch of the profile of optical density as a function of position x assuming x-ray film is used at the detector. Which part of the film is more transparent?

5.21 An X-ray imaging system is set up as shown in Figure P5.3. All length units in the figure and in the equations below are in centimeters. Assume that the overall system response can be modeled as

$$I_d(x_d, y_d) = Ks(x_d/m, y_d/m) * t(x_d/M, y_d/M),$$

where x_d, y_d denote coordinates in the detector plane and that m and M are the source and object magnification, respectively. K is a constant, $s(x, y)$ denotes the x-ray source distribution, and $t(x, y)$ is the spatial transmission function of the object, whose thickness is ignored. Assume the x-ray source can be modeled as a 1-D Gaussian distribution as

$$s(x, y) = S_0 e^{-x^2} \delta(y),$$

where S_0 is a constant.

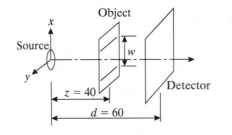

Figure P5.3
Diagram for Problem 5.21.

(a) Compute the object magnification M.

(b) Compute the source magnification m.

An ideal line phantom is to be imaged by the system, whose transmission function is given by

$$t(x, y) = \delta\left(x - \frac{w}{2}\right) + \delta\left(x + \frac{w}{2}\right).$$

(c) Determine the image of the phantom on the detector plane.

(d) Determine the minimal value of w such that the images of the two lines on the detector plane can be distinguished from each other.

5.22 A slab of soft tissue with one blood vessel running in the middle is imaged under an X-ray imaging system, as shown in Figure P5.4. For ease of computation, assume the tissue and the vessel both have square-shaped cross-sections; the dimensions are shown in the figure. Assume that the x-ray source produces $N_i = 4 \times 10^6$ photons at either 15 keV or 40 keV, and the photons are uniformly shed upon the side of the tissue.

The linear attenuation coefficients μ of the soft tissue, blood, and a radiographic contrast agent at two energy levels are given in the table below.

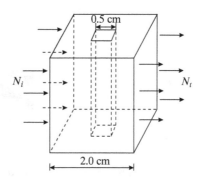

	Linear Attenuation Coefficient (cm^{-1})		
Energy(keV)	Soft Tissue	Blood	Contrast Agent
15	4.0	3.0	5.0
40	0.4	0.2	20

Ignoring photon noise, detector efficiency, beam divergence, Compton scattering, path length, and magnification effects,

(a) Determine the <u>total number</u> of exiting photons at the two energy levels respectively. At which energy level are more photons absorbed?

(b) Calculate the <u>local contrast</u> of the blood vessel (target) at each energy level. At which energy level is the local contrast of the blood vessel better?

(c) Suppose the contrast agent whose linear attenuation coefficient is given in the table were injected into the blood vessel. Would you expect there to be much difference in the local contrast at 15 keV after injection? How about at 40 keV? (No calculations, please, just mathematical reasoning.)

(d) Explain why it is expected that the linear attenuation coefficient of the contrast agent is much larger at 40 keV than at 15 keV?

5.23 An x-ray imaging system is shown in Figure P5.5 to image a bar phantom. The detector is placed on the $z = 0$ plane. The bar phantom with two dark bars is placed on the $z = z_0$ plane, and the x-ray source is placed on the $z = 3z_0$ plane. The two dark bars on the phantom have widths w and are separated by a distance of w. The X-ray source fires monochromatic X-ray beams that are uniformly shed upon the phantom. Suppose the dark bars on the phantom absorb 75 percent of the photons passing through the phantom, and the rest of the phantom let all photons go through. Ignore Compton scattering, inverse square law, obliquity, and image noise.

(a) Assume an ideal point source and ignore the thickness of the phantom, sketch the intensity profile on the detector as a function of position x for $y = 0$. Assume the intensity is 1 at $x = 0$, $y = 0$. Carefully label the axes.

(b) What is the contrast of the image on the detector? Assume the dark bars are the targets.

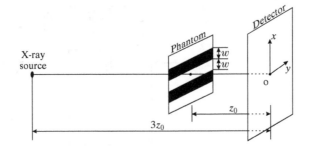

Figure P5.5
A medical imaging system
and a bar phantom.

(c) The image on the detector will be digitized. Suppose the film/screen combination blurs the image by the following point spread function prior to point sampling

$$h(x, y) = \text{sinc}(\alpha x)\text{sinc}(\beta y), \qquad \alpha > 0, \ \beta > 0.$$

What are the largest sampling periods (in x- and y- directions) that will give an aliasing-free sampling?

(d) Now assume the phantom is a slab of a certain thickness. The dark bars attenuate the X-ray photons by the linear attenuation coefficient

$$\mu(E) = \ln\left(\frac{640(\text{keV})}{E(\text{keV})}\right)\text{cm}^{-1}, \qquad 100 \text{ keV} \le E \le 160 \text{ keV},$$

which is a function of energy, E. The X-ray source fires N photons at $E = 160$ keV per unit area during a burst of a unit time. How thick is the phantom such that the dark bars attenuate 75 percent of the photons? Ignore Compton scattering, inverse square law, obliquity, the path length effect, and image noise.

5.24 Consider an x-ray imaging system using a film-screen receptor. Due to the limited dynamic range of film, an essentially constant x-ray intensity is required for the film to be properly exposed. Assume for simplicity that this means a *constant number of photons per unit area* is incident on the image receptor.

(a) Assuming all other factors are constant, if the x-ray tube kilovoltage were increased and the exposure time was appropriately adjusted to produce an optimal film exposure, what would be the effect on SNR? Explain.

(b) Assuming the same conditions as in (a), what would be the effect on the patient dose? Explain.

(c) For each of the following changes, indicate the effect on the subject contrast using this notation: I=increase, D=decrease, N=no effect:
- Increase in patient thickness
- Reduction in kilovoltage
- Increase in detector efficiency
- Reduction in the x-ray field size
- Use of a high atomic number contrast media

(d) Under what conditions is an "air gap" (space between the patient and the image receptor) most effective in reducing the scatter fraction?

(e) What ultimately limits the contrast sensitivity of an x-ray imaging system: geometric unsharpness, the display gamma, the image noise, or the scatter fraction?

(f) The attenuation of a polychromatic x-ray spectrum can be approximately determined using its average photon energy. For a given tube, the average photon energy is mainly determined by: the beam current, the Fourier transform of the x-ray spectrum, the material in the beam path, or the scatter fraction?

5.25 Consider the x-ray projection radiography system shown in Figure P5.6. Relevant dimensions in the figure are $L = 1$ m, $D = 4$ m, $w = 1$ cm, $h = 3$ cm, $R = 0.1$ cm, and $D_{td} = 10$ cm.

A contrast agent is used to enhance the image of the tumor. Assume a 35 keV (monochromatic) x-ray source and linear attenuation coefficients given in the table below.

	Linear Attenuation Coefficient (cm^{-1})		
Energy(keV)	Tissue	Tumor	Tumor with Contrast Agent
35	1	0.75	10

(a) The K-shell energies of iodine and barium are 33.2 keV and 37.4 keV, respectively. Assuming that either agent could be made into a compound that would go to the tumor, what would be the best agent to use and why?

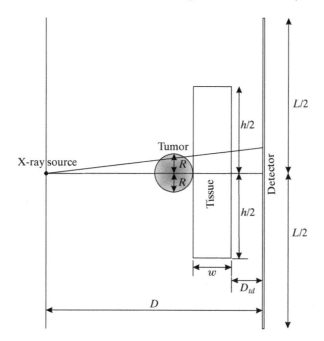

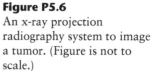

Figure P5.6
An x-ray projection radiography system to image a tumor. (Figure is not to scale.)

(b) Let the local contrast of the tumor be defined as $C = (I_t - I_b)/I_b$, where I_t is the intensity at the center of the tumor and I_b is the intensity just outside the edge of the tumor. Neglecting Compton scattering, find both the local contrast C_1 of the tumor before contrast is applied and the local contrast C_2 of the tumor after contrast is applied. In the next parts, assume that Compton scattering events are taking place throughout the tissue and, for simplicity, assume that $w \approx 0$.

(c) Make a sketch of the photon path that yields a photon hitting the detector with the lowest possible energy. (Assume that only single scatter Compton events can occur.)

(d) Find the energy of the Compton scattered photon in part (c) (that has the lowest possible energy to be detected given the assumed imaging geometry).

Computed Tomography

6.1 Introduction

A *tomogram* is an image of a plane or slice within the body. A "conventional x-ray tomogram" is generated using *motion tomography*, a procedure in which the x-ray source and detector screen are moved in opposite directions in order to keep one plane in focus, as shown in Figure 6.1. However, although only one plane is in focus, the intervening tissue still affects the film density. Therefore, the images still suffer from blurring and overlaying of out-of-plane features, artifacts that are inherent to conventional radiographs.

X-ray computed tomography (CT) has almost completely replaced motion tomography, although it has poorer in-plane spatial resolution. CT is popular because it completely eliminates artifacts from overlaying tissues. One way to think about the basic mechanism of CT is to imagine taking a series of conventional chest x-rays, where the patient is rotated slightly around an axis running from head to foot between each exposure. Each developed film contains a 2-D projection of a 3-D body, but because they are each exposed at a different angle, there is different information in each one. In fact, each horizontal line within a film gives a 1-D projection of a 2-D axial cross-section of the body at that angle. Thus, a collection of horizontal lines, one from each film at the same height, contains information about only one axial cross-section. These *projection data* are then used to *reconstruct* cross-sectional images. A typical CT scanner is shown in Figure 6.2.

What can we say about the tissues in an axial cross-section from a collection of 1-D projections taken at different angles? Fortunately, we can say a great deal, and this is the basic premise of CT. The transformation that takes 1-D projections of a 2-D object over many angles is called the *2-D Radon transform*, and it has an inverse. That is, given a collection of projections we can, in theory, reconstruct the axial cross-section exactly. Practical implementation of the *inverse 2-D Radon transform* is accomplished in every CT scanner in one fashion or another. Therefore, after studying how a CT scanner acquires these projections (after all, taking a

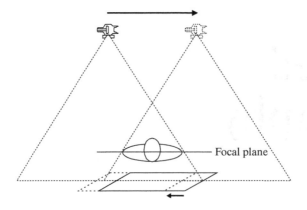

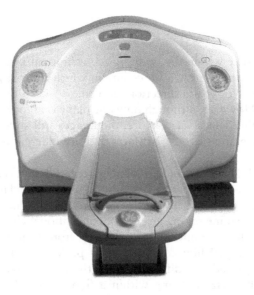

Figure 6.1
Motion tomography.

Figure 6.2
A typical CT scanner. (Used with permission of GE Healthcare.)

collection of chest x-rays is hardly practical), we will study in detail the 2-D Radon transform and its theoretical and practical inverse.

The advent of x-ray CT in the mid-1960s and its development through the 1970s and 1980s produced a profound change in the role of diagnostic imaging in medicine. A wide variety of medical conditions is visible in CT images that are not visible in conventional radiographs. Thus, diagnosis and monitoring of many diseases are possible through CT alone; this has reduced the amount of exploratory surgery previously required for these diseases. As a measuring device, the x-ray CT scanner has also come a long way since its origins. Now, through standard calibration procedures, CT scanners measure so-called *CT numbers* in *Hounsfield units*, which are constant from scan-to-scan and across different scanners. Therefore, quantitative analysis is possible, and is in fact in common practice today.

Because of the growing importance of CT during the early years, manufacturers put considerable effort toward lowering both the scanning and image reconstruction times. Accordingly, the time required to scan a patient and generate an image has dropped from several hours with the earliest clinical machines to under 100 milliseconds for some current machines. Multiple cross sections can now be collected rapidly using *helical CT* and *multislice CT*; hence, 3-D data sets are readily available. A very important area of current research and development in both the academic and industrial arenas involves the display and analysis of 3-D CT data sets.

In this chapter, we present the fundamentals of x-ray CT imaging. Since the properties of x-rays—their generation, detection, and interaction with the human body—have been discussed in Chapter 5, the focus of this chapter is on the various geometries of CT scanners and on a mathematical description of the algorithms necessary to reconstruct images. The mathematical abstraction is necessary since it leads to the actual methods, implemented in software and hardware, used in modern-day clinical scanners. As well, the same mathematics and methods can be used in SPECT, PET, and MRI.

6.2 CT Instrumentation

The fundamental measurement required by a CT scanner is the measurement of x-ray attenuation along a line between an x-ray source and an x-ray detector. In order to reconstruct an image of a 2-D cross-section, a collection of such measurements are required along all lines within the cross-section. Seven *generations* of basic CT designs have been developed to obtain these fundamental data. Roughly speaking, the seven generations were developed in sequence to improve overall performance, mostly toward faster acquisition of both cross-sectional and volumetric data. Although first and second generation scanners are no longer used in medical imaging, it is still important to understand their geometry. In particular, the parallel ray geometry of first generation scanners is used to explain the reconstruction formulas that underly all CT scanners. As we describe the CT generations, it may be helpful to refer to Table 6.1, which compares various properties of CT scanners across the generations.

6.2.1 CT Generations

First-generation (1G) scanners are no longer manufactured for medical imaging, but their geometry is still used to explain the theoretical ideas underlying image reconstruction. The geometry of a 1G scanner is shown in Figure 6.3. It consists of a single source, collimated (meaning that its beam is restricted) to a thin line, and a single detector that move in unison along a linear path tangent to a circle that contains the patient. After making a linear scan, the source and detector apparatus are rotated so that a linear scan at a different angle can be made, and so on. The 1G geometry has the advantage that an arbitrary number of *rays*—i.e., paths between source and detector—can be measured within a given projection and an arbitrary number of angular projections may be measured. It has the further advantage that scattered radiation goes mostly undetected (because it misses the detector), and therefore, the measured attenuation of the beam is almost certainly due to tissues along the ray.

TABLE 6.1

Comparison of CT Generations

Generation	Source	Source Collimation	Detector	Detector Collimation	Source-Detector Movement	Advantages	Disadvantages
1G	Single x-ray tube	Pencil beam	Single	None	Move linearly and rotate in unison	Scattered energy is undetected	Slow
2G	Single x-ray tube	Fan beam, not enough to cover FOV	Multiple	Collimated to source direction	Move linearly and rotate in unison	Faster than 1G	Lower efficiency and larger noise because of the collimation in detectors
3G	Single x-ray tube	Fan beam, enough to cover FOV	Many	Collimated to source direction	Rotate in synchrony	Faster than 2G, continuous rotation using a slip ring	More expensive than 2G, low efficiency
4G	Single x-ray tube	Fan beam covers FOV	Stationary ring of detectors	Cannot collimate detectors	Detectors are fixed, source rotates	Higher efficiency than 3G	High scattering since detectors are not collimated
5G (EBCT)	Many tungsten anodes in single large tube	Fan beam	Stationary ring of detectors	Cannot collimate detectors	No moving parts	Extremely fast, capable of stop-action imaging of beating heart	High cost, difficult to calibrate
6G (Spiral CT)	3G/4G	3G/4G	3G/4G	3G/4G	3G/4G plus linear patient table motion	Fast 3D images	A bit more expensive
7G (Multislice CT)	Single x-ray tube	Cone beam	Multiple arrays of detectors	Collimated to source direction	3G/4G/6G motion	Fast 3D images	Expensive

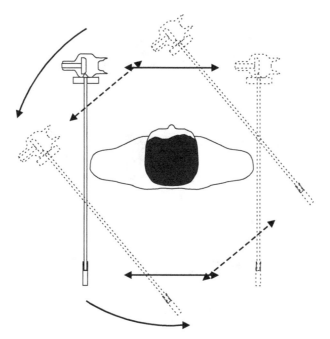

Figure 6.3
The geometry of a
first-generation (1G) CT
scanner.

A *second-generation (2G) scanner*, as illustrated in Figure 6.4, has additional detectors, forming a detector array, arranged along a line or a circle. As in the 1G scanner, the source and detector array move linearly in unison, to cover the field of view (FOV). However, unlike the 1G scanner, the source in a 2G scanner is collimated as a fan beam, so that energy is kept within the slice but spread over the detector array. Therefore, while the source and central detector scan the same projection as the 1-D projection, the additional detectors are simultaneously obtaining additional projections from different angles.

With the 2G scanner geometry, we can make a larger rotation after each linear scan and thereby complete a full scan in less time, making the 2G scanner faster. Scattering effects must be considered, however, in the design of a 2G scanner, since radiation scattered from one ray could end up in another detector. Because of this phenomenon, the detectors in a 2G scanner are usually collimated so that they can receive radiation only from the correct direction. Collimation on detection, however, reduces efficiency and increases noise for a given dose. Hence, for a given dose and with an identical number of projections and measurements per projection, the 1G scanner will produce a better image. Still, the payoff in scan time is so great that slightly increasing the dose to counter the deleterious effects from collimation is worth it.

EXAMPLE 6.1
Consider a 1G or 2G scanner whose source-detector apparatus can move linearly at a speed of 1.0 m/sec and that the field of view has a diameter of 0.5 m. Suppose further that 360 projections over 180° are required and that it takes 0.5 sec for the source-detector apparatus to rotate one angular increment, regardless of the angle.

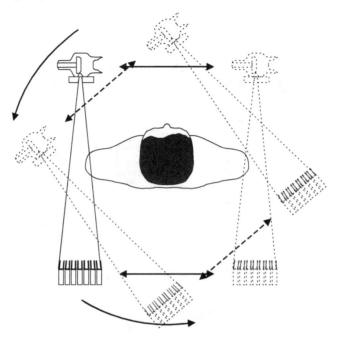

Figure 6.4
The geometry of a second-generation (2G) CT scanner.

Question What is the scan time for a 1G scanner? What is the scan time for a 2G scanner having 9 detectors spaced 0.5° apart?

Answer It takes $0.5/1.0 = 0.5$ sec to measure one projection. So the time used for making measurements is 360×0.5 s $= 180$ sec, and 360 rotations are needed, which take 360×0.5 s $= 180$ s. So the total scan time is $180 + 180 = 360$ sec, or 6 minutes.

For the 2G scanner, 9 projections are acquired simultaneously. (The required angular increment is $180°/360 = 0.5°$, which agrees with the detector separation.) Therefore, only $360/9 = 40$ rotations are required, which takes 40×0.5 s $= 20$ s. Ignoring the small amount of overscanning required to cover the fan angle itself, the time for each linear scan is still 0.5 sec. Therefore, the total scan time is $40 \times 0.5 + 20 = 40$ sec. ∎

The geometry of a *third-generation (3G) scanner*, as shown in Figure 6.5, has a fan-beam that covers the image region with the source held in a single position. Therefore, the source and detector array need not perform a linear scan; instead they simply rotate in synchrony. To obtain a sufficient number of samples per (fan-beam) projection, there must be a large number of detectors, and this pushes the cost of a 3G scanner beyond that of a 2G scanner. The simple rotational motion and highly parallel detection capability, however, allows for a dramatic decrease in scan time. A typical scanner acquires 1,000 projections with a fan-beam angle of 30 to 60 degrees incident upon 500 to 700 detectors and does this in 1 to 20 seconds. As in 2G systems, the detectors of a 3G system are collimated to reduce scattering artifacts, so there is some loss of detector collection efficiency. Furthermore, since the detectors must be very small, they cannot be as efficient as those in 1G and 2G

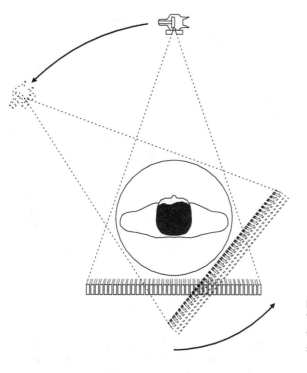

Figure 6.5
The geometry of a third-generation (3G) CT scanner.

systems. Therefore, once again, to obtain the same image quality with a 3G scanner as that obtainable in 1G and 2G systems, the dose must be slightly greater.

A *fourth-generation (4G) scanner*, as shown in Figure 6.6, has a single rotating source with a larger ring of stationary detectors. A variation on this theme has the source outside the detectors with slight gaps between the detectors through which the x-rays can be fired. Collimation cannot be used in this geometry since a detector must receive energy from a source that moves through many positions. Because the detectors are not collimated and because they can be physically large since they lie on a large ring, the detection efficiency of a 4G scanner is higher than that of a 3G scanner. However, scattering is a major problem with 4G systems; because the detectors cannot be collimated, they can also receive scattered radiation. These two factors compromise the performance, so that image quality in 4G systems is comparable to that of 3G systems.

One major design feature shared by the first four generations of CT scanners is the presence of only one x-ray tube. This feature is primarily due to the fact that x-ray tubes are expensive and bulky and that they must be constantly calibrated. The need to rotate a bulky object—x-ray tube and/or detectors—limits the overall imaging speed of the scanner. *Fifth-generation (5G) scanners* are designed to solve this problem. Fifth generation CT scanners, also known as *electron beam computed tomography* (EBCT) scanners, are commercially available at this time but are not in widespread clinical use because of their high cost. EBCT uses a flying electron beam, steered electromagnetically, to hit one of four tungsten anode strips that encircle the patient. X-rays are generated when the electron beam strikes the tungsten

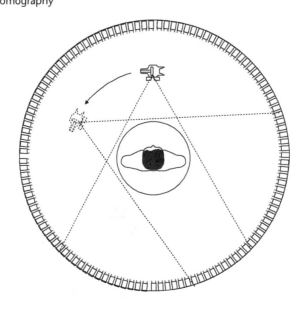

Figure 6.6
The geometry of a
fourth-generation (4G) CT
scanner.

anode; the resultant radiation is collimated into a fan-beam, which passes through the patient and is detected on the other side by a stationary ring of detectors, as in 4G CT systems. Since the anodes and detectors are stationary, no moving parts are required, and this allows a full set of fan-beam projection data to be acquired in about 50 milliseconds. EBCT is an expensive design, but because of the extremely small scan-time it is the only commercially available CT method that can capture stop-action images of a beating heart without electrocardiographic (ECG) gating.

Sometimes considered the *sixth-generation (6G) scanner*, the *helical CT* was developed in the late 1980s to address the need for rapid volumetric data acquisition. Today, about 15 percent of the CT scanners in the United States are capable of helical CT; and the changeover is accelerating. The reason is simple: With helical CT, full 60 cm torso scan can be obtained in about 30 seconds, a full 24 cm lung study in 12 seconds, and a detailed 15 cm angiography study in 30 seconds. With these kinds of speeds, motion artifacts can be reduced or eliminated (using breath-holds), and critically ill patients can be scanned quickly. For entry-level scanners, the equipment may cost about US$500,000 for a helical CT unit versus about US$350,000 for a conventional CT unit. A top-of-the-line helical CT unit may reach the $1 million mark.

A helical CT scanner consists of a conventional arrangement of the x-ray source and detectors (as in 3G and 4G systems) which can continuously rotate. While the tube is rotating and acquiring projection data, the patient table is set into motion, sliding the patient through the source-detector plane. It is easy to see that with this geometry and movement, the position of the source carves out a helix with respect to the patient. Continuous rotation of the large mass comprising the x-ray source and detectors requires what is called *slip ring* technology in order to communicate with the controlling stationary hardware. In particular, power is provided using rings and brushes, while data is passed using optical links. Rotation periods are typically 1 to 2 seconds per revolution.

A *seventh-generation (7G) scanner* is emerging with the advent and growth of *multislice* CT scanners. In these scanners, a "thick" fan-beam is used, and multiple (axial) parallel rows of detectors are used to collect the x-rays within this thick fan. Essentially, multiple (up to 64) 1-D projections are collected at the same time using this geometry. This leads to more economical use of the x-ray tube, since less of the generated beam must be absorbed by lead. Also, provided that the x-ray tube and detectors are far enough apart, the simultaneous data that are collected from the parallel rows of detectors corresponds approximately to parallel planes. This means that a "slab" of cross-sectional images can be imaged simultaneously. When a multiple detector array is combined with a helical scanner, the pitch of the helix can be larger, and full 3-D scans can be even faster.

The advent of helical and multislice CT has made the requirement for new developments in data processing even more critical. In particular, while a conventional CT might have reconstructed 40 slices over a region of interest, with helical CT there is nothing to stop a clinician from acquiring 80 to 120 slices over the same region in less time. Some extensive studies may comprise 500 slices—e.g., a 50 cm scan at 1 mm intervals might be used for imaging the colon. The burden is now to read these slices, which takes time. A recent study was conducted in which helical CT images were used to detect liver lesions. In this study, it was found that reviewers could detect 86 to 92 percent of the lesions using all the data, but only 82 to 85 percent of the lesions using every other slice. Most clinicians are looking to image processing and computer vision techniques—e.g., virtual reality—to reduce the amount of time it takes to read this enormous volume of data. At the same time, it is believed that considerably improved diagnostic accuracy can be achieved.

6.2.2 X-Ray Source and Collimation

Nearly all commercial medical CT scanners use just one x-ray source due to the initial cost and maintenance costs associated with x-ray tubes. In most respects, the x-ray tubes in CT scanners are just like those in projection radiography systems. All use the rotating anode design and are oil-cooled to reduce heat damage to the anode. Some CT scanners operate in pulse mode, but most continuously excite the x-ray tube during data acquisition. Some scanners use a pulse mode that excites the tube at alternating kVp's—e.g., 80 kVp and 140 kVp—so that dual energy studies can be conducted. In most cases, cool-down periods are required between scans to avoid heat damage to the x-ray tube. Even so, with 20 to 30 exposures per examination and 20 to 30 examinations per day, the x-ray tubes in CT scanners often wear out in less than one year. Maintaining the x-ray tubes through regular calibration and replacing them quite often adds considerably to the cost of operating a CT scanner.

The x-rays generated by a CT x-ray tube require collimation and filtration that is somewhat different than that in projection radiography systems. For one, CT typically requires a fan-beam geometry (3G, 4G, and 5G systems) rather than the cone-beam geometry used in projection radiography. Collimation into a fan, typically between 30 to 60 degrees in fan angle, is accomplished using two pieces of lead that form a slit between them. This collimator is placed as close to the patient as possible, just outside a protective plexiglass tube, so that the geometry of the fan has the most constant thickness possible through the patient. A motor controls how wide the slit is, so that the fan thickness (or height)—typically 1 to 10 mm—can be selected by the operator. In multislice CT systems, the thickness of the fan is typically larger than in single-slice systems in order to cover the width (perhaps 20 to 30 mm) of the multiple detectors, but otherwise the principle is the same. The x-rays pass through a mirror, which can be optically illuminated from below in order to create a thin line on the patient showing the position and thickness of the x-ray collimation to be used during the scan.

As we will see in Section 6.3, CT image formation requires the x-ray source to be approximated as a monoenergetic source. In order to make the actual CT x-ray source more monoenergetic in practice, more filtering is employed than is typical in projection radiography systems. Generally, copper followed by aluminum is used to narrow the energy spectrum ("harden the beam") of the x-rays entering the patient. Such filtration was discussed in Section 5.2.2.

6.2.3 CT Detectors

Most modern scanners use so-called *solid-state detectors*, as shown in Figure 6.7(a). These detectors contain a *scintillation crystal* in the first stage, typically a cadmium tungstate, sodium iodide, bismuth germanate, yttrium-based, or cesium iodide crystal. X-rays interact with the crystal mainly by photoelectric effect, producing photoelectrons, similar to what happens to phosphor in an intensifying screen. These electrons are excited and emit visible light when they spontaneously de-excite. This *scintillation* process results in a burst of light. The light is then converted to electric current using a solid-state photo-diode which is attached tightly to the scintillator. (Some 4G and 5G scanners use photomultiplier tubes to convert light to electricity; see Section 8.2.3 for a description of such tubes.) For 1G, 2G, and 3G systems, which use line collimators in front of the detectors, the efficiency of each detector may be increased by using thick crystals. However, thinner crystals must be used in 4G systems to allow the detectors to detect off-axis incident photons (as is required by the scanner geometry); unfortunately, thinner crystals have lower efficiency.

In 3G scanners, very small and highly directional detectors are required. Either solid-state detectors, as described above, or xenon gas detectors, as shown in

Figure 6.7
(a) Solid-state detectors, (b) xenon gas detectors, and (c) multiple (solid-state) detector array.

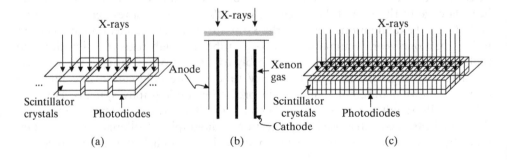

Figure 6.7(b), can be designed to satisfy these requirements. Xenon gas detectors use compressed xenon gas in long, thin tubes, which when ionized generate a current between an anode and cathode (maintained at a high potential difference). These detectors are typically less efficient than solid-state detectors, but they are highly directional, and therefore appropriate for the geometry of 3G scanners. Solid-state detectors must be accompanied by external collimation in order to provide necessary directivity (i.e., scatter rejection). In order to provide both good resolution and high efficiency, in a single-slice scanner each crystal would typically be 1.0 mm × 15 mm in size.

An important concept related to detector dimensions is *slice thickness*. In a typical CT scanner, the axial response is a rect function whose width is the slice thickness. The maximum slice thickness is equal to the detector height, 15 mm in the typical case. Smaller slice thicknesses are created by further restricting the beam using movable blades in the x-ray tube collimator. Slice thickness is controlled at the console and is typically set in millimeter increments, starting from 1 mm.

In a multiple detector array system, as shown in Figure 6.7(c), the individual solid-state detectors would typically be 1.0 mm × 1.25 mm in size. In these systems, slice thickness is controlled by the detector height; in particular, slice thickness must be in multiples of the detector height. For example, if there are 16 rows of detectors, each 1.25 mm in height, we could acquire 16 simultaneous projections, each 1.25 mm thick. Alternatively, we could acquire 8 simultaneous projections, each 2.5 mm thick, and so on. We will see later that there is a tradeoff in image quality that must be addressed in selecting slice thickness, so it is not always a good idea to simply acquire the thinnest possible slices.

6.2.4 Gantry, Slip Ring, and Patient Table

The *gantry* of a CT system holds the x-ray tube and detectors so that they can be rotated around the patient rapidly and repeatably. The fan angle, size of the detector array, and separation between the x-ray tube and detector array must be capable of imaging a 50 cm (typical) field of view. A full 2-D scan in under 1 sec is required. The gantry can typically be tilted in order to facilitate acquisition of nonaxial slices.

Most modern scanners are capable of continuous rotation without requiring a "rewinding" of the source and detectors. Since there are large voltages that must go to the x-ray tube and hundreds of signals that must be transmitted out of the detectors, continuous rotation makes special demands on the mechanical and electrical design, which the so-called *slip ring* solves. A slip ring comprises a large cylinder with grooves on the outside so that brushes can make continuous electrical contact with the rotating cylinder. The x-ray tube and detectors are mounted inside the cylinder and are in continuous electrical contact with the (stationary) controlling and data processing electronic hardware.

The patient table is more than just a place to put the patient. In helical scanners, it is an integral part of the data acquisition hardware, since it must be moved smoothly and precisely in synchrony with the source and detector rotation. Even in single-slice scanners, the table's positioning capabilities must be quite flexible. Typically, the table can be extended well out from the gantry and lowered in order to facilitate patient transfer from a hospital bed. It must be capable of "docking" with the scanner and locked into place so that motor-controlled positioning of the patient can be accomplished. Typically, there are positioning lights that can illuminate the patient to show the technician where the slice(s) will be acquired. The table can then be moved to position the patient under the guide lights in order to acquire the desired image(s).

6.3 Image Formation

6.3.1 Line Integrals

In a CT system, the x-ray tube makes a short burst of x-rays that propagate (generally in a fan-beam geometry) through a cross-section of the patient. The detectors detect the exit beam intensity integrated along a line between the x-ray source and each detector. As presented in Chapter 5 [see (5.3)], the (integrated) x-ray intensity at any given detector is given by

$$I_d = \int_0^{E_{\max}} S_0(E)E \exp\left[-\int_0^d \mu(s; E)ds\right] dE, \tag{6.1}$$

where $S_0(E)$ is the x-ray spectrum and $\mu(s; E)$ is the linear attenuation coefficient along the line between the source and detector.

Unfortunately, the integration over energy in (6.1), while physically correct, is in a mathematically intractable form for CT image reconstruction (to be discussed in Section 6.3.3). To get around this, we use the concept of *effective energy*, $\overline{E}$, which is defined as the energy that, in a given material, will produce the same measured intensity from a monoenergetic source as is measured using the actual polyenergetic source. This concept was first presented in Example 4.4. Given this concept, it is correct to say that

$$I_d = I_0 \exp\left[-\int_0^d \mu(s; \overline{E})\, ds\right]. \tag{6.2}$$

Given a measurement of I_d and knowledge of I_0, (6.2) can be rearranged to yield a basic projection measurement g_d,

$$g_d = -\ln\left(\frac{I_d}{I_0}\right), \tag{6.3}$$

$$= \int_0^d \mu(s; \overline{E})\, ds. \tag{6.4}$$

Thus, we can make a very important observation about the function of a CT scanner: The basic measurement of a CT scanner is a line integral of the linear attenuation coefficient at the effective energy of the scanner.

In an actual CT system, the *reference intensity* I_0 must be measured for each detector; this is a calibration step. In fan-beam systems, an auxiliary measurement typically is made using a detector positioned at the end of the array so that there is always air between the source and this detector. Using this measurement, the reference intensities for the other detectors can be determined using prior calibration data of all detectors taken in air.

6.3.2 CT Numbers

A CT scanner reconstructs the value of μ at each pixel within a cross-section, using a process that we will describe in great detail in the following sections. Different CT scanners, however, have different x-ray tubes, which in turn have different effective energies. Thus, the exact same object will produce different numerical values of μ on different scanners. Worse, since the x-ray tube on a busy CT scanner may need to be replaced about once every year, the same CT scanner will produce a different scan of the same object in successive years. This is clearly not a desirable situation.

In order to compare data from different scanners, which may have different x-ray sources and hence different effective energies, CT numbers are computed from the measured linear attenuation coefficients at each pixel. The *CT number* is defined as

$$ h = 1000 \times \frac{\mu - \mu_{\text{water}}}{\mu_{\text{water}}}, \tag{6.5} $$

and it is said to contain Hounsfield units (HU). Clearly, $h = 0$ HU for water; and since $\mu = 0$ in air, we find that $h = -1000$ HU for air. The largest CT numbers typically found naturally in the body are for bone, where $h \approx 1000$ HU for average bone, although CT numbers can surpass $h \approx 3000$ HU for metal and contrast agents. Usually, CT numbers are rounded or truncated to the nearest integer; they are typically reproducible to about ± 2 HU between scans and across scanners.

6.3.3 Parallel-Ray Reconstruction

We have learned that the basic CT measurement is a line integral of the effective linear attenuation coefficient within a cross-section. But line integrals are not what we desire; we really want a picture of μ, or equivalently its CT number h, over the entire cross-section. Therefore, the important question is this: Can we reconstruct a picture of μ given a collection of its line integrals? The answer is "yes" to this question, and in this section we explore the theory and practice of reconstruction from projections for parallel-ray geometry.

Geometry Let x and y be rectilinear coordinates in the plane. A line in the plane is given by

$$ L(\ell, \theta) = \{(x, y) \mid x \cos \theta + y \sin \theta = \ell\}, \tag{6.6} $$

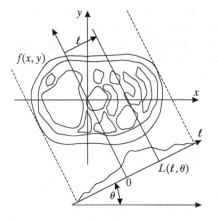

Figure 6.8
The geometry of lines and projections.

where ℓ is the lateral position of the line and θ is the angle of a unit normal to the line, as first introduced in (2.10) and shown in Figure 6.8. The line integral of function $f(x, y)$ is given by

$$g(\ell, \theta) = \int_{-\infty}^{\infty} f(x(s), y(s)) \, ds \,, \tag{6.7}$$

where

$$x(s) = \ell \cos \theta - s \sin \theta \,, \tag{6.8}$$

$$y(s) = \ell \sin \theta + s \cos \theta \,. \tag{6.9}$$

This is a parametric form of the line integral; an alternate expression is given by

$$g(\ell, \theta) = \int_{-\infty}^{\infty} \int_{-\infty}^{\infty} f(x, y) \delta(x \cos \theta + y \sin \theta - \ell) \, dx dy \,. \tag{6.10}$$

Here, the sifting property of the 1-D impulse function $\delta(\cdot)$ [see (2.6)] causes the integrand to be zero everywhere except on the line $L(\ell, \theta)$. The integral acts on the delta function by integrating the values of $f(x, y)$ only along the line—hence, it takes a line integral.

For a fixed θ, $g(\ell, \theta)$ is called a *projection*; for all ℓ and θ, $g(\ell, \theta)$ is called the *2-D Radon transform* of $f(x, y)$. The relationship of a projection to the object $f(x, y)$ is shown in Figure 6.8. If we make the identifications

$$f(x, y) = \mu(x, y; \overline{E}) \,, \tag{6.11}$$

$$g(\ell, \theta) = -\ln \left(\frac{I_d}{I_0} \right) \,, \tag{6.12}$$

we see that this mathematical abstraction exactly characterizes the CT measurement situation. In what follows, we assume that $g(\ell, \theta)$ corresponds to the measurements and $f(x, y)$ corresponds to the underlying unknown function or object that we wish to reconstruct. Notice that our definition of a projection corresponds to a collection of line integrals for parallel lines. Hence, these are called *parallel-ray projections*, and they correspond in geometry to 1G CT scanners only. We will find in Section 6.3.4, however, that the formalism and methods developed in this section lead directly to a reconstruction approach for fan-beam projections.

EXAMPLE 6.2
Consider the unit disk given by

$$f(x, y) = \begin{cases} 1 & x^2 + y^2 \leq 1 \\ 0 & \text{otherwise} \end{cases}.$$

Question What is its 2-D Radon transform?

Answer This function is circularly symmetric, so its projections are independent of the angle. It is therefore sufficient to calculate the projection at $\theta = 0°$ where the lateral displacement ℓ is horizontal and line integrals are vertical. (See Figure 6.9.)

Accordingly, we have

$$g(\ell, \theta) = \int_{-\infty}^{\infty} f(\ell, y) \, dy.$$

Here, $g(\ell, \theta) = 0$ when $|\ell| > 1$. When $|\ell| \leq 1$, the integral must integrate the function value on the unit disk, just the number 1, from the bottom of the unit circle to the top. Accordingly, in this range we can write

$$g(\ell, \theta) = \int_{-\sqrt{1-\ell^2}}^{\sqrt{1-\ell^2}} dy.$$

Performing the integration and supplying the known value outside this range yields

$$g(\ell, \theta) = \begin{cases} 2\sqrt{1 - \ell^2} & |\ell| \leq 1 \\ 0 & \text{otherwise} \end{cases},$$

which is the desired result.　　　　　　　　　　　　　　　　　　　■

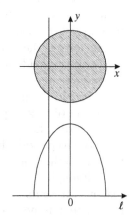

Figure 6.9
Projection of a disk.

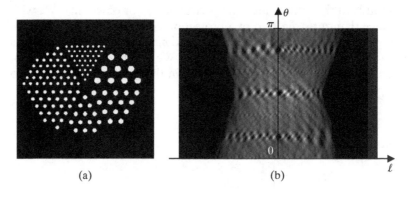

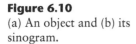

Figure 6.10
(a) An object and (b) its
sinogram.

Sinogram   An image of $g(\ell, \theta)$ with ℓ and θ as rectilinear coordinates is called a *sinogram*. It is a pictorial representation of the Radon transform of $f(x, y)$ and represents the data that are necessary to reconstruct $f(x, y)$.

An object and its sinogram are shown in Figure 6.10. It is important to observe several features about this sinogram. First, by convention in this book, the bottommost row of the sinogram corresponds to the projection of the object at $\theta = 0$. With reference to Figure 6.8, we see that this is the projection comprising vertical integrals of the object. Since the object in Figure 6.10(a) is widest in the horizontal direction, this gives the widest projection, which is evident in Figure 6.10(b).

Moving up from the bottommost row in Figure 6.10(b), the projections are increasingly thinner, and become thinnest at the angle $\theta = \pi/2$. This corresponds to the projection comprising horizontal integrals. The topmost row of the sinogram corresponds to an angle just "shy" of $\theta = \pi$, returning nearly to a projection comprising vertical lines. There is no need to proceed further in angle, since these projections are redundant.

There are other interesting features in Figure 6.10(b). It is possible to see sweeping features that run vertically through the sinogram. These features are actually sinusoidal in character, and if the sinogram were "turned on its side," they would be described by sinusoidal functions. This is the origin of the name *sinogram*. It is straightforward to show that these features correspond to individual objects in the image; the Radon transform of small points in the image corresponds to (vertical) sinusoids in the sinogram. There are also accumulations of bright white spots in certain rows of the sinogram in Figure 6.10(b). Upon careful inspection of the image in Figure 6.10(a), it is apparent that these bright spots correspond to the alignment of bright spots on parallel lines in the projection at these specific angles. In other words, the regular alignment of features in the image along certain directions gives rise to periodic bright spots in the projections, which are revealed as organized features in the sinogram.

Although the sinogram is certainly "understandable," it could never be argued that it represents an image of the object. Clearly, it is necessary to take the data represented in the sinogram and process them so that they look like the cross-section from which they came. This is the process of *reconstruction*, which we now begin to develop.

Backprojection Let $g(\ell, \theta)$ be the 2-D Radon transform of $f(x, y)$, and consider the projection at $\theta = \theta_0$. In general, there are an infinite number of functions $f(x, y)$ that could give rise to this projection; hence, we cannot determine $f(x, y)$ uniquely from a single projection. Intuition tells us that if $g(\ell, \theta_0)$ takes on a large value at $\ell = \ell_0$, then $f(x, y)$ must be large over the line (or somewhere on the line) $L(\ell_0, \theta_0)$. One way to create an image with this property is to simply assign every point on $L(\ell_0, \theta_0)$ the value $g(\ell_0, \theta_0)$. When we repeat this for all ℓ, the resulting function is called a *backprojection image* and is given formally by

$$b_\theta(x, y) = g(x \cos \theta + y \sin \theta, \theta) \,. \tag{6.13}$$

The backprojection image $b_{30°}(x, y)$ for the example in Figure 6.10 is shown in Figure 6.11(a).

Loosely speaking, the backprojection image at angle θ_0 is consistent with the projection at angle θ_0, but its values are assigned with no prior information about the distribution of the image intensities.[1] To incorporate information about the projections at other angles, we can simply add up (integrate) their backprojection images, yielding the so-called *backprojection summation image*

$$f_b(x, y) = \int_0^\pi b_\theta(x, y) \, d\theta \,, \tag{6.14}$$

also called a *laminogram*. An example of a backprojection summation image is shown in Figure 6.11(b). The early scanners did this using a discrete approximation to the integral. However, it was soon shown analytically that this is the wrong thing to do; this is also revealed from the blurriness of Figure 6.11(b). The concept is useful, however, because the correct procedure uses backprojection also, except that it applies to filtered versions of the projections, as we shall see.

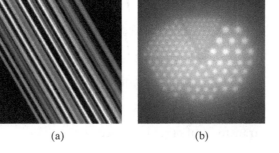

(a)   (b)

Figure 6.11
(a) A backprojection image and (b) a backprojection summation image.

---

[1]Actually, a backprojection image has infinite energy (if its domain is considered to be the infinite plane), and its projection at angle θ_0 is infinite. Therefore, it is not really consistent with the projection.

EXAMPLE 6.3

Question What is the backprojection image $b_\theta(x, y)$ at $\theta = 45°$ given the projection $g(\ell, 45°) = \operatorname{sgn}(\ell)$ where $\operatorname{sgn}(\ell)$ is -1 when $\ell \le 0$ and is $+1$ when $\ell > 0$.

Answer We have

$$b_{45°}(x, y) = g(x \cos 45° + y \sin 45°, 45°),$$

$$= g\left(x\frac{\sqrt{2}}{2} + y\frac{\sqrt{2}}{2}, 45°\right),$$

$$= g\left(\frac{\sqrt{2}}{2}(x + y), 45°\right).$$

When $x + y$ is negative, g is -1; when $x + y$ is positive, g is $+1$. Mathematically, this is

$$b_{45°}(x, y) = \begin{cases} -1 & x + y \le 0 \\ +1 & x + y > 0 \end{cases},$$

which is the desired result and is shown in Figure 6.12. ∎

Projection-Slice Theorem In this section, we develop a very important relationship between the 1-D Fourier transform of a projection and the 2-D Fourier transform of the object.

We begin by taking the 1-D Fourier transform of a projection with respect to ℓ:

$$G(\varrho, \theta) = \mathcal{F}_{1D}\{g(\ell, \theta)\} = \int_{-\infty}^{\infty} g(\ell, \theta) e^{-j2\pi\varrho\ell} \, d\ell, \tag{6.15}$$

where ϱ denotes spatial frequency (like u or v, except in an arbitrary direction). Now we substitute the analytic expression for $g(\ell, \theta)$ given in (6.10) and manipulate this expression:

$$G(\varrho, \theta) = \int_{-\infty}^{\infty} \int_{-\infty}^{\infty} \int_{-\infty}^{\infty} f(x, y)\delta(x \cos\theta + y \sin\theta - \ell) e^{-j2\pi\varrho\ell} \, dx\, dy\, d\ell,$$

$$= \int_{-\infty}^{\infty} \int_{-\infty}^{\infty} f(x, y) \int_{-\infty}^{\infty} \delta(x \cos\theta + y \sin\theta - \ell) e^{-j2\pi\varrho\ell} \, d\ell\, dx\, dy,$$

$$= \int_{-\infty}^{\infty} \int_{-\infty}^{\infty} f(x, y) e^{-j2\pi\varrho(x \cos\theta + y \sin\theta)} \, dx\, dy, \tag{6.16}$$

where the last step followed from the sifting property of the delta function [see (2.6)].

The final expression for $G(\varrho, \theta)$ in (6.16) is reminiscent of the 2-D Fourier transform of $f(x, y)$, defined as

$$F(u, v) = \int_{-\infty}^{\infty} \int_{-\infty}^{\infty} f(x, y) e^{-j2\pi(xu+yv)} \, dx\, dy, \tag{6.17}$$

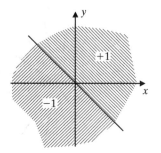

Figure 6.12
A backprojection image.

where here the variables u and v are the frequency variables in the x and y directions, respectively. In fact, making the identification $u = \varrho \cos\theta$ and $v = \varrho \sin\theta$ leads directly to the equivalence, and to the very important relationship

$$G(\varrho, \theta) = F(\varrho \cos\theta, \varrho \sin\theta). \qquad (6.18)$$

Equation (6.18) is known as the *projection-slice theorem*, and it is the basis of three important reconstruction methods. The relationship states that the 1-D Fourier transform of a projection is a *slice* of the 2-D Fourier transform of the object. Said another way, the 1-D Fourier transform of the projection equals a line passing through the origin of the 2-D Fourier transform of the object at that angle corresponding to the projection. A graphical interpretation is shown in Figure 6.13. We see that ϱ and θ may be interpreted as the polar coordinates of the 2-D Fourier transform.

EXAMPLE 6.4
The projection-slice theorem helps us to understand how angular sampling can influence reconstructions. Suppose that only eight projections are acquired at angles $\theta_i = \pi(i + 0.5)/8$, $i = 0, \ldots, 7$. Figure 6.14 shows the locations of the acquired Fourier data corresponding to these projections.

Question Show that the function $f(x, y) = \cos x$ is invisible at these angles—that is, it will produce projections that are identically zero at these angles.

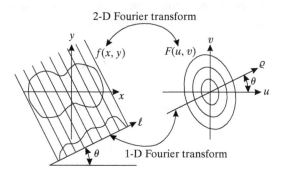

Figure 6.13
The projection-slice theorem in graphical form.

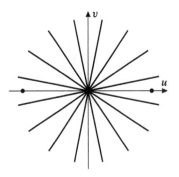

Figure 6.14
Figure for Example 6.4

Answer The 2-D Fourier transform of $\cos x$ is

$$F(u,v) = \pi \left[\delta(2\pi u - 1) + \delta(2\pi u + 1) \right] \delta(v),$$

which has two 2-D impulse functions on the u-axis, as shown in Figure 6.14. On the other hand, the projection-slice theorem tells us that, given the available observations, we will know the Fourier transform of f only on the lines $\{(\varrho \cos \theta_i, \varrho \sin \theta_i) \mid -\infty < \varrho < \infty\}$ for $i = 0, \ldots, 7$. Notice that although these lines are equally spaced over $[0, \pi]$, they do not include the u-axis. Therefore, they will be zero, and the function $\cos x$ will be invisible. ∎

The Fourier Method A conceptually simple reconstruction method, called the *Fourier method*, follows immediately from the projection-slice theorem. One simply takes the 1-D Fourier transform of each projection, inserts it with the corresponding correct angular orientation into the correct slice of the 2-D Fourier plane, and takes the inverse 2-D Fourier transform of the result. Accordingly, we have

$$f(x, y) = \mathcal{F}_{2D}^{-1}\{G(\varrho, \theta)\}. \tag{6.19}$$

The Fourier method is not widely used in CT due to the practical problem of interpolating polar data onto a Cartesian grid, and the need to use the relatively time-consuming 2-D inverse Fourier transform. Further analytic manipulation of (6.19), however, leads to improvements, as we shall see below.

Filtered Backprojection The inverse Fourier transform of $F(u, v)$ [see (6.17)] can be written in polar coordinates as

$$f(x, y) = \int_0^{2\pi} \int_0^{\infty} F(\varrho \cos \theta, \varrho \sin \theta) e^{j2\pi \varrho (x \cos \theta + y \sin \theta)} \varrho \, d\varrho \, d\theta. \tag{6.20}$$

Using the projection-slice theorem, we have

$$f(x, y) = \int_0^{2\pi} \int_0^{\infty} G(\varrho, \theta) e^{j2\pi \varrho (x \cos \theta + y \sin \theta)} \varrho \, d\varrho \, d\theta, \tag{6.21}$$

from which it follows that

$$f(x, y) = \int_0^\pi \int_{-\infty}^\infty |\varrho| G(\varrho, \theta) e^{j2\pi\varrho(x\cos\theta + y\sin\theta)} \, d\varrho d\theta, \qquad (6.22)$$

after some work and use of the fact that $g(\ell, \theta) = g(-\ell, \theta + \pi)$.

From the point of view of integration over ϱ, the term $(x\cos\theta + y\sin\theta)$ in (6.22) is a constant, say ℓ. Hence, (6.22) may be written as

$$f(x, y) = \int_0^\pi \left[\int_{-\infty}^\infty |\varrho| G(\varrho, \theta) e^{j2\pi\varrho\ell} \, d\varrho \right]_{\ell = x\cos\theta + y\sin\theta} d\theta. \qquad (6.23)$$

We recognize the inner integral in (6.23) as an inverse 1-D Fourier transform [see (2.64)]. The added term $|\varrho|$ makes (6.23) a *filtering* equation. Here, the Fourier transform of the projection $g(\ell, \theta)$ is multiplied by the frequency filter $|\varrho|$ and inverse-transformed. After the inverse transform, the filtered projection is backprojected (this is accomplished by replacing ℓ with $x\cos\theta + y\sin\theta$), which is followed by a "summation" of all filtered projections. This reconstruction approach is appropriately called *filtered backprojection*, and it is a considerable improvement over the Fourier method in speed and flexibility. The term $|\varrho|$ is known as the *ramp filter* because of its appearance in Fourier space.

An intuitive understanding of the role of the ramp filter can be found from inspection of Figure 6.14. Here, the straight application of the Fourier method leads to sampling that is inversely proportional to ϱ. Additional area or "weight" is required to compensate for the sparser sampling at higher frequencies.

Convolution Backprojection From the convolution theorem of Fourier transforms [see (2.92)], we may write (6.23) as

$$f(x, y) = \int_0^\pi \left[\mathcal{F}_{1D}^{-1}\{|\varrho|\} * g(\ell, \theta) \right]_{\ell = x\cos\theta + y\sin\theta} d\theta. \qquad (6.24)$$

Then, defining $c(\ell) = \mathcal{F}_{1D}^{-1}\{|\varrho|\}$, we have

$$f(x, y) = \int_0^\pi [c(\ell) * g(\ell, \theta)]_{\ell = x\cos\theta + y\sin\theta} \, d\theta, \qquad (6.25)$$

$$= \int_0^\pi \int_{-\infty}^\infty g(\ell, \theta) c(x\cos\theta + y\sin\theta - \ell) \, d\ell d\theta, \qquad (6.26)$$

where (6.26) results from substitution of the convolution integral and is the equation for *convolution backprojection*. Performing a convolution rather than a filtering operation (a pair of Fourier transforms with a multiplication in-between) is generally more efficient if the impulse response is narrow. It is so in this case, and generally, most CT scanners perform (some form of) convolution backprojection rather than filtered backprojection.

Unfortunately, $c(\ell)$ does not exist, since $|\varrho|$ is not integrable (and therefore its inverse Fourier transform is undefined). However, various formal expressions involving generalized functions (e.g., delta functions and their derivatives) or limits of functions have been developed. While these expressions are useful in a theoretical setting, we require a practical approach to design an impulse response that can be used in an actual CT scanner. This is accomplished by *windowing* $|\varrho|$ with a suitable windowing function $W(\varrho)$, such as a square, Hamming, or cosine window. Here, windowing denotes the use of a filter, in addition to the ramp filter, that modifies the observed projection. In practice, therefore, convolution backprojection algorithms use the approximate impulse response

$$\tilde{c}(\ell) = \mathcal{F}_{1D}^{-1}\{|\varrho|W(\varrho)\}. \tag{6.27}$$

It is also customary to set the filter value at $\varrho = 0$ to be nonzero, in order to produce the correct reconstructed average image value.

Reconstruction in Three Steps   Filtered backprojection (6.23) uses three basic steps to reconstruct an image from a sinogram: (1) filtering, (2) backprojection, and (3) summation. Convolution backprojection (6.24) uses the same three steps except that the filtering operation is implemented using a convolution. The reconstructed images resulting from these two approaches will be identical except for round-off errors due to numerical implementation issues.

Figure 6.15 illustrates the filtering (or convolution) step. Each row of the sinogram in Figure 6.15(a) is filtered using a windowed ramp filter to yield the filtered sinogram in Figure 6.15(b). The ramp filter is essentially a high-pass filter, and its value at $f = 0$ is 0. As a result, high frequency detail is accentuated, the

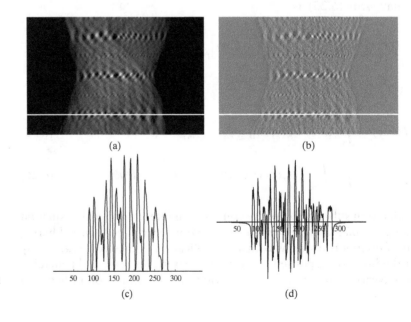

Figure 6.15
Convolution step:
(a) Original sinogram;
(b) filtered sinogram;
(c) profile of sinogram row [white line in (a)]; and
(d) profile of filtered sinogram row [white line in (b)].

background value is now zero (gray in the filtered sinogram), and there are negative values (dark values in the filtered sinogram). A projection and its corresponding filtered projection are also shown in this figure for comparison.

We have already seen the backprojection operation. Here, backprojection is applied to each filtered projection, rather than the raw projection. An example is shown in Figure 6.16. Since the filtered projections have both positive and negative values, backprojection images also have positive and negative values. It can be understood on an intuitive level that the presence of negative values will permit the reconstruction of a zero value outside the region of support of the object; of course, this requires the summation of many backprojection images for this to happen.

The last step in reconstruction is the summation step, a process that is illustrated in Figure 6.17. Notice that the summation process is accomplished using an "accumulator" concept. Starting with a "zero" image, the first (filtered) backprojection image is added. Then, the second backprojection image is added, and so forth.

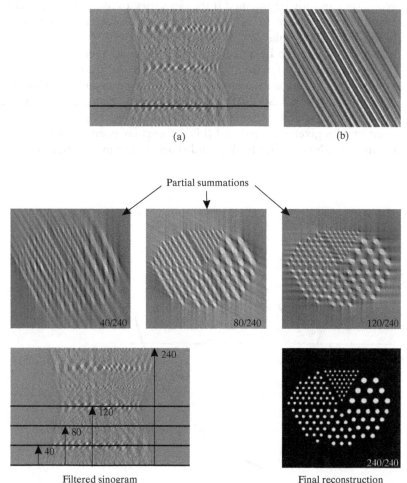

(a)　　　　　　　(b)

Figure 6.16
Backprojection step.

Partial summations

40/240　　　　80/240　　　　120/240

Filtered sinogram　　　　Final reconstruction

240/240

Figure 6.17
Summation step.

There is no need to store hundreds of backprojection images. The overall process of filtered backprojection is very time and memory efficient, which accounts for its great popularity over the past two or three decades.

6.3.4 Fan-Beam Reconstruction

As discussed in Section 6.2, all modern commercial scanners have a fan-beam source-to-detector arrangement. Therefore, in practice it is necessary to reconstruct images from fan-beam projections rather than parallel-ray projections. In this section, we develop a convolution backprojection reconstruction algorithm that can be applied to a fan-beam geometry.

Geometry   There are three basic fan-beam geometries to consider: (1) those that have equal angles between the measured ray-paths, (2) those that have equal detector spacing, and (3) those that have equal angles and equal detector spacing. The third geometry can only be satisfied if the detectors are placed along a circular arc whose center is at the source. In this section, we consider the equal-angle case with detectors positioned on a circular arc. A reconstruction formula for the case of equal detectors on a straight line can be developed using analogous means.

We consider the fan-beam geometry shown in Figure 6.18. This is a 3G geometry in which the source and detector rotate together around the laboratory origin, also called the rotational *isocenter*. A fan-beam projection, measured at each rotational position of the source, is denoted $p(\gamma, \beta)$ [in analogy to $g(\ell, \theta)$], where γ is the angular position of a given detector and β is the angular position of the source, as shown in Figure 6.18. Notice that both γ and β are angles measured in the counterclockwise

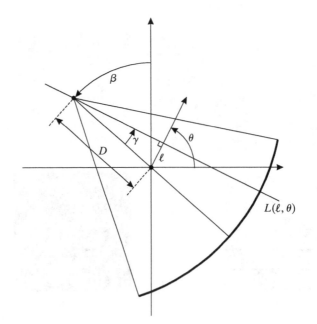

Figure 6.18
Fan-beam geometry.

direction; however, their origins are different and so are their reference directions (physical orientations when taking on zero values). In particular, β measures the source position around the *isocenter*, and it is zero when the source is resting on the $+y$-axis. On the other hand, γ is measured around the source position, and it is zero for rays that pass through the center of the detector array.

It is easiest to derive the fan-beam convolution backprojection formula by starting with what we already know about parallel-ray reconstruction. Within the fan-beam projection we have identified a line $L(\ell, \theta)$ in Figure 6.8, where ℓ is the lateral displacement and θ is the angular orientation, exactly as before. We can see from the geometry that the parameters of this line, ℓ and θ, are related to the fan-beam parameters by

$$\theta = \beta + \gamma, \tag{6.28}$$

$$\ell = D \sin \gamma, \tag{6.29}$$

where D is the distance from the source to the origin (isocenter).

Fan-Beam Reconstruction Formula Our development of a fan-beam reconstruction formula begins with the equation for parallel-ray convolution backprojection [(6.26)], which we repeat here for convenience:

$$f(x, y) = \int_0^\pi \int_{-\infty}^\infty g(\ell, \theta)c(x \cos \theta + y \sin \theta - \ell) \, d\ell \, d\theta. \tag{6.30}$$

From the geometry of parallel-ray tomography,

$$f(x, y) = \frac{1}{2} \int_0^{2\pi} \int_{-T}^T g(\ell, \theta)c(x \cos \theta + y \sin \theta - \ell) \, d\ell \, d\theta, \tag{6.31}$$

where we have assumed that $g(\ell, \theta) = 0$ for $|\ell| > T$. Thus, in (6.31) we consider rotation over the full circle and objects that are zero outside the disk with radius T, centered at the origin.

Let (r, ϕ) be polar coordinates in the plane. Then $x = r \cos \phi$, $y = r \sin \phi$, and $x \cos \theta + y \sin \theta = r \cos \phi \cos \theta + r \sin \phi \sin \theta = r \cos(\theta - \phi)$. Substituting this last expression into (6.31) gives

$$f(r, \phi) = \frac{1}{2} \int_0^{2\pi} \int_{-T}^T g(\ell, \theta)c(r \cos(\theta - \phi) - \ell) \, d\ell \, d\theta, \tag{6.32}$$

which is a parallel-ray reconstruction formula for f written in polar coordinates.

We now want to integrate over γ and β instead of ℓ and θ. This is a transformation of coordinates, where the transformation is given in (6.28) and (6.29). The Jacobian of this transformation is $D \cos \gamma$, and direct substitution yields

$$f(r, \phi) = \frac{1}{2} \int_{-\gamma}^{2\pi - \gamma} \int_{\sin^{-1} \frac{-T}{D}}^{\sin^{-1} \frac{T}{D}} g(D \sin \gamma, \beta + \gamma)$$

$$\times c(r \cos(\beta + \gamma - \phi) - D \sin \gamma)D \cos \gamma \, d\gamma \, d\beta. \tag{6.33}$$

The following simplifications to this integral can be made:

- The limits of the outer integral can be replaced by 0 (lower limit) and 2π (upper limit), since the functions are periodic in β with period 2π.
- The expression $\sin^{-1}\frac{T}{D}$ represents the largest angle γ_m that needs to be considered given that the object is contained in a disk of radius T. Thus, we can replace the upper and lower limits on the inner integral by γ_m and $-\gamma_m$, respectively.
- We recognize the fan-beam projection as

$$p(\gamma, \beta) = g(D\sin\gamma, \beta + \gamma).\qquad(6.34)$$

This gives the basic fan-beam reconstruction formula:

$$f(r, \phi) = \frac{1}{2}\int_0^{2\pi}\int_{-\gamma_m}^{\gamma_m} p(\gamma, \beta)c\left(r\cos(\beta + \gamma - \phi) - D\sin\gamma\right)D\cos\gamma\,d\gamma\,d\beta.\qquad(6.35)$$

Fan-Beam Convolution Backprojection In this section, we manipulate the fan-beam reconstruction formula of (6.35) into a form that more closely resembles ordinary convolution backprojection.

Consider an arbitrary point in the cross-section, given by the polar coordinates (r, ϕ), as shown in Figure 6.19. Its position can also be defined relative to the source/detector positions using the angle γ' and radius D', as shown in the figure. The argument of $c(\cdot)$ can be written in simpler form using these coordinates (after some trigonometric manipulation) as

$$r\cos(\beta + \gamma - \phi) - D\sin\gamma = D'\sin(\gamma' - \gamma),\qquad(6.36)$$

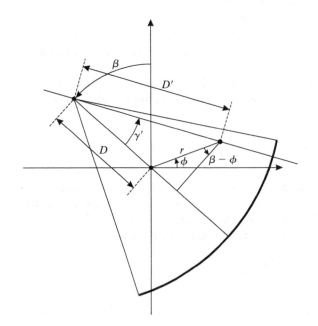

Figure 6.19
Rotating polar coordinates
for fan-beam geometry.

where γ is an arbitrary angle as depicted in Figure 6.18. This yields

$$f(r,\phi) = \frac{1}{2} \int_0^{2\pi} \int_{-\gamma_m}^{\gamma_m} p(\gamma,\beta)c(D'\sin[\gamma'-\gamma])D\cos\gamma \, d\gamma \, d\beta, \qquad (6.37)$$

where D' and γ' are determined by r and ϕ.

It can be shown (see Problem 6.18) that

$$c(D'\sin\gamma) = \left(\frac{\gamma}{D'\sin\gamma}\right)^2 c(\gamma). \qquad (6.38)$$

Accordingly, looking back at (6.37), we see that it is useful to define

$$c_f(\gamma) = \frac{1}{2}D\left(\frac{\gamma}{\sin\gamma}\right)^2 c(\gamma). \qquad (6.39)$$

Combining (6.37) and (6.38) with this definition, we find that

$$f(r,\phi) = \int_0^{2\pi} \frac{1}{(D')^2} \int_{-\gamma_m}^{\gamma_m} \tilde{p}(\gamma,\beta)c_f(\gamma'-\gamma) \, d\gamma \, d\beta, \qquad (6.40)$$

where $\tilde{p}(\gamma,\beta) = \cos\gamma \, p(\gamma,\beta)$, which is our final fan-beam reconstruction formula.

To more clearly identify the "convolution" part and the "backprojection" part, we split the reconstruction formula of (6.40) into two pieces:

$$q(\gamma,\beta) = \tilde{p}(\gamma,\beta) * c_f(\gamma), \qquad (6.41)$$

$$f(r,\phi) = \int_0^{2\pi} \frac{1}{(D')^2} q(\gamma',\beta) \, d\beta. \qquad (6.42)$$

Here, $q(\cdot)$ is simply a filtered projection where the impulse response $c_f(\cdot)$ is a weighted (or windowed) version of $c(\cdot)$, the usual (approximate) ramp filter [as in (6.27)]. However, the backprojection operator is a bit different, since D' depends on the image coordinate (r,ϕ) or, equivalently, (x,y). Therefore, each filtered projection backprojects along the ray-paths of integration, but does so with a weighting that depends on the distance from the source. Thus, one should think of the phrase *convolution weighted-backprojection* when the projections are fan-beams.

6.3.5 Helical CT Reconstruction

Consider the helical CT scenario, in which there is a single detector array and the patient table is continuously moved through the gantry. In this case, the nature of the acquired data is different than in single-slice CT. Since x-rays move at the speed of light, any acquired projection still corresponds to a single plane within the patient; however, because of the continuous movement of the patient, no two projections will correspond to the same plane. A given (fan-beam) projection is therefore identified uniquely by its longitudinal position z_j, measured relative to the patient. The index j, in fact, not only identifies the position z_j of the projection but also determines its angle β_j.

Suppose there are M angles acquired over 360 degrees. Then the acquired angles repeat such that $\beta_j - \beta_{j+M} = 0$. How far has the patient moved when the angle repeats? This is determined by the *pitch* of the helix, which is given by

$$\zeta = z_{j+M} - z_j, \tag{6.43}$$

which is in turn determined by the speed of gantry rotation and the table speed.

Suppose that we want to reconstruct an image at the longitudinal position z_0. In conventional (single-slice) tomography, we would need many projections taken at a collection of angles within the plane $z = z_0$. In the helical geometry, however, we are not guaranteed even to have one such projection (because z_0 may not correspond to z_n for any n). It is standard practice to "create" the needed projections using linear interpolation of the projections that are measured in nearby slices. For example, suppose we would like the projection corresponding to the angle β_j. We know that the projection angle β_j is acquired at all linear positions z_{j+kM} for $k = 0, 1, 2, \ldots$. It is straightforward to find k such that $z_{j+kM} \leq z_0 < z_{j+[k+1]M}$. We know that the scanner acquires projections at angle β_j at both positions z_{j+kM} and $z_{j+[k+1]M}$. Linear interpolation between these two projections yields an estimate of the desired projection

$$\hat{p}_{z_0}(\gamma, \beta_j) = \frac{z_0 - z_{j+kM}}{\zeta} p_{z_{j+[k+1]M}}(\gamma, \beta_j)$$

$$+ \frac{z_{j+[k+1]M} - z_0}{\zeta} p_{z_{j+kM}}(\gamma, \beta_j), \tag{6.44}$$

which is valid for any position γ within the projection. This procedure can be repeated for all $j = 1, \ldots, M$, noting that the choice of k depends on both z_0 and j. In this way, a complete set of projections for fan beam reconstruction of the 2-D slice at z_0 is created. A fan-beam reconstruction formula can then be used to reconstruct a picture of the slice.

Many factors will affect the quality of a helical CT image. In particular, the quality of interpolated data will be affected by the pitch of the helix. The pitch is determined by the rate of rotation of the gantry and the speed of the table feed. If the pitch is too big, while the fan is too thin, then aliasing artifacts will appear. To compensate, the fan could be made thicker, which provides a natural analog low-pass filter. Then, the images will be free from aliasing but will be blurry. In clinical scanners today, protocols are preset into the scanner for certain types of studies. These protocols balance the conflicting requirements of image quality with field-of-view requirements and speed of acquisition.

6.3.6 Cone Beam CT

With the goal of rapid coverage and reconstruction of 3-D volumes, it is natural to consider the use of a cone beam radiation pattern and an area detector (rather than a fan-beam and a linear detector). A standard fluoroscope, for example, can rotate the cone-beam source and 2-D detector in a circle around the body, electronically recording 2-D projection images at a collection of different angles. The data within the plane of the circular source path are exactly what one achieves with a conventional fan-beam CT scanner; thus, an image can be reconstructed by conventional means. It turns out, however, that there is not sufficient coverage to yield mathematically correct reconstructions in any other plane through the object. (This situation is analogous to that given in Example 6.4.) The situation is further complicated by the fact that projections are often *truncated*; i.e., the body extends beyond the x-ray cone.

Despite these complications, so-called *cone beam reconstruction algorithms* have been developed to produce passable 3-D reconstructions. The Feldkamp algorithm is the most well known, and has been shown to produce the best reconstructions possible for the geometry described above.

6.4 Image Quality in CT

While the theory presented in the previous sections captures the ideal behavior of CT systems, in practice there are limitations to what one can achieve. It is not possible to achieve arbitrarily high resolution, since detectors have finite width and there are sampling limitations, both in the number of projections and in the number of ray-paths per projection. As well, since the patient will absorb x-rays, we cannot use an arbitrarily large dose, and this limits the attainable signal-to-noise ratio (SNR). A third problem is beam-hardening, which is a phenomenon caused by energy selective absorption of x-rays by human tissues (as discussed in Section 5.2.2). In this section, we describe these phenomena, and indicate how they affect the appearance of reconstructed CT images.

6.4.1 Resolution

Filtered backprojection [see (6.23)] is an exact formula for the inverse Radon transform. As noted in (6.27), however, the ideal (ramp) filter $|\varrho|$ is not realizable, so an approximate filter $W(\varrho)|\varrho|$ using a window function $W(\varrho)$ [see (6.27)] must be used in practice. In addition, the line integrals themselves cannot be imaged exactly due to the finite size of the detectors. In fact, CT detectors are *area detectors* that locally integrate the underlying true signal, an effect that can be modeled using an additional filter $S(\varrho)$ [see (2.129)], which is the Fourier transform of the indicator function $s(\ell)$ defining a detector. Combining these effects, the filtered backprojection formula becomes

$$\hat{f}(x, y) = \int_0^\pi \left[\int_{-\infty}^\infty G(\varrho, \theta) S(\varrho) W(\varrho) |\varrho| e^{j2\pi\varrho\ell} \, d\varrho \right]_{\ell = x\cos\theta + y\sin\theta} d\theta . \qquad (6.45)$$

The reconstruction in (6.45) is approximate, and the resulting function is a blurred version of $f(x, y)$. We can discover a relation between $\hat{f}(x, y)$ and $f(x, y)$ by manipulating (6.45).

First, we recognize that (6.45) is precisely the inverse Radon transform of the function

$$\hat{g}(\ell, \theta) = \mathcal{F}^{-1}\{G(\varrho, \theta)S(\varrho)W(\varrho)\}.$$ (6.46)

By the 1-D convolution theorem,

$$\hat{g}(\ell, \theta) = g(\ell, \theta) * \tilde{h}(\ell),$$ (6.47)

where

$$\tilde{h}(\ell) = s(\ell) * w(\ell).$$ (6.48)

The result of filtered backprojection can therefore be interpreted as the inverse Radon transform of the blurry projections $\hat{g}(\ell, \theta)$.

In order to determine how the blurry projections cause the image of the object to be blurred, it is convenient to draw upon the *convolution property* of the Radon transform. In particular, it can be shown that the Radon transform of the convolution of two functions is the convolution of their Radon transforms [see Problem 6.9]:

$$\mathcal{R}\{f * h\} = \mathcal{R}\{f\} * \mathcal{R}\{h\}.$$ (6.49)

Notice that the convolution on the left in (6.49) is two-dimensional, while that on the right is one-dimensional. Comparing (6.47) with (6.49) allows us to identify

$$\mathcal{R}\{h\} = \tilde{h}(\ell),$$ (6.50)

and therefore, upon taking the inverse Radon transform of both sides,

$$h(x, y) = \mathcal{R}^{-1}\{\tilde{h}(\ell)\}.$$ (6.51)

From the convolution property of the Radon transform (6.49), we can conclude that the reconstructed object is given by

$$\hat{f}(x, y) = f(x, y) * \mathcal{R}^{-1}\{\tilde{h}(\ell)\}.$$ (6.52)

Therefore, the reconstructed estimate $\hat{f}(x, y)$ is a blurry version of the truth, $f(x, y)$, and the resolution of the system can be characterized by the point spread function $\mathcal{R}^{-1}\{\tilde{h}(\ell)\}$.

To find the point spread function $h(x, y)$, we use the projection slice theorem (6.18). The Fourier transform of $\tilde{h}(\ell)$ is given by

$$\tilde{H}(\varrho) = S(\varrho)W(\varrho),$$ (6.53)

which is independent of θ. Accordingly, the 2-D Fourier transform of $h(x, y)$ must be circularly symmetric and given by

$$H(q) = S(q)W(q). \tag{6.54}$$

It follows that the point spread function is also circularly symmetric and given by the inverse Hankel transform [see (2.113)]

$$h(r) = \mathcal{H}^{-1}\{S(\varrho)W(\varrho)\}. \tag{6.55}$$

Finally, the reconstructed image is given by

$$\hat{f}(x, y) = f(x, y) * h(r), \tag{6.56}$$

where $r = \sqrt{x^2 + y^2}$. The PSF is circularly symmetric and can be characterized by a FWHM given the detector width, which then determines $S(\varrho)$, and the ramp filter window function $W(\varrho)$.

EXAMPLE 6.5

Suppose a CT system uses rectangular detectors having width d and a rectangular window function with highest frequency $\varrho_0 \gg 1/d$.

Question   What is the approximate PSF of this CT system?

Answer   Rectangular detectors with width d blur the underlying projection with impulse response function

$$s(\ell) = \text{rect}\left(\frac{\ell}{d}\right).$$

In the frequency domain, this is represented by the filter

$$S(\varrho) = d \, \text{sinc}(d\varrho).$$

The ramp filter window function can be written as

$$W(\varrho) = \text{rect}\left(\frac{\varrho}{2\varrho_0}\right).$$

The first zero of $S(\varrho)$ is at frequency $\varrho = 1/d$. Since $\varrho_0 \gg 1/d$, the frequency cutoff defined by the detectors is much less than that of the ramp filter, and the ramp filter window function can be ignored. Therefore, the (circularly symmetric) 2-D transfer function is (approximately) given by

$$H(q) \approx d \, \text{sinc}(dq).$$

From Table 2.3,

$$\mathcal{H}\{\text{sinc}(r)\} = \frac{2 \, \text{rect}(q)}{\pi\sqrt{1 - 4q^2}}.$$

Using this fact, the fact that the forward and inverse Hankel transforms are the same, and the fact that [see (2.121)]

$$\mathcal{H}\{f(ar)\} = \frac{1}{a^2} F(q/a),$$

we can work out the following (approximate) impulse response:

$$h(r) = \mathcal{H}^{-1}\{d \operatorname{sinc}(d\varrho)\},$$

$$= d \frac{1}{d^2} \frac{2 \operatorname{rect}(r/d)}{\pi \sqrt{1 - 4(r/d)^2}},$$

$$= \frac{2 \operatorname{rect}(r/d)}{\pi \sqrt{d^2 - 4r^2}}. \qquad \blacksquare$$

6.4.2 Noise

Measurement Statistics   A CT detector measures the intensity of the x-ray pulse after losses in the patient. There is intrinsic noise in such a measurement, coming from the Poisson nature of x-rays [see (3.52)]. In addition, body attenuation and limited detector efficiency reduce the measured number of photons, further increasing noise.[2] We will make many approximations in this section in arriving at a useful approximate expression for the noise in a CT image. The first approximation is that we will assume that the CT system is monoenergetic, operating at the effective energy $\overline{E}$ of the scanner. In this case, the statistical behavior of the scanner can be understood by converting the observed intensity at a given detector I_d to an observed photon count as $N_d = I_d/\overline{E}$.

We denote the basic CT measurement for the ith detector and jth angle by

$$g_{ij} = -\ln\left(\frac{N_{ij}}{N_0}\right), \qquad (6.57)$$

where N_0 is the incident photon count and N_{ij} is the number of detected photons. Unfortunately, because the absorption of photons is a statistical phenomenon, the same experiment repeated again would probably yield a different measurement. In fact, N_{ij} is a Poisson random variable with mean

$$\overline{N}_{ij} = N_0 \exp\left(-\int_{L_{ij}} \mu(s)\, ds\right), \qquad (6.58)$$

where L_{ij} is the ray-path from the source to detector i for the jth projection angle. Equation (6.58) is just another way of writing the line integral introduced in (6.1). Notice that the mean depends on the number of incident photons and on the line

---

[2]The x-ray beam itself arises from a random process, and the number of x-rays in the beam is a Poisson random variable. Body attenuation and detector efficiency produce *random deletions* (in statistical jargon) of these x-rays, and the number of x-rays in the detected beam is also a Poisson random variable.

integral of the linear attenuation coefficient along the ray-path between the source and detector. In fact, the mean is the desired measurement.

The basic measurement g_{ij} is a transformation of the random variable N_{ij}, however, not a transformation of the mean of N_{ij}. Therefore, g_{ij} is a random variable and, assuming N_0 is large, its mean $\bar{g}_{ij}$ and variance $\text{var}(g_{ij})$ are given by

$$\bar{g}_{ij} \approx -\ln\left(\frac{\bar{N}_{ij}}{N_0}\right) \tag{6.59}$$

$$\text{var}(g_{ij}) \approx \frac{1}{\bar{N}_{ij}} \tag{6.60}$$

We assume that the random variables N_{ij} are independent, and it follows that the random variables g_{ij} are also independent. Given this second-order characterization of the basic CT measurements, we are now in a position to determine the second-order statistics of the images that we create from these measurements. These second-order statistics will permit us to define a signal-to-noise ratio of reconstructed images, which is a measure of the quality of reconstructions.

Image Statistics In this section, we develop an expression for the mean and variance of the reconstructed linear attenuation coefficients when (discrete) convolution backprojection (CBP) is used. For simplicity, we assume parallel-ray geometry.

A discrete approximation to the CBP integral in (6.26) yields the discrete CBP formula

$$\hat{\mu}(x, y) = \frac{\pi T}{M} \sum_{j=1}^{M} \sum_{i=-N/2}^{N/2} g_{\theta_j}(iT)\, \tilde{c}\left(x \cos \theta_j + y \sin \theta_j - iT\right), \tag{6.61}$$

where M is the number of projections taken over the range $[0, \pi)$, $N+1$ (odd) is the number of ray-paths per projection, T is the physical spacing between detectors, and $g_{ij} = g_{\theta_j}(iT)$. Notice that $\tilde{c}(\cdot)$ is a realizable approximation to the ramp filter. The mean of the reconstructed image is therefore

$$\text{mean}[\hat{\mu}(x, y)] = \frac{\pi T}{M} \sum_{j=1}^{M} \sum_{i=-N/2}^{N/2} \bar{g}_{ij}\, \tilde{c}\left(x \cos \theta_j + y \sin \theta_j - iT\right), \tag{6.62}$$

which for large N_0, M, and N is exactly what we want. Therefore, CBP has the desirable property that the mean of the reconstructed image approaches exact reconstruction as the quality of measurement increases.

Since g_{ij} are assumed to be independent random variables, the variance of the sum is the sum of the variances, given by

$$\sigma^2(x, y) = \text{var}[\hat{\mu}(x, y)], \tag{6.63}$$

$$= \frac{\pi^2 T^2}{M^2} \sum_{j=1}^{M} \sum_{i=-N/2}^{N/2} \text{var}[g_{ij}] \left[c \left(x \cos \theta_j + y \sin \theta_j - iT \right) \right]^2 , \quad (6.64)$$

$$= \frac{\pi^2 T^2}{M^2} \sum_{j=1}^{M} \sum_{i=-N/2}^{N/2} \frac{1}{\overline{N}_{ij}} \left[c \left(x \cos \theta_j + y \sin \theta_j - iT \right) \right]^2 . \quad (6.65)$$

To proceed, we make a rather drastic approximation: $\overline{N}_{ij} \approx \overline{N}$. That the mean number of detected photons is a constant is clearly false for nearly all objects; however, by using this approximation, we can develop some important relationships that are otherwise obscured in the summations.

Using the approximation $\overline{N}_{ij} \approx \overline{N}$,

$$\sigma^2(x, y) = \sigma_\mu^2 = \frac{\pi^2 T^2}{M^2 \overline{N}} \sum_{j=1}^{M} \sum_{i=-N/2}^{N/2} \left[c \left(x \cos \theta_j + y \sin \theta_j - iT \right) \right]^2 . \quad (6.66)$$

For large N and M, we may make the approximation

$$\frac{\pi}{M} \sum_{j=1}^{M} T \sum_{i=1}^{N} \left[c \left(x \cos \theta_j + y \sin \theta_j - iT \right) \right]^2$$

$$\approx \int_0^\pi \int_{-\infty}^\infty \left[c \left(x \cos \theta + y \sin \theta - \ell \right) \right]^2 d\ell d\theta , \quad (6.67)$$

$$= \pi \int_{-\infty}^\infty \left[c \left(\ell \right) \right]^2 d\ell , \quad (6.68)$$

$$= \pi \int_{-\infty}^\infty \left| C(\varrho) \right|^2 d\varrho , \quad (6.69)$$

where the last equality follows from Parseval's theorem [see (2.97)]. For a rectangular window with bandwidth ϱ_0 applied to the ramp filter,

$$\pi \int_{-\infty}^\infty \left| C(\varrho) \right|^2 d\varrho = \pi \int_{-\varrho_0}^{\varrho_0} \varrho^2 \, d\varrho = \frac{2\pi \varrho_0^3}{3} . \quad (6.70)$$

Thus, for a rectangularly windowed ramp filter and the approximation $\overline{N}_{ij} = \overline{N}$, the reconstructed image variance is independent of (x, y), and is given by

$$\sigma_\mu^2 \approx \frac{\pi T}{M} \frac{1}{\overline{N}} \frac{2\pi \varrho_0^3}{3} , \quad (6.71)$$

$$\approx \frac{2\pi^2}{3} \varrho_0^3 \frac{1}{M} \frac{1}{\overline{N}/T} . \quad (6.72)$$

We now interpret this expression for image noise variance with the following observations:

- If we increase the bandwidth of the rectangular window, the image variance also increases. This follows one's natural intuition that noise has high frequency components, while the image bandwidth tends to diminish at high frequencies. We will see in nuclear medicine applications (Chapter 9) that the design of windowing functions that optimally balance the increase in noise contributions with the loss of image sharpness is very important.

- If we decrease T, the spacing between detectors, the variance decreases. Notice, however, that if the size of each detector is forced to decrease in order to do this, or the efficiency decreases, then the decreased spacing will be offset by a lower $\overline{N}$.

- If we increase $\overline{N}$, the variance decreases. In general, one increases N_0 to do this. One could also increase $\overline{N}$ by increasing the incident x-ray energy, which lowers the amount of x-ray absorption. However, one would then have a loss of contrast, a subject that is considered in the following section on SNR.

- If we increase M, the variance decreases. Thus, the more angles the better. This assumes constant acquisition time per angle—it would not be true with constant total scan time. (However, increased M for constant total scan time could still provide benefits from improved angular sampling.)

- The fraction $\overline{N}/T$ indicates the average number of photons per unit distance along the detector array. In fan-beam geometries, we can increase this ratio, and hence decrease the image variance, by increasing N_0.

Image SNR Image variance is an important concept when we are trying to measure the linear attenuation coefficient. However, in radiology we must also consider the contrast between tissues, since this is what creates the visual effect of objects or structures with boundaries in a medical image. Therefore, any useful definition of SNR must contain both contrast and noise. We define SNR here as

$$\mathrm{SNR} = \frac{C\overline{\mu}}{\sigma_\mu}, \tag{6.73}$$

where C is the fractional change in μ from $\overline{\mu}$, $\overline{\mu}$ is the mean linear attenuation coefficient, and σ_μ is the standard deviation of the measurement (the square root of the variance). Equation (6.73) is thus the CT version of the differential SNR introduced in (3.69). Combining (6.72) with (6.73) gives

$$\mathrm{SNR} = \frac{C\overline{\mu}}{\pi}\varrho_0^{-3/2}\sqrt{\frac{3}{2}(\overline{N}/T)M}. \tag{6.74}$$

In a good CT scanner design, we set $\varrho_0 \approx k/d$, where $k \approx 1$ and d is the width of a detector. This gives

$$\mathrm{SNR} \approx 0.4kC\overline{\mu}d^{3/2}\sqrt{(\overline{N}/T)M}. \tag{6.75}$$

If, in fact, $d = T$ (as in a 3G scanner), then

$$\text{SNR} \approx 0.4k\overline{C\mu}d\sqrt{\overline{N}M}. \tag{6.76}$$

In the fan-beam case,

$$\overline{N}_f : \text{mean photon count per fan-beam,}$$

$$D : \text{number of detectors,}$$

$$L : \text{length of detector array,}$$

and, therefore, $\overline{N} = \overline{N}_f/D$ and $d = L/D$, which yields

$$\text{SNR} \approx 0.4k\overline{C\mu}LD^{-3/2}\sqrt{\overline{N}_fM}. \tag{6.77}$$

Therefore, in the fan-beam case we have the odd situation that increasing the number of detectors actually *lowers* the SNR. The reason, it turns out, is that convolution of the projections with the ramp filter couples the noise between detectors, and does so to a greater extent as the number of detectors increases. Why shouldn't we reduce the number of detectors to, say, three or one? Although such a system would give excellent SNR, we know from Section 6.4.1 that the resolution would be abysmal.

EXAMPLE 6.6

Consider a fan-beam CT system with one source, D detectors, M angles, and J by J reconstructed images, where $D = M = J = 256$. Assume that the width of each detector is $d = 0.25$ cm and the ramp filter uses a rectangular window with cutoff $\varrho_0 = 1/d$. The scanner is used to image a lesion with contrast $C = 0.005$ embedded in water ($\overline{\mu} = 0.15$ cm^{-1}).

Question   We require the image to have a signal-to-noise ratio of at least 20 dB. What is the minimum number of photons per projection at the detectors that is required in order to meet this SNR constraint?

Answer   Since $\text{SNR(dB)} = 20\log_{10}\text{SNR}$, we have

$$\text{SNR} = 10^{20\text{dB}/20} = 10.$$

Since $\text{SNR} = C\overline{\mu}/\sigma_\mu = 10$, we have

$$\sigma_\mu = \frac{0.005}{10} \times 0.15 \text{ cm}^{-1} = 7.5 \times 10^{-5} \text{ cm}^{-1}.$$

Thus,

$$\sigma_\mu^2 = 5.625 \times 10^{-9} \text{ cm}^{-2} = \frac{2\pi^2}{3}\frac{\varrho_0^3 T}{M\overline{N}}.$$

The number of photons per projection is

$$P_p = \overline{N}D = \frac{2\pi^2}{3}\frac{\varrho_0^3 T}{M\sigma_\mu^2}D.$$

Since $\varrho_0 = 1/d$, $T = d = 0.25$ cm, and $D = M$, we have

$$P_p = \frac{2\pi^2}{3} \frac{1}{0.25^2} \frac{1}{\sigma_\mu^2} = 1.87 \times 10^{10} \text{ minimum}.$$

∎

6.4.3 Artifacts

This section identifies and briefly explains some of the artifacts that can appear in CT images. Several of these are depicted in Figure 3.12.

Aliasing   The projections will be aliased if they are undersampled. The effect on a projection is that higher frequency information will appear as lower frequency artifacts, as discussed in Section 2.8. These artifacts will continue through the process of reconstruction (CBP, Fourier method, etc.), and appear as artifacts in the image. The most readily apparent of these artifacts are *streak artifacts* which appear to emanate, in particular, from small bright objects within the image. An insufficient number of projections also causes aliasing. These artifacts also appear in the reconstructed images, often as streaks emanating from object boundaries at points which have small radii of curvature. These streaks may be either dark or bright, depending on the precise location of the objects and edges within the field-of-view.

Aliasing may be eliminated by sampling at a high enough (spatial sampling) rate, assuming the object is band limited, or by low-pass filtering prior to sampling. Both of these goals are achieved to some degree in current clinical scanners. High sampling rates are achieved by providing numerous detectors within the fan-beam and by sampling at numerous angles. Ultimately, the number of detectors is limited by physical size, efficiency, and the fact that smaller detectors receive fewer photons and, hence, have higher measurement variance. The number of angles is limited by storage capacity time (both to scan and reconstruct), and dose. Low-pass filtering (along each projection) is achieved by the physical aperture of each detector, which acts as a boxcar filter. There is no effective low-pass filtering in angle.

Another potential source for artifacts is in the backprojection-summation process, which must supply filtered projection values for a finite set of points (x_i, y_j) (pixels) in the plane. The critical issues are how many pixels will be used over the field-of-view and what will be the method of interpolation. Too few, and we produce an aliased version of the correct image, showing moiré patterns that appear as a structured textures that do not really exist. For such a coarse image, the projections should be low-pass filtered before backprojection.

A general rule of thumb for designing CT systems is that the number of detectors should be approximately equal to the number of projections that should be approximately equal to the number of points on the side of a reconstructed image. A typical 3G system has around 700 detectors, acquires 1000 projections, and reconstructs a 512×512 image.

Beam Hardening   *Beam hardening* is caused by energy-selective attenuation of the x-rays, as discussed in Section 5.2.2, and refers to the phenomenon in which the mean energy of the x-ray spectrum increases while propagating through the body.

In the human body, the linear attenuation coefficient decreases with increasing E, as shown in Figure 4.8. Therefore, low-energy photons are preferentially absorbed. Hence, since all CT x-ray sources have a distribution (spectrum) of energies, the propagating beam becomes richer in high-energy photons.

We had concluded in Section 6.2 that the concept of effective energy could be used for polyenergetic sources. However, this concept assumes that the exit spectrum is the same as the entering spectrum. When this is not true, and standard reconstruction methods are used, the so-called *interpetrous lucency artifact* may appear. (It is called "interpetrous" because it often occurs in bone.) This artifact, which is particularly apparent in head scans (because of the outer ring of bone), creates a kind of halo effect around the brain parenchyma. In other parts of the body, streak artifacts, particularly at the tips of bones and at metal pieces within the body (shrapnel, surgical clips, screws, etc.) may also appear. Methods exist for compensating for or eliminating these artifacts, including preprocessing projection data, postprocessing images, and dual-energy imaging. In general, radiologists simply know to expect them and interpret the images accordingly.

Other Artifacts   Three other sources of artifacts in CT images are worth mentioning: (1) electronic or system drift, (2) x-ray scatter, and (3) motion. The electronic or system artifact of particular interest is that which is caused by miscalibration or gain drift. The most dramatic effect is caused by the complete failure of a detector. In a 3G scanner, this situation (usually) causes the detector to report zero photons, and thus 100 percent attenuation at all angular positions. The resulting reconstruction contains a ring artifact whose radius is equal to the position of the defective detector from the detector array's center.

X-ray scatter generally causes a convolutional blurring of each projection.[3] A blurred projection creates a blurry image. It is possible to prevent scatter by collimation and/or selective energy detection, but this comes at the expense of lower efficiency. It is also possible to compensate for scatter by deconvolution, but this has the tendency to further increase the noise in the projections. One way that has been used to measure the amount of scattering present is off-plane detectors.

Finally, motion is a difficult problem. Since a complete scan takes (typically) between 1 and 10 seconds, the heart is almost certainly going to have significant motion during the scan. Breathing is going to be a problem on the longer scans when patients cannot hold their breath. It is possible to gate the data acquisition so that data will be taken only at a certain stage in the cardiac cycle and/or breathing cycle. The data acquired over successive cycles can then be pieced together and treated as if they were taken simultaneously. These methods are used on some machines in certain clinical situations where deemed necessary.

6.5   Summary and Key Points

Computed tomography solves the problem of loss of contrast present in projection radiography that is caused by superposition of overlying structures, by focusing

---

[3]This may have the desirable property of low-pass filtering the projection if aliasing is a problem.

directly (and exclusively) on the cross-sectional image slice of interest. This strategy significantly improves contrast, at the expense of spatial resolution, but the gain in contrast is so important that CT is ubiquitous in medical imaging practice. In this chapter, we presented the following key concepts that you should now understand:

1. A *tomogram* is an image of a cross-sectional plane or slice within or through the body.
2. *X-ray computed tomography* (CT) produces tomograms of the distribution of linear attenuation coefficients, expressed in *Hounsfield units.*
3. There are currently seven *generations* of CT scanner design, which depend on the relation between the x-ray source and detectors, and the extent and motion of the detectors (and patient bed).
4. The *basic imaging equation* is identical to that for projection radiography; the difference is that the ensemble of projections is used to reconstruct cross-sectional images.
5. The most common reconstruction algorithm is *filtered backprojection*, which arises from the *projection slice theorem.*
6. In practice, the reconstruction algorithm must consider the *geometry* of the scanner—parallel-beam, fan-beam, helical-scan, or cone-beam.
7. As in projection radiography, *noise* limits an image's signal-to-noise ratio.
8. Other *artifacts* include *aliasing*, *beam hardening*, and—as in projection radiography—*inclusion of Compton scattered photons.*

Bibliography

Kak, A. C., and Slaney, M. *Principles of Computerized Tomographic Imaging.* New York: IEEE Press, 1988.

Macovski, A. "Physical Problems of Computerized Tomography." *Proceedings of the IEEE* 71, 1983: 373–78.

Macovski, A. *Medical Imaging Systems.* Englewood Cliffs, NJ: Prentice Hall, 1983.

Wolbarst, A. B. *Physics of Radiology.* Norwalk, CT: Appleton and Lange, 1993.

Problems

Instrumentation

6.1 A CT calibration experiment measures two Hounsfield numbers: $h_m^W = 10$ for water and $h_m^A = -1100$ for air.

(a) Find expressions for a and b so that

$$a h_m^W + b = h^W$$

$$a h_m^A + b = h^A \,.$$

(b) What are the correct CT numbers h^W and h^A for water and air, respectively?

(c) What are the values of a and b for this calibration experiment?

6.2 Suppose a 6G CT scanner has a patient table that moves at a speed of 2 cm/s. The x-ray source detector apparatus rotates at a speed of $4\pi/s$. Also assume that it takes 1 ms to measure a projection.

(a) What is the pitch of the helix?

(b) How many projections does the system measure over a 2π angle?

(c) How long does it take to do a 60 cm torso scan?

Radon Transform

6.3 Show that the Radon transform is a linear operator.

6.4 Show that the Radon transform of $f(x - x_0, y - y_0)$ is $g(\ell - x_0 \cos\theta - y_0 \sin\theta, \theta)$.

6.5 Find the 2-D Radon transform $g(\ell, \theta)$ of $f(x, y) = \exp(-x^2 - y^2)$. (*Hint*: Use the rotational symmetry of $f(x, y)$ to simplify your integration.)

6.6 Show that the function $h_\ell(\ell)h_\theta(\theta)$ cannot represent a 2-D Radon transform $g(\ell, \theta)$ unless $h_\theta(\theta)$ is a constant.

6.7 (a) Find the 2-D Radon transform $g(\ell, \theta)$ of a 2-D function $f(x, y) = \cos 2\pi f_0 x$. Show that *filtered* backprojection produces the correct reconstruction.

(b) Repeat part (a) for $f(x, y) = \cos 2\pi ax + \cos 2\pi by$ and $f(x, y) = \cos 2\pi (ax + by)$.

6.8 A first-generation CT scanner is used to image a unit-square shaped object (i.e, length of each side = 1). The object is surrounded by air and has a constant linear attenuation coefficient of μ_0. The coordinate system is set up such that the origin is at the object center, and the x- and y- axes are parallel to the sides of the object.

(a) Write a mathematical expression for the linear attenuation function $\mu(x, y)$. (*Hint*: Use the rect function.)

(b) What is the Fourier transform of $\mu(x, y)$?

(c) Write a mathematical relationship between the projection $g(\ell, \theta)$ computed using the observed x-ray intensities and $\mu(x, y)$.

(d) Using the *projection-slice theorem*, find $G(\varrho, \theta)$.

(e) Take the inverse Fourier transform of $G(\varrho, \theta)$ to find an expression for $g(\ell, \theta)$.

(f) Sketch $g(\ell, 30°)$ (include axis labels) and sketch its backprojection image $b_{30°}(x, y)$.

6.9 Prove the convolution property of the Radon transform:

$$\mathcal{R}\{f * h\} = \mathcal{R}\{f\} * \mathcal{R}\{h\}.$$

CT Reconstruction

6.10 The "unit square function" is given by

$$s(x, y) = \begin{cases} 1 & -1 \leq x \leq 1, -1 \leq y \leq 1 \\ 0 & \text{otherwise} \end{cases}$$

Let $g_s(\ell, \theta)$ be the 2-D Radon transform of $s(x, y)$.
(a) Show that $g_s(\ell, \theta + \pi/2) = g_s(\ell, \theta)$.
(b) Show that $g_s(\ell, -\theta) = g_s(\ell, \theta)$.
(c) Write an expression for $g_s(\ell, \theta)$, $-\infty < \ell < +\infty$, $0 \leq \theta < \pi$ given knowledge of $g_s(\ell, \theta)$ only in the range $-\infty < \ell < +\infty$, $0 \leq \theta < \pi/4$.
(d) Sketch $g_s(\ell, \theta)$ at $\theta = 0$, $\theta = \pi/8$, and $\theta = \pi/4$.
(e) Determine an expression for $g_s(\ell, \theta)$, $0 \leq \theta < \pi/4$.

6.11 The *mass* of an object is defined as

$$m = \int_{-\infty}^{\infty} \int_{-\infty}^{\infty} f(x, y) \, dx \, dy,$$

and its *center of mass* is defined as $\mathbf{c} = (c_x, c_y)$ where

$$c_x = \frac{1}{m} \int_{-\infty}^{\infty} \int_{-\infty}^{\infty} x f(x, y) \, dx \, dy,$$

$$c_y = \frac{1}{m} \int_{-\infty}^{\infty} \int_{-\infty}^{\infty} y f(x, y) \, dx \, dy.$$

The mass of a projection is defined as

$$m_p(\theta) = \int_{-\infty}^{\infty} g(\ell, \theta) \, d\ell,$$

and its center of mass is given by

$$c_p(\theta) = \frac{1}{m_p(\theta)} \int_{-\infty}^{\infty} \ell g(\ell, \theta) \, d\ell.$$

(a) Show that $m_p(\theta) = m$.
(b) Show that $c_p(\theta) = c_x \cos \theta + c_y \sin \theta$.
(c) Define the *triangle function* as

$$f(x, y) = \begin{cases} 1 & 0 \leq y \leq 1 - |x| \\ 0 & \text{otherwise} \end{cases}.$$

Find $m_p\left(\frac{\pi}{4}\right)$ and $c_p\left(\frac{\pi}{4}\right)$ for the *triangle function*.

6.12 Consider an object comprising two small metal pellets located at $(x, y) = (2, 0)$ and $(2, 2)$ and a piece of wire stretched straight between $(0, -2)$ and $(0, 0)$.

(a) Sketch this object. Assume N photons are fired at each lateral position ℓ in a parallel-ray configuration. For simplicity, assume that each metal object stops $1/2$ the photons that are incident upon it no matter what angle it is hit.

(b) Sketch the number of photons you would expect to see as a function of ℓ for $\theta = 0°$ and $\theta = 90°$.

(c) Draw the projections you would see at $\theta = 0°$ and $\theta = 90°$.

(d) Sketch the backprojection image you would get at $\theta = 0°$ (without filtering).

6.13 Consider the object in Figure P6.1 that is an equilateral triangle centered at the origin, each side of which is of length a and one side of which is parallel to the x axis. Suppose the object has constant linear attenuation coefficient μ and is being imaged in a 1G CT scanner. Assume $\mu = 1$ and $a = 6$.

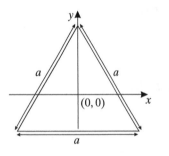

Figure P6.1
Problem 6.13.

(a) Calculate a formula for the projection $g(\ell, 60°)$ and sketch it.

(b) What is the value of $b_{60°}(0, a/4)$ for $g(\ell, 60°)$ calculated in the above.

(c) Find $F(\varrho \cos 60°, \varrho \sin 60°)$.

6.14 In finding an approximation $\tilde{c}(\ell)$ for the "rho-filter" in filtered backprojection, you decide to multiply $|\varrho|$ by the triangular window of width ϱ_0 given by

$$
W(\varrho) = \begin{cases} 1 - \dfrac{|\varrho|}{\varrho_0} & |\varrho| \leq \varrho_0 \\ 0 & \text{otherwise} \end{cases}.
$$

(a) Derive an expression for the approximate filter $\tilde{c}(\ell)$.

(b) Describe what happens as $\varrho_0 \to \infty$.

6.15 Show that

$$
\int_0^{2\pi} \int_0^\infty \varrho\, G_\theta(\varrho) e^{j2\pi \varrho \omega \cdot x}\, d\varrho\, d\theta = \int_0^\pi \int_{-\infty}^\infty |\varrho| G_\theta(\varrho) e^{j2\pi \varrho \omega \cdot x}\, d\varrho\, d\theta
$$

6.16 In this problem, we want to figure out what is "wrong" with the backprojection summation image given by

$$f_b(x, y) = \int_0^\pi g(x \cos\theta + y \sin\theta, \theta) \, d\theta ,$$

and to see how to fix it.

(a) Show that the 2-D Radon transform of the 2-D delta function $\delta(x, y)$ is $\delta(\ell)$. (*Hint*: Use the projection slice theorem.)

(b) Show that the backprojection summation image of $\delta(x, y)$ is $\dfrac{1}{\sqrt{x^2 + y^2}}$.

(*Hint*: Write (x, y) in polar coordinates (r, ϕ); use the fact that $\sin\theta \approx \theta$ for small θ; and the fact that $\delta(a\ell) = \delta(\ell)/|a|$.)

(c) Show that the backprojection summation image of $\delta(x - x_0, y - y_0)$ is $1/\sqrt{(x - x_0)^2 + (y - y_0)^2}$—i.e., that the backprojection summation operator is shift invariant.

(d) Show that $f_b(x, y) = f(x, y) * (1/\sqrt{x^2 + y^2})$.

(e) Describe how you could, in principle, recover $f(x, y)$ from $f_b(x, y)$. Describe any problems you may foresee with your approach.

6.17 With each measured projection $g(\ell, \theta)$ (viewed as a function of ℓ) a backprojection image $b_\theta(x, y)$ can be generated.

(a) Write down a formula for $b_\theta(x, y)$. Now assume that we know only one projection: $g(\ell, 30^\circ) = e^{-|\ell|}$.

(b) What is the value of $b_{30^\circ}(1, 2)$?

(c) Can you determine $b_{45^\circ}(1, 2)$? If yes, what is it? If no, why not?

(d) Can you determine $b_{210^\circ}(1, 2)$? If yes, what is it? If no, why not?

(e) Sketch $b_{30^\circ}(x, y)$. Suppose $g(\ell, \theta)$ is sampled so that only the following values are known: $g(nT, \theta)$, $-\infty < n < \infty$.

(f) Can you determine $b_{30^\circ}(1, 2)$ from the sampled projection $g(n, 30^\circ)$ (i.e., $T = 1$)? If yes, what is it? If no, say why not and suggest an approximate value.

(g) Can you determine $b_{30^\circ}(2, 1)$ from the sampled projection $g(n, 30^\circ)$? If yes, what is it? If no, say why not and suggest an approximate value.

6.18 Prove the following relationship satisfied by the ramp filter $c(\cdot)$:

$$c(D' \sin\gamma) = \left(\frac{\gamma}{D' \sin\gamma} \right)^2 c(\gamma) .$$

Image Quality

6.19 In a computed tomography system, each projection is actually obtained using a uniform scanning beam of width W instead of an infinitesimal pencil beam. Find the resultant estimate $\hat{f}(x, y)$ of the function $f(x, y)$ when convolution backprojection is used.

6.20 Find the signal-to-noise ratio of the computed tomography reconstruction of a lesion immersed in a 20-cm (diameter) cylinder of water whose attenuation coefficient is 5 percent different than that of the water. A scanned source is used providing 100 projections at 0.1R (roentgen) per projection. The detector and beam dimensions are 2.0 × 2.0 mm. Make appropriate assumptions about the reconstruction filter. Assume $\bar{\mu} = 0.15$ cm^{-1} and that the detectors are touching each other. Also assume that there are 2.5×10^{10} photons/cm^2 per roentgen.

6.21 Joe is designing a new CT scanner. It will be a fan-beam design, with one source and D detectors. He will use M angles, reconstruct images with N by N pixels, and strictly enforce the design rule of thumb $D = M = N$. The width of each detector d will be selected to be as large as possible in order to completely fill a 1 meter detector array with D detectors. The reconstruction rho-filter will be designed using a rectangular window with bandwidth $\varrho_0 = 1/d$.

    (a) Joe requires that a lesion with contrast $C = 0.005$ embedded in water ($\bar{\mu} = 0.15$ cm^{-1}) will have a signal-to-noise ratio of at least 20 dB. Suppose $D = M = N = 300$. What is the minimum number of photons per projection at the detectors that is required in order to meet this SNR constraint.

    (b) Assume the patient has 0.125 m^2 cross-sectional area (lying completely in each fan beam). Assume that the patient is exposed to 1 roentgen when 2.5×10^{10} photons pass through 1 cm^2 of tissue, and assume that the dose to the patient in rads is equal to the exposure in roentgens. What is the largest $D(= M = N)$ that can be selected which will satisfy the SNR requirement of part (a) and will keep the total dose for the entire cross-sectional scan less than or equal to 2 rads. (Neglect attenuation due to the patient and assume the detectors have 100 percent efficiency.)

6.22 Your problem is to determine the number of detectors D to fit into an array of fixed length L in order to optimize the SNR of a third generation CT scanner.

    (a) Assume the approximate rho-filter is $\tilde{c}(\ell) = \mathcal{F}^{-1}\{|\varrho|W(\varrho)\}$, where $W(\varrho)$ is a rectangular window with (single-sided) bandwidth ϱ_0. Assume that $\varrho_0 = \min\{d^{-1}, \varrho_{max}\}$, where d is the width of a detector and ϱ_{max} is a fixed constant. Assume that the number of projections M is given by $1.5D$. Find an expression for the SNR as a function of D.

    (b) To avoid aliasing a $J \times J$ pixel reconstructed image, it is decided to choose $\varrho_{max} = J/(2L)$. We also restrict the number of detectors to lie in the range $1 \leq D \leq J$. What D gives the best SNR?

6.23 Suppose we have a first-generation CT scanner with a mechanical problem. For each projection angle, it measures zero at $\ell = 0$. Now we use the scanner to image a uniform disk.

    (a) Sketch the sinogram.

    (b) What would the reconstructed image look like?

    (c) If the scanner always skips measurement at $\ell = \ell_0$, what would the reconstructed image look like?

Applications and Advanced Topics

6.24 Suppose we know that $f(x, y) = \sum_{j=1}^{n} f_j \phi_j(x, y)$. Define the ith line integral as $g_i = \int_{L_i} f(x, y) ds$ $i = 1, \ldots, m$ and the ith measured line integral as $y_i = g_i + v_i$ $i = 1, \ldots, m$, where v_i is noise. Now define the following vectors $f = [f_1 \cdots f_n]^T$, $y = [y_1 \cdots y_m]^T$, and $v = [v_1 \cdots v_m]^T$. Then in matrix notation, our measurements are given by $y = Hf + v$, where H is an $m \times n$ matrix.

(a) Determine H_{ij} the (i, j)th entry in H.

(b) In an ideal world $v = 0$, $m = n$, and H^{-1} exits; then $f = H^{-1}y$ gives an exact reconstruction of f (and we never even spoke of the Radon transform!). Discuss the meaning of the following situations:

1. H^{-1} exists but $v \neq 0$.

2. $v = 0$ but $m < n$.

3. $v = 0$ but $m > n$.

(c) Suppose $m > n$ and assume that the rank of H is n. Find an expression for $\hat{f}$, the vector that minimizes $E = (y - Hf)^T(y - Hf)$. (*Hint*: Multiply out E and take the derivative with respect to each component of f and set each to zero. Organize your answer using matrix notation and solve for f.)

(d) In a typical system we may have 256×256 pixels in a reconstructed image and 360×512 line integral measurements. Given these sizes, explain the main difficulty with the above approach to image reconstruction.

6.25 (a) Find the 2-D Radon transform $g_\theta(\ell)$, for $0 \leq \theta \leq \pi/4$ only, of the unit square indicator function:

$$f(x, y) = \begin{cases} 1 & -1/2 \leq x, y \leq 1/2 \\ 0 & \text{elsewhere} \end{cases}$$

(b) Sketch $g_\theta(\ell)$ for some θ, where $0 < \theta < \pi/4$.

(c) Find $\int_{-\infty}^{\infty} g_\theta(\ell) \, d\ell$.

(d) Suppose only two projections, for $\theta = 0$ and $\theta = \pi/2$, are available. Find an expression for the approximate summation backprojection image $\hat{f}_b(x, y)$ (given only these two images) and make a sketch of the result.

(e) In general, is it possible to reconstruct $f(x, y)$ perfectly given only a *finite* number of projections? Explain your answer.

6.26 A first-generation (1G) CT scanner is used to image the object shown in Figure P6.2 (surrounded by air). Suppose for $E > 100$ keV

$$\mu_1(E) = 1.0 \exp -E[\text{keV}]/100[\text{keV}] \, \text{cm}^{-1} \quad \text{and}$$

$$\mu_2(E) = 2.0 \exp -E[\text{keV}]/100[\text{keV}] \, \text{cm}^{-1}.$$

Suppose that each incident photon burst is polychromatic but has only two photon energies, and that there are 10^6 photons at 100 keV, and 0.5×10^6 at 140 keV.

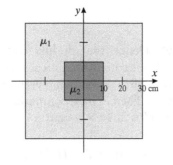

Figure P6.2
Problem 6.26.

(a) Calculate the incident intensity of the x-ray burst.

(b) Ignore the dimension of the detector in this part. Find the intensity I_d as a function of x for the measured projection at $\theta = 0°$. Sketch it.

(c) Calculate the local contrast of the middle square given the measured projection $I_d(x)$. What is the local contrast of the projection $g(\ell, 0)$?

(d) Suppose the detector has a width of 1 cm. Find the measured projection $I_d(x)$ (x being a continuous variable) in this case, and sketch it. Will this change affect the local contrast? Explain.

6.27 A CT system is designed to acquire M parallel-ray projections of $f(x, y)$ at angles $\theta_i = \pi i/M$, $i = 0, \ldots, M - 1$. The collection of projections measured by this system (modeled to be continuous in ℓ) is defined to be the *M-projection Radon transform*:

$$\mathcal{D}_M f = \{g(\ell, \theta_i)| \ i = 0, \ldots, M - 1\},$$

where $g(\ell, \theta_i) = \iint f(x, y)\delta(x \cos \theta_i + y \sin \theta_i - \ell) \, dx \, dy$ is the projection at angle θ_i.

(a) Draw a picture showing the four positions of the detector array required to determine $\mathcal{D}_4 f$. Label which projection is acquired at each position.

(b) Knowledge of $\mathcal{D}_M f$ gives partial knowledge of the Fourier transform $F(u, v)$ of $f(x, y)$. Draw a picture showing the set of points in the u, v-plane where $F(u, v)$ is known given $\mathcal{D}_4 f$. Indicate which projection gave rise to each subset.

(c) Now let $G_1 = \mathcal{D}_M f_1$. Define $G_2 = \mathcal{D}_M(f_1 + f_2)$, where $f_2 = \cos 2\pi f_x x \cos 2\pi f_y y$, $f_x = \cos(3\pi/16)$ and $f_y = \sin(3\pi/16)$. Show that $G_2 = G_1$ for $M = 4$.

(d) For f_2 given above, find the minimum value of M in order to guarantee that $G_2 \neq G_1$.

(e) The result of part (c) means that you can add a function to the original without changing its M-projection Radon transform. In this context, such functions are called *ghost functions*. In theory, is it possible to make M large enough so that there are no ghost functions of $\mathcal{D}_M f$? Explain your answer. What happens in practice?

(f) Suppose one used large detectors, effectively low-pass filtering each projection upon observation. Would f_2 still be a ghost function? Explain.

6.28 Consider an object comprising three squares of the same size with width 20 cm, as shown in Figure P6.3. The origin of the coordinate system is located at the center of the middle square. The linear attenuation coefficients in these three regions are $\mu_1 = 0.1$ cm^{-1}, $\mu_2 = 0.2$ cm^{-1}, and $\mu_3 = 0.3$ cm^{-1}, respectively. Assume that *parallel ray geometry* is used.

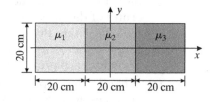

Figure P6.3
Problem 6.28.

(a) Find $g(\ell, 0°)$.
(b) Find $g(\ell, 90°)$.
(c) Find $g(\ell, 45°)$. Sketch it and carefully label the axes.
(d) Sketch $b_{45°}(x, y)$. Find the value $b_{45°}(1, 1)$.
    Now, assume we are using a third-generation CT scanner. The source-detector distance is 1.5 m.
(e) What is the smallest possible circular field of view (FOV) to image the entire object shown in the figure? What is the shortest length of the detector array that will "cover" the FOV?
(f) Using the CT "rule of thumb," determine how many angles should be acquired if the detector array has 256 elements. What is the pixel size of the reconstructed image (assuming the image covers the entire FOV.)

6.29 A square object with side-width of 40 cm is imaged by a first-generation CT scanner.
(a) What type of collimators should be used for the source and the detector?
(b) What is the smallest circular FOV to cover the entire object?
(c) Suppose the smallest angular increment of the scanner is 0.25 degree and we strictly obey the CT "rule of thumb." How many line integrals should we measure for each angle? If the reconstructed image is just large enough to cover the FOV, what is the size of the pixels?
(d) We found that the reconstructed image was noisy. Is it possible to double the SNR given that we cannot change the 0.25 degree angular increment and will not violate the rule of thumb? Explain.
(e) Suppose the object has a constant linear attenuation coefficient of 0.1 cm^{-1}. Sketch $g(\ell, 45°)$ and carefully label the axes.
(f) Sketch $b_{45°}(x, y)$. Find the value $b_{45°}(10, 10)$.

6.30 You are designing a CT scanner. You always use the rules $M = D = J$ and $\varrho_0 = 1/d$. You want a circular FOV having a 60 cm diameter, so your image will be a square with 60 cm sides.

(a) You want your system to be able to resolve two point sources separated by 1 mm, so that at least one square pixel can always be put between them no matter what their orientation relative to the pixel area. What is the minimum number of detectors you must use?

(b) You decide to use 925 detectors, each 0.8 mm wide and fit side-by-side on a line. You decide to place the x-ray source 180 cm away from the center of the detector array (orthogonal to the array, of course). Will your FOV fit within the fan? Explain your answer. Suppose $\bar{\mu} = 0.2$ cm^{-1} and suppose there are an average of 1.5×10^{11} photons hitting the detector array for each projection.

(c) What is the approximate SNR in dB of a tumor mass whose linear attenuation coefficient is 0.25 cm^{-1}.

6.31 You have the good fortune to acquire a used third-generation (3G) CT scanner that has slip ring technology (continuously rotating 360-degree x-ray tube and detector array). Everything is in good working order except that only one detector channel is operational. Fortunately, you are able to switch this channel between detectors so that you can acquire data from any detector, but from only one detector for each pulse of the x-ray tube. The tube can be pulsed continuously at a rate of one pulse per millisecond. The gantry can rotate at one revolution per second. The geometry of the fan-beam is shown in Figure P6.4.

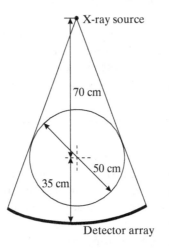

Figure P6.4
Problem 6.31.

(a) Determine the fan angle from the depicted geometry.

(b) Determine the spacing between the 703 detectors that are packed into the array depicted.

(c) Suppose data from only the central detector is acquired (as fast as possible) over a single counterclockwise rotation of the gantry (from x-ray source on top back to x-ray source on the top). What line integrals are acquired? Draw the trajectory of data acquired over this one acquisition cycle on a sinogram diagram. Are any line integrals repeated?

(d) Now suppose that data from only the bottom-left detector (with reference to the diagram above) is acquired, again over a single counterclockwise rotation of the gantry with the fastest possible rotation and x-ray tube pulsing. What line integrals are acquired this time? Draw the trajectory on a sinogram diagram and explain whether there are repeated line integrals or not.

(e) Devise a strategy based on the ideas in (c) and (d) that will scan a sinogram without redundancy. In order to approximately obey the CT rule of thumb, how many rotations will be required to scan the sinogram? Explain.

(f) What is so bad about this CT scanner that it should never be used to scan patients?

6.32 Object motion can cause artifacts in CT images. This happens in part because object motion produces an observation that is not a legitimate Radon transform of any object. Assume parallel-ray geometry.

(a) Prove that the Radon transform of $\delta(x, y)$ is $\delta(\ell)$. Sketch its sinogram.

(b) Prove the following theorem:

$$\mathcal{R}f(x - x_0, y - y_0) = g(\ell - x_0 \cos \theta - y_0 \sin \theta, \theta).$$

(c) Find the Radon transform of $\delta(x - 1, y)$. Sketch its sinogram.

We now assume that while the scanner acquires data over $\theta \in [0, \pi/2]$ the object is $\delta(x, y)$ and while the scanner acquired data over $\theta \in [\pi/2, \pi)$ the object is $\delta(x - 1, y)$. In other words, the object *moved* during the scan from $(0, 0)$ to $(1, 0)$.

(d) Sketch the sinogram that is acquired by this scan.

(e) Prove the following:

$$\int_{-\infty}^{\infty} \ell g(\ell, \theta) \, d\ell = q_x \cos \theta + q_y \sin \theta,$$

where

$$q_x = \int_{-\infty}^{\infty} \int_{-\infty}^{\infty} x f(x, y) \, dx \, dy \quad \text{and}$$

$$q_y = \int_{-\infty}^{\infty} \int_{-\infty}^{\infty} y f(x, y) \, dx \, dy.$$

(f) Show that the sinogram acquired by the scanner when the object moved cannot be the Radon transform of *any* object.

6.33 The plot in Figure P6.5(a) shows the input energy spectrum of a polychromatic x-ray source and the linear attenuation coefficient of a material as a function of photon energy.

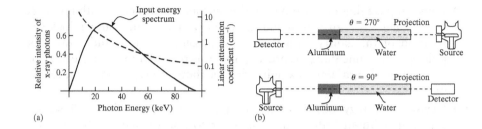

Figure P6.5
Problem 6.33.

(a) Sketch the energy spectrum of the x-ray beam after it passes through the material. Explain your sketch.

(b) Suppose that an object is imaged using a first-generation CT scanner. In Figure P6.5(b) there are two different positions of the source and detector while imaging this same object. Explain why the two measurements should ideally be the same but are actually different in practice.

(c) The fact that the two measurements are different causes artifacts in CT reconstruction. One way to reduce these artifacts is to acquire two measurements with two different input x-ray spectra, one having higher mean energy than the other. Describe how you might obtain these two measurements using the same x-ray tube.

(d) This so-called *dual energy CT* setup requires calibration. Four measurements are made on two test objects (Figure P6.6) using the two input spectra, yielding measurements given in the table below. The values in the table are *line integrals* (not x-ray intensities or numbers of photons). Explain why g_{c1}^L is greater than g_{c1}^H.

Figure P6.6
Problem 6.33.

	Configuration 1	Configuration 2
Higher photon energy (keV)	$g_{c1}^H = 2.2$	$g_{c2}^H = 4.3$
Lower photon energy (keV)	$g_{c1}^L = 3.16$	$g_{c2}^L = 6.79$

(e) Suppose you had access to a 75 keV (monoenergetic) x-ray source. Given $\mu(\text{aluminum}, 75 \text{ keV}) = 0.7 \text{ cm}^{-1}$ and $\mu(\text{water}, 75 \text{ keV}) = 0.1866 \text{ cm}^{-1}$, find the line integrals, g_{c1}^{75} and g_{c2}^{75}, for each of the two configurations.

(f) For calibration, coefficients are to be determined so that measurements using the high and low spectra can be combined in a linear equation to yield what is expected from a monoenergetic 75 keV x-ray source. Given your result from part (e), find the coefficients a^L and a^H such that

$$g_{c1}^{75} = a^L g_{c1}^L + a^H g_{c1}^H$$

$$g_{c2}^{75} = a^L g_{c2}^L + a^H g_{c2}^H$$

(g) For an arbitrary object, your dual energy CT scanner will obtain two sinograms: $g^L(\ell, \theta)$ and $g^H(\ell, \theta)$. Write a mathematical expression for the reconstructed object, $\mu(\text{object}, 75 \text{ keV})$, given these sinograms.

Nuclear Medicine Imaging

Overview

In Part II of this book, we considered projection radiography and computed tomography, two imaging modalities that rely on the *transmission* of photons through the body to form images. We now turn our attention in Part III to nuclear medicine, an imaging modality that relies on the *emission* of photons from within the body.

In contrast to projection radiography and computed tomography, the biological behavior of a substance's biodistribution in the body is of interest in nuclear medicine. Each molecule of the substance is labeled with a radioactive atom. Here, the ionizing radiation emitted when the radioactive atom undergoes radioactive decay is used to determine the location of the molecule within the body; the ionizing radiation is of no medical interest per se. Since the biodistribution of the radiolabeled substance—the *radiotracer*—is determined by the body's physiological and biochemical functioning, nuclear medicine is considered a *functional* imaging modality. Figure III.1 makes this point. Figure III.1(a) shows a projection radiograph of the arm, shoulder, and ribs. Here, image intensity reflects the varying absorption of transmitted x-rays through the bone. Figure III.1(b) shows a nuclear medicine "bone scan" of the corresponding area. Here, image intensity reflects the metabolic activity of the bone.

Nuclear medicine is used whenever a physician needs information on physiologic or biochemical function. For example, the two most common nuclear medicine procedures are bone scanning and myocardial perfusion imaging. Bone scanning, as shown in Figure I.1(b) and Figure III.1(b), looks at the metabolic activity of bones; this is complementary to the structural (anatomical) information from a projection radiograph [e.g., Figure III.1(a)]. A projection radiograph can show a fracture; a bone scan can show active metabolism during the healing process. Similarly, coronary angiography, which depicts the anatomy of the coronary arteries, shows the vessels that supply blood to the heart muscle, whereas myocardial perfusion imaging shows the distribution of blood flow in the muscle. Figure III.2 illustrates the use of nuclear medicine to depict myocardial perfusion. The images show

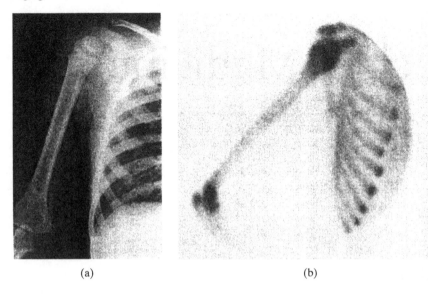

(a)   (b)

Figure III.1
Representative
(a) transmission image and
(b) nuclear medicine
emission image of the arm,
shoulder, and ribs.

the distribution of thallium chloride, which—as a potassium analog—distributes in heart muscle according to blood flow. Panel (a) is an anatomic orientation guide; panel (b) shows a normal heart; panel (c) shows a heart with an apical inferoseptal myocardial infarction ("heart attack"); and panel (d) shows a heart with an inferolateral myocardial infarction [i.e., a different portion of the heart muscle is diseased compared with (c)]. Notice the regions of decreased image intensity—identified with arrows—in (c) and (d) compared with (a).

In thinking about Figure III.1(a) versus (b), and the blood vessels in Figure II.1(d) versus perfusion in Figure III.2, we hope the distinction between anatomy or structure and physiology or function becomes clearer. Structural and functional images show anatomy and its functional consequences, respectively—that is why both types of imaging modalities are important.

In general, a radiotracer is injected into a peripheral arm vein of the patient, or the patient inhales or ingests the radiotracer. Specialized instrumentation produces images of the internal distribution of radioactivity, which is assumed to mirror the distribution of the compound of interest. These images are compared with known distributions in different disease states. Because there are hundreds of different radiotracers routinely available in nuclear medicine, there are literally hundreds of different nuclear medicine studies, each of which assesses the function of a different physiologic process or organ system within the body. The technology of nuclear medicine permits direct measurements to be made of body processes in humans that in the past could be examined only in experimental animals. In order to most accurately depict the biodistribution of radiotracers, instrumentation that emphasizes high image quality and quantitative accuracy has been developed. Digital image processing plays an important role in nuclear medicine, not only in image enhancement but also in extraction of quantitative information about physiological function.

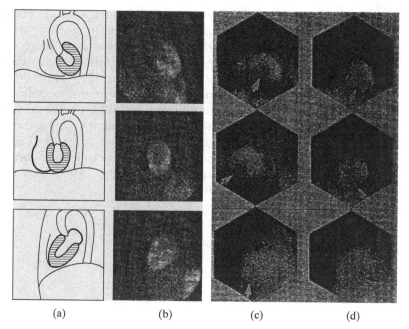

(a)   (b)   (c)   (d)

Figure III.2
Myocardial perfusion imaging. *Top row*: Anterior projection. *Middle row*: Left anterior oblique projection. *Bottom row*: Left lateral projection. (a) Anatomic orientation. (b) Normal heart. (c) Heart with apical inferoseptal myocardial infarct. (d) Heart with inferolateral infarct. (M. P. Sandler, et al., Diagnostic Nuclear Medicine, 3/e, Lippincott, Williams, and Wilkin, 1996.)

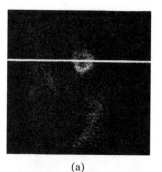

(a)

(c)

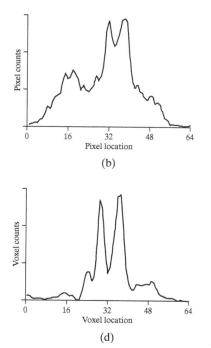

(b)

(d)

Figure III.3
Myocardial perfusion imaging. *Top row*: Planar imaging. *Bottom row*: Tomographic imaging. Parts (a) and (c) are representative images; (b) and (d) are count profiles—plots of counts as a function of pixel location along the line depicted in the image—corresponding to each image. Notice how the contrast (or difference in counts) between the left ventricular myocardial wall (the two central peaks in the count profile) and the cavity (the valley in-between the two peaks) is greater in the tomographic image than in the planar image. (Reprinted from Nuclear Medicine and PET, Fifth Edition, by Christian, Bernier, and Langan, page 245, Figure 9.2, © 2004, with permission from Elsevier.)

The images in Figure III.2 are projection images, which in nuclear medicine are often called *planar* images or *planar scintigraphy*. In the same way that computed tomography is the 3-D tomographic modality based on projection radiography, *emission computed tomography* is the 3-D nuclear medicine modality based on planar scintigraphy. Figure III.3 shows a planar (2-D projection) image and an emission tomographic (slice) image; in this case, the tomographic image is from single photon emission computed tomography (SPECT). The corresponding "count profiles"—which plot image intensity as a function of location along the line depicted in white over each image—are also shown. Notice the improved contrast in the tomographic image (i.e., higher peaks and lower valleys). Figure I.4(c) also shows an emission computed tomographic image; in this case, it is a positron emission tomography (PET) scan of the brain.

The Physics of Nuclear Medicine

7.1 Introduction

Nuclear medicine relies on radiotracers introduced into the body to produce a spatial distribution of measurable radiation. External imaging devices, such as *scintillation cameras*, record the emissions coming from the patient, and produce either a planar, two-dimensional image or cross-sectional images akin to those from computed tomography.

Depending on the specific radiotracer, different physiological or biochemical functions are being imaged. This is strikingly different than projection radiography or computed tomography. In those two transmission imaging modalities, while the specific characteristics of the signal depend on specific imaging and instrument parameters, the basic type of information does not change from image to image. In nuclear medicine, each different radiotracer produces a depiction of a completely different function, so the basic information itself is different.

In this chapter, we consider the basic physical processes that give rise to radionuclides, radioactive decay, and the emissions that form the basis of nuclear medicine imaging.

7.2 Nomenclature

Recall from Chapter 4 that an atom consists of a nucleus of protons and neutrons, which together are called *nucleons*, surrounded by orbiting electrons. The *atomic number* Z is equal to the number of protons in the nucleus, and defines the element. The *mass number* A is equal to the number of nucleons in the nucleus. The term *nuclide* refers to any unique combination of protons and neutrons that forms a nucleus. If a particular nuclide is radioactive (i.e., it can undergo radioactive decay), it is termed a *radionuclide*. Nuclides are typically denoted by either $^A_Z X$ or X-A, where X is the element symbol.

Atoms with the same atomic number but different mass number (i.e., different numbers of neutrons) are called *isotopes*. For example, carbon-11 is an isotope of carbon. It decays by positron decay and is used in positron emission tomography, as discussed in Chapter 9. Because isotopes have the same number of protons (and hence electrons), they are chemically identical. Atoms with the same mass number but different atomic numbers are called *isobars*. Carbon-11 decays to boron-11; the two are isobars. Atoms with the same number of neutrons are called *isotones*. Atoms with the same atomic and mass number (i.e., the same nuclide) but with different energy levels are called *isomers*. Technetium-99m decays to technetium-99; the two are isomers.

This nomenclature becomes important when we consider the radionuclides used in nuclear medicine. Radioactive isotopes of certain elements serve as the source of the ionizing radiation we image in nuclear medicine. As discussed below, we are particularly interested in obtaining certain radiations, which drives our choice of nuclides.

7.3 Radioactive Decay

7.3.1 Mass Defect and Binding Energy

The sum of the masses of the constituents of an atom (i.e., the protons, neutrons, and electrons) is greater than the atom's actual mass. The difference between the sum of the masses of the atom's constituents and the actual mass is called the *mass defect*.

As an example, consider stable carbon-12. An atom of carbon-12 has 6 protons, 6 neutrons, and 6 electrons. On the atomic scale, it is common to express mass in *unified atomic mass units* (u), where 1 u is exactly 1/12 the mass of a carbon-12 atom. The mass of a proton is 1.007276 u, the mass of a neutron is 1.008665 u, and the mass of an electron is 0.000548 u. The mass defect of carbon-12 is therefore $6 \times 1.008665 + 6 \times 1.007276 + 6 \times 0.000548 - 12 = 0.098934$ u.

Einstein's famous relationship, $E = mc^2$, states that mass and energy are related, and that matter and energy cannot be created or destroyed, only transformed from one form into the other. Since mass and energy are related by Einstein's equation, there is an amount of energy "missing" from the atom (equivalent to the mass defect) that would otherwise be predicted to be present based on the constituents of that atom. This "missing" energy is termed the *binding energy* and can be computed from $E = \Delta mc^2$, where Δm is the mass defect and c is the speed of light (3×10^8 m/s). The unit we have been using throughout the book is the *electron volt*, eV, which is defined as the amount of energy an electron gains when it is accelerated across a voltage potential of 1 volt. Based on $E = mc^2$, one u is equivalent to 931 MeV.

In general, a more massive nuclide is more likely to have a larger mass defect, and therefore a larger binding energy. Thus, it is more appropriate to consider the binding energy per nucleon, rather than the total binding energy per se. As an example, we can again consider carbon-12. Since the mass defect of a carbon-12 atom is 0.098934 u, the binding energy is $0.098934 \times 931 = 92.1$ MeV. Since carbon-12 has 12 nucleons, the binding energy/nucleon = $92.1/12 = 7.67$ MeV/nucleon. A graph of the binding energy per nucleon as a function of mass number is shown in Figure 7.1.

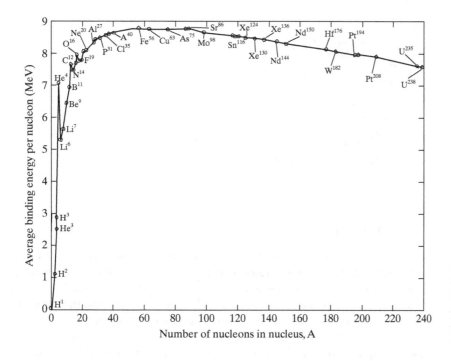

Figure 7.1
Average binding energy per nucleon as a function of mass number. Adapted from Kraushaar and Ristinen, 1984.

Binding energy applies to both the protons and neutrons in the nucleus, and to the orbiting electrons. An electrostatic attractive force exists between particles with opposite charge (e.g., positively charged protons in the nucleus and negatively charged orbiting electrons). An electron in an inner orbit is attracted to the positively charged nucleus with a force greater than is an electron farther away. The energy required to remove an electron completely from an atom is the *electron binding energy*, which is greater for electrons in orbits closer to the nucleus, because of the greater attractive force of the nucleus for inner electrons. Normally, an electrostatic repulsive force exists between particles with the same charge; protons repel each other when separated by a distance greater than the diameter of a nucleus. Within the nucleus, however, an attractive force, the *nuclear* or *strong force*, is responsible for holding neutrons and protons together. The energy required to separate the constituent protons and neutrons in a nucleus is the *nuclear binding energy*.

Radioactive decay is the process by which an atom rearranges its constituent protons and neutrons to end up with lower inherent energy. Radioactive decay occurs spontaneously, and energy is released in the process. The result of radioactive decay is an atom (the *daughter*) with less inherent energy than that which preceded it (the radioactive *parent* atom). This change in energy is reflected in the nuclear binding energy discussed above. Since binding energy is the amount of energy "missing" from an atom, the daughter atom has a higher binding energy/nucleon than the parent atom.

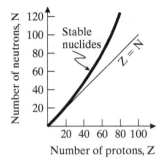

Figure 7.2
Stable elements having proton/neutron pairings that lie on the *line of stability*.

7.3.2 Line of Stability

If the different unique combinations of protons and neutrons as a nucleus that are found in nature are catalogued, they can be separated into two groups: non-radioactive nuclides, which are stable, and radioactive nuclides, which are called *radionuclides*. In general, the total number of nucleons, and the ratio of neutrons to protons, determine whether a nuclide is stable or radioactive.

A graph of the number of neutrons as a function of the number of protons for each unique stable nuclide produces a curve that follows the line of identity at low to intermediate atomic numbers (i.e., the nucleus has the same number of protons as neutrons), and then diverges up, so that for stable nuclides at higher atomic numbers there are more neutrons than protons in the nucleus. This curve, shown in Figure 7.2, is called the *line of stability*, because it depicts the data from all the stable nuclides. (It is called a "line" in practice even though it is not a straight line.) When plotted on the same graph, radionuclides fall off this line of stability. One way of conceptualizing radioactive decay is as the attempt of any radioactive atom that is "off" of this line of stability to reach the line of stability. In order to move toward the line of stability, an atom changes its proton to neutron ratio in the process of radioactive decay.

The greater the binding energy per nucleon, the more stable the atom. Thus, a second way to conceptualize radioactive decay is as a process an atom undergoes to increase its binding energy per nucleon. When a parent undergoes radioactive decay to produce a daughter, the daughter will always have a higher binding energy per nucleon than the parent. Therefore, Figure 7.1 should also be helpful in visualizing the relationships between parents and daughters in radioactive decay.

7.3.3 Radioactivity

The term *radioactivity* describes how many radioactive atoms are undergoing radioactive decay every second. It does not reflect what type of radiation is being emitted, or the energy of that radiation. The common unit for radioactivity is the *curie*, abbreviated Ci, where 1 Ci $= 3.7 \times 10^{10}$ disintegrations per second (dps). A *disintegration* is an atom undergoing radioactive decay. The SI unit for radioactivity is the *becquerel*, abbreviated Bq, where 1 Bq = 1 dps. Clearly,

$$1 \text{ Ci} = 3.7 \times 10^{10} \text{ Bq}. \tag{7.1}$$

Radioactivity in the range of mCi or MBq is common in nuclear medicine procedures.

EXAMPLE 7.1

The intensity of radiation incident on a detector at range r from a radioactive source is given by

$$I = \frac{AE}{4\pi r^2},$$

where A is the radioactivity of the material and E is the energy of each photon.

Question For technetium-99m with radioactivity of 1 mCi, what is the intensity at a distance of 20 cm?

Answer From Table 7.1 (see page 247), we know that the photon energy for technetium-99m is

$$E = 140 \text{ keV}$$

The radioactivity is $A = 1$ mCi $= 3.7 \times 10^7$ Bq. So the intensity at a distance 20 cm from the source is

$$I = \frac{3.7 \times 10^7 \text{ Bq} \times 140 \text{ keV}}{4\pi (0.2 \text{ m})^2} = 1.03 \times 10^{10} \frac{\text{keV}}{\text{sec} \cdot \text{m}^2}.$$

■

7.3.4 Radioactive Decay Law

The radioactive decay law states that the loss of atoms in a radioactive source per unit time is proportional to the number of radioactive atoms. Thus, letting N represent the number of radioactive atoms in the source (and treating it as a continuous quantity), we have

$$-\frac{dN}{dt} = \lambda N, \tag{7.2}$$

where λ, the constant of proportionality, is called the *decay constant*. The decay constant has units of inverse time, and it is a constant for a given radionuclide. Assuming there are N_0 atoms at time $t = 0$, this expression can be integrated to determine the number of atoms N_t at time t, yielding

$$N_t = N_0 e^{-\lambda t}. \tag{7.3}$$

The *radioactivity A* of a source is defined as the number of atoms disintegrating per unit time. (Do not confuse this A with the mass number A of a nuclide.) Combining this fact with (7.2) and (7.3),

$$A = -\frac{dN}{dt} = \lambda N, \tag{7.4}$$

and

$$A_t = A_0 e^{-\lambda t}. \tag{7.5}$$

Either (7.3) or (7.5) is commonly referred to as the *radioactive decay law*. The factor that multiplies the initial value (N_0 or A_0) is called the *decay factor* (DF), and is given by

$$DF = e^{-\lambda t}. \tag{7.6}$$

Recall that an exponential curve never goes to zero, so some radioactivity always remains. The *half-life* $t_{1/2}$ is the time it takes for the radioactivity (or the number of radioactive atoms) to decrease by a factor of 2. By definition,

$$\frac{A_{t_{1/2}}}{A_0} = \frac{1}{2} = e^{-\lambda t_{1/2}}. \tag{7.7}$$

Taking the natural logarithm of both sides and rearranging yields

$$t_{1/2} = \frac{0.693}{\lambda}. \tag{7.8}$$

Thus, the half-life and decay constant have a fixed relationship across radionuclides. The half-life is a constant for a given radionuclide (but differs across radionuclides).

EXAMPLE 7.2

Consider two radionuclides P and Q. Suppose the half-life of P is twice that of Q, $t_{1/2}^P = 2t_{1/2}^Q$. At $t = 0$, we have N_0 atoms of both radionuclides.

Question   When will the radioactivities of two radionuclides be equal?

Answer   Since $t_{1/2}^P = 2t_{1/2}^Q$, the decay constants of two radionuclides have relation $\lambda_P = \lambda_Q/2$. So at $t = 0$, the radioactivities for P and Q are

$$A_0^P = \lambda_P N_0 \quad \text{and} \quad A_0^Q = \lambda_Q N_0 = 2A_0^P.$$

Based on the radioactive decay law, we have

$$A_t^P = A_0^P e^{-\lambda_P t} \quad \text{and} \quad A_t^Q = A_0^Q e^{-\lambda_Q t} = 2A_0^P e^{-2\lambda_P t}.$$

By equating A_t^P and A_t^Q, we have the following equation for t:

$$e^{-\lambda_P t} = 2e^{-2\lambda_P t}.$$

The solution is $t = \frac{\ln 2}{\lambda_P} = t_{1/2}^P$. So at $t = t_{1/2}^P$, the radioactivities of two radionuclides are equal.   ∎

7.4 Modes of Decay

It is important to understand the different modes of radioactive decay, as these govern the different types of ionizing radiation produced. There are four main modes of decay: (1) *alpha decay*, which results in emission of an alpha particle; (2) *beta decay*, which results in emission of a beta particle; (3) *positron decay*, which results in emission of a positron; and (4) *isomeric transition*, which results in emission of a gamma ray. These ionizing radiations fall into two classes: (1) *particulate radiation* and (2) *electromagnetic radiation*. The main particulate ionizing radiations resulting from radioactive decay are alphas, betas, and positrons. Alpha particles consist of two protons and two neutrons; beta particles are like electrons; and positrons are antimatter electrons. The electromagnetic ionizing radiations resulting from radioactive decay are gamma rays. Of these, we are only concerned in medical imaging with *positrons* (used for positron emission tomography in Chapter 9) and *gamma rays* (used for planar scintigraphy in Chapter 8 and single photon emission computed tomography in Chapter 9).

7.4.1 Positron Decay and Electron Capture

A nuclide with a ratio of neutrons to protons too low for stability can undergo electron capture or positron decay. These processes can be thought of as the transformation of a proton into a neutron, although that is not what literally occurs. A *positron* β^+ is emitted in the process of positron decay:

$$p \rightarrow n + \beta^+ + \nu,$$

where p is a proton, n is a neutron, and ν is a neutrino. For example, carbon-11 decays by positron emission to boron-11:

$$^{11}_{6}C \rightarrow {}^{11}_{5}B + \beta^+ + \nu.$$

A positron is actually an antimatter electron, having a +1 charge and the same rest mass as an electron. The neutrino, emitted with the positron, is a massless, chargeless subatomic particle. For a given radionuclide undergoing positron decay, the total kinetic energy of the positron plus the neutrino is a constant; however, the division of this total energy among the positron and neutrino varies from emission to emission.

Since a positron is an antimatter electron, an "unusual" atomic process occurs when a positron and electron meet. As a positron is emitted, it travels several millimeters in the material in which it is emitted, depositing its kinetic energy. (Its kinetic energy is transferred to the material by both collisional and radiative transfer, just as with a beta particle or energetic electron.) It then meets a free electron in the tissue, and—because it is an antimatter electron—mutual annihilation occurs. From conservation of energy, two 511 keV annihilation photons appear (511 keV is the energy equivalent to the rest mass of an electron or positron from $E = mc^2$); from conservation of momentum, they are emitted 180-degrees back-to-back.

Atoms that undergo positron decay have an alternative decay mode—the nucleus can actually capture an electron:

$$p + e^- \rightarrow n + \nu.$$

Typically, an inner shell (K or L shell) electron is captured, as these are in the closest physical proximity to the nucleus. A given radionuclide source will undergo electron capture a certain fraction of the time, and positron decay the other, but a given atom will (of course) only undergo decay once, by one of the two possible modes.

7.4.2 Isomeric Transition

A radionuclide may decay to a more stable nuclide that has the same atomic and mass numbers; both the parent and daughter are not only the same element but the same isotope of that element. In such a case, the parent usually represents a transient, *metastable* state with higher energy. This excess energy is released in the form of gamma rays. Recall from Chapter 4 that gamma rays are in practice indistinguishable from the x-rays that are now familiar to you. (Some textbooks indicate that the distinction between x-rays and gamma rays is based on wavelength or energy, but it is actually based on site of origin: gamma rays from the nucleus and x-rays from the electron cloud.)

An example of an isomeric transition is

$$\text{Cs-137} \rightarrow \text{Ba-137m} \rightarrow \text{Ba-137} + \gamma, \tag{7.9}$$

where γ represents the 662 keV gamma ray photon released when Ba-137m decays to Ba-137. Here, the "m" in Ba-137m represents the metastable state.

7.5   Statistics of Decay

Radioactive decay is a random process. If you conduct two experiments, in which you start with exactly N_0 atoms of the same radionuclide in each experiment, and then count the number of decays over a period of exactly one thousandth of a half-life, you are very likely to get two different counts. The difference you observe is due not to experimental error but to the inherently random nature of radioactive decay. This tells us immediately that the radioactive decay law described in the previous section [(7.3) and (7.5)] must be considered as a description of average behavior; the radioactive decay law predicts the *mean* number of radioactive atoms and activity, not the exact instantaneous number and amount.

For large numbers of radioactive atoms and for periods of time much smaller than their half-lives, the random behavior of radioactive decay is governed by the Poisson distribution, which was first introduced in Section 3.4.3 and is given by

$$\Pr[N = k] = \frac{a^k e^{-a}}{k!}, \tag{7.10}$$

where a is a parameter of the distribution. This is a *probability mass function* describing the probability that random variable N will equal k in a given experiment. It can be shown that both the mean and variance of a Poisson random variable are equal to a, the only parameter of the distribution, as noted in (3.53) and (3.54).

As an example, suppose that we have N_0 identical radioactive atoms having a long half-life. Over a short period of time Δt, only a tiny fraction of these atoms will disintegrate; therefore, we can assume that N_0 is constant over this time. From the radioactive decay law [see in particular (7.2)], we would expect to see exactly $\Delta N = \lambda N_0 \Delta t$ disintegrations over this short time period. In fact, over repeated measurements we would actually see a variable number of disintegrations. We would also notice that the *average* number of disintegrations would tend toward this predicted value. This value turns out to be the mean of a Poisson distribution, which describes the true random nature of radioactive decay. That is, for this experiment we have

$$a = \lambda N_0 \Delta t,$$

and the Poisson distribution governing the decay of these radioactive atoms is

$$\Pr[\Delta N = k] = \frac{(\lambda N_0 \Delta t)^k e^{-\lambda N_0 \Delta t}}{k!}. \qquad (7.11)$$

Let us consider the probability that there are no disintegrations in an extremely small time interval. Evaluating (7.11) for $k = 0$ yields

$$\Pr[\Delta N = 0] = e^{-\lambda N_0 \Delta t}, \qquad (7.12)$$

which for very small Δt is approximated by

$$\Pr[\Delta N = 0] \approx 1 - \lambda N_0 \Delta t. \qquad (7.13)$$

Since for an extremely small interval, there is going to be only one disintegration, or none at all, and since probabilities must sum to one, we can interpret $\lambda N_0 \Delta t$ as the probability of having one disintegration from all N_0 radioactive atoms in the time interval Δt. This gives us another way to interpret the radioactive decay constant λ: It can be thought of as the probability of radioactive decay per radioactive atom per unit (small) time.

If Δt is treated as a parameter and allowed to vary, then (7.11) characterizes a time-varying process $\Delta N(\Delta t)$, which is known as a *Poisson counting process*. Such a process counts the number of events (disintegrations, in this case) that have occurred over a certain period of time. In this case, the quantity λN_0 is known as the *Poisson rate* (see Example 3.7). It has units of events (disintegrations) per second and can be thought of as a type of *intensity* of the Poisson process. It is also a measure of the activity A of the decay process. However, it must be reemphasized that the Poisson process behavior takes place over time periods much smaller than the half-life of a radionuclide. On the other hand, the radioactive decay describes the change in A over much longer periods of time.

EXAMPLE 7.3

A patient study needs to be completed in no more than 10 minutes with at least 3.5 million counts of photons to achieve desired image quality.

Question   Suppose we detect 6 K photons in the first second. What is the minimal half-life of the radionuclide for the study to be successful?

Answer   During the first second, the number of photons detected is

$$\Delta N = \int_0^1 \lambda N_0 e^{-\lambda t} dt = N_0 (1 - e^{-\lambda}) = 6 \text{ K}.$$

We need at least 3.5 million counts in 10 minutes, so

$$\Delta N = \int_0^{600} \lambda N_0 e^{-\lambda t} dt = N_0 (1 - e^{-600\lambda}) \geq 3500 \text{ K}.$$

The above two equations give us

$$\frac{1 - e^{-600\lambda}}{1 - e^{-\lambda}} \geq \frac{3500}{6}.$$

Solving the inequality equation, we have

$$\lambda \leq 9.45 \times 10^{-5} \text{ sec}^{-1}.$$

The minimal half-life is

$$t_{1/2} = \frac{0.693}{9.45 \times 10^{-5} \text{Sec}^{-1}} = 7333 \text{ sec} \approx 2 \text{ hr}.$$
∎

7.6   Radiotracers

There are about 1,500 known radionuclides, of which 200 or so can be purchased. Out of these, however, only a dozen or so are suitable for nuclear medicine, for several reasons. First, we desire "clean" gamma ray emitters—i.e., ones that do not also emit alpha and beta particles (because these particles would only contribute to patient radiation dose without usefully contributing to image formation). Positron emitters are also suitable because the positrons rapidly annihilate with electrons to produce gamma rays. Second, unlike radiography and CT, where we require attenuation of the radiation to produce image contrast, in nuclear medicine we would prefer that there would be no attenuation of the radiation. This is because in nuclear medicine we are trying to determine the location of the emitters; attenuation simultaneously adds to the patient dose and reduces the signal we can detect. This requirement suggests that the energy of the gamma rays should be high, so that they leave the body with little attenuation. On the other hand, we must be able to actually detect the radiation once it leaves the body, and the higher the energy the less likely the gamma rays will interact in the detector. For these reasons (and others, as we will see later), the gamma emitters used in nuclear medicine have energies in the range 70–511 keV.

Another important property of a radiotracer is its half-life. We need to be able to form images in a matter of minutes in general, not seconds or hours. If the radionuclide decays in seconds, there is hardly time enough to administer and metabolize the compound before the source is lost. On the other hand, if it takes hours to create an image, patient motion will be a serious problem and natural metabolic processes will change the distribution of the radiotracer in the body over the period of the exam. So the half-life of the radionuclide should be short, but not too short. Typical times are on the order of minutes to several hours. Because of this relatively short time requirement, some radiotracers are made on-site in generators or cyclotrons, while others are ordered from a nearby radiopharmacy. A list of the more commonly used radionuclides and some of their properties is given in Table 7.1.

TABLE 7.1

Common Radionuclides in Nuclear Medicine			
	Gamma Emitters		
Z	Nuclide	Half-life	Photon Energy (keV)
24	Chromium-51	28d	320
31	Gallium-67	79.2h	92, 184, 296
34	Selenium-75	120d	265
38	Strontium-87m	2.8h	388
43	Technetium-99m	6h	140
49	Indium-111	2.8d	173, 247
	Indium-113m	1.73h	393
53	Iodine-123	13.3h	159
	Iodine-125	60d	35, 27
	Iodine-131	8.04d	364
54	Xenon-133	5.3d	81
80	Mercury-197	2.7d	77
81	Thallium-201	73h	135, 167
	Positron Emitters		
Z	Nuclide	Half-life	Positron Energy (keV)
6	Carbon-11	20.3min	326
7	Nitrogen-13	10.0min	432
8	Oxygen-15	2.1min	696
9	Fluorine-18	110min	202
29	Copper-64	12.7h	656
31	Gallium-68	68min	1900
33	Arsenic-72	26h	3340
35	Bromine-76	16.1h	3600
37	Rubidium-82	1.3min	3150
53	Iodine-122	3.5min	3100

Source: Wolbarst, 1993.

An important characteristic of a radionuclide is that it must be useful and safe to "trace" within the body, either by itself or attached to a compound. Some of the radionuclides in Table 7.1 are more useful than others in this respect. Iodine, for example is a naturally occurring substance in the body, accumulating in the thyroid gland. Iodine-123 or I-131 in a sodium salt can be administered orally and measured in the thyroid a day later to assess thyroid function. Technetium-99m can be used to label DTPA, which is filtered by the kidneys, and serial images of the kidneys can be used to assess renal function. Gaseous O_2 in which one oxygen atom has been replaced by oxygen-15 is used to measure blood flow and to assess oxygen metabolism with positron emission tomography. Fluorodeoxyglucose (FDG) is used by the body like glucose, except that the (labeled) fluorine-18 atoms remain where the molecule is first used. Imaging the uptake of FDG in the brain, for example, is thought to reveal aspects of the mental processes involved in motor, perceptual, and cognitive tasks.

A final and somewhat subtle desirable characteristic for gamma emitters is that they should have monoenergetic (single energy) decay. The reason for this is that with monoenergetic emissions, energy-sensitive detection can be used to discriminate the primary photons from those that have been Compton scattered. It can be seen from Table 7.1 that most of these common radionuclides are monoenergetic. Even those that are not monoenergetic, such as Tl-201 and Ga-67, emit only two or three different energy photons, in contrast to a typical polyenergetic x-ray beam, which has a continuous spectrum of energies.

Technetium-99m (Tc-99m) is by far the most commonly used gamma ray producing radionuclide in nuclear medicine. It is easily generated on-site using a so-called *cow* containing molybdenum-99, whose radioactive decay (half-life is 67 hours) produces Tc-99m. Tc-99m can in turn be tagged to a variety of molecules, including sulphur colloids, glucoheptonate, albumin macroaggregates, and phosphates, which can be used to track a variety of physiological processes in the body. Tc-99m has a half-life of 6 hours and monoenergetic gamma ray energy of 140 keV, which are both ideal in terms of balancing the various radionuclide requirements. As we now proceed in the following chapters to describe the instrumentation related to planar scintigraphy and SPECT, it will be useful to think of Tc-99m as the radioactive source, and to work out examples with this source as well.

EXAMPLE 7.4
Consider an experiment that uses dual radionuclides, technetium-99m and thallium-201. At $t = 0$, we have equal number of atoms of both radionuclides.

Question When will the count rate of technetium-99m be smaller than 20 percent of the total count rate?

Answer The average number of photons from technetium-99m is given by

$$\Delta N_{Tc}(t) = \lambda_{Tc} N_0 e^{-\lambda_{Tc} t}.$$

Similarly, for thallium-201, we have

$$\Delta N_{Tl}(t) = \lambda_{Tl} N_0 e^{-\lambda_{Tl} t}.$$

When the counts of technetium-99m are smaller than 20 percent of the total counts, we must have

$$4\Delta N_{Tc}(t) \leq \Delta N_{Tl}(t).$$

From Table 7.1, we know $\lambda_{Tc} = 0.693/6\ \mathrm{hr}^{-1}$, and $\lambda_{Tl} = 0.693/73\ \mathrm{hr}^{-1}$. Solving above equation, we have

$$t = 36.65\ \mathrm{hr}. \qquad \blacksquare$$

7.7   Summary and Key Concepts

Nuclear medicine makes use of radiotracers (compounds that are chemically labeled with radioactive atoms), which depend on the radioactive decay of radionuclides. The choice and behavior of specific radionuclides significantly influences the utility of the radiotracer. In this chapter, we presented the following key concepts that you should now understand:

1. *Nuclear medicine* produces images that depict the distribution of a *radiotracer*; this distribution is governed by body function, not structure.
2. Nuclear medicine makes use of *radionuclides*; these are unique nuclear species representing radioactive atoms, which emit ionizing radiation upon (spontaneous) decay.
3. A given radionuclide is characterized by its *decay mode* (which indicates the type of ionizing radiation emitted) and *half-life* (the time for half the radioactive atoms to decay, on average).
4. *Radioactive decay* is a random process (governed by the *Poisson distribution*); it is described by a nonexponential relation and is a function of time and the *decay constant* (the probability of decay), which itself is a characteristic of a given radionuclide.
5. Radiotracers make use of radionuclides that emit radiation of appropriate type and energy, have half-lives that are appropriate, and are chemically inert.

Bibliography

Attix, F. H. *Introduction to Radiological Physics and Radiation Dosimetry*. New York: Wiley, 1986.

Bushberg, J. T., Seibert, J. A., Leidholdt Jr., E. M., and Boone, J. M. *The Essential Physics of Medical Imaging*, 2nd ed., Philadelphia, PA: Lippincott Williams & Wilkins, 2002.

Carlton, R. R., and Adler, A. M. *Principles of Radiographic Imaging: An Art and a Science*, 3rd ed. Albany, New York: Delmar, 2001.

Cember, H. *Introduction to Health Physics*, 2nd ed. New York: Pergamon Press, 1983.

Early, P. J., and Sodee, D. B., *Principles and Practice of Nuclear Medicine*, 2nd ed. St. Louis, MO: Mosby, 1995.

Johns, E. J., and Cunningham, J. R. *The Physics of Radiology*, 4th ed. Springfield, IL: Charles C Thomas Publisher, 1983.

Kraushaar, J. J., and Ristinen, R. A. *Energy and Problems of a Technical Society*. New York: Wiley, 1984.

Sorenson, J. A., and Phelps, M. E. *Physics in Nuclear Medicine*, 2nd ed. Philadelphia, PA: W. B. Saunders Company, 1987. Wang, C. H., Willis, D. L., and Loveland, W. D. *Radiotracer Methodology in* *the Biological, Environmental, and Physical Sciences*. Englewood Cliffs, NJ: Prentice Hall, 1975. Wolbarst, A. B. *Physics of Radiology*. Norwalk, CT: Appleton & Lange, 1993.

Problems

Fundamentals of Atoms

7.1 Show that 1 u is equivalent to 931 MeV.

7.2 A deuteron has 1 proton and 1 neutron and has a mass of 2.01355 u. Compute its mass defect and binding energy.

Radioactive Decay and Its Statistics

7.3 Compute the mean and variance of a Poisson distribution with parameter a.

7.4 Suppose there are 1×10^9 radioactive atoms in a given sample of iodine-123 (^{123}I), which has a half-life of 13 hours.

(a) What is the radioactivity of the original sample?

(b) How many radioactive atoms can be expected to be present after 24 hours?

(c) What is the probability that there will be 1×10^8 radioactive atoms left after 24 hours?

7.5 Consider the problem in Example 7.1, suppose we have 1×10^{12} atoms of technetium-99m to start with. What is the intensity measured at $t = 0$ and at $t = 1$ hour?

7.6 (a) How long will it take for a radioactive sample with activity 1 Ci to decay to activity 1 Bq if the half-life is τ.

(b) What approximate order of magnitude would you want the half-life of a radioactive tracer used in nuclear medicine to be? Milliseconds? Seconds? Minutes? Hours? Days? Weeks? Years? Explain your answer.

7.7 Consider the imaging system shown in Figure P7.1. The $D \times D$ square detector is on the x-y plane. The center of the detector coincides with the origin of the x-y plane. A point source, S, is on the z-axis a distance R away from the detector plane. The intensity on the detector is defined as the energy of photons detected on a unit area of detector in unit time. If the radioactivity of the source at time $t = 0$ is A_0.

(a) What is the intensity at each point on the detector plane as a function of time?

(b) What is the average intensity on the plane as a function of time?

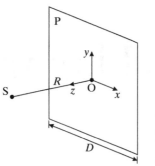

Figure P7.1
A simple imaging system with a radioactive source and a planar detector.

7.8 (a) Show that the decay factor DF is related to the half-life by

$$DF = e^{-0.693t/T_{1/2}}$$

(b) The *average lifetime* τ of a radioactive sample is given by $\tau = 1/\lambda$. Find an expression for τ in terms of the half-life.

7.9 A vial containing ^{99m}Tc is labeled "2 mCi/ml @ 8 a.m."

(a) What is the activity of the sample at 4 p.m. on the same day?

(b) What volume should be withdrawn at 4 p.m. on the same day to prepare an injection of 1.5 mCi for a patient?

CHAPTER

8

Planar Scintigraphy

8.1 Introduction

As in diagnostic x-ray imaging, nuclear medicine imaging evolved from projection imaging to tomographic imaging. Unlike diagnostic x-ray imaging, however, projection studies in nuclear medicine, called *planar scintigraphy*, have always used the *Anger scintillation camera*, a type of electronic detection instrumentation. The corresponding tomographic imaging method, called *single photon emission computed tomography* (SPECT), uses a rotating Anger camera to obtain projection data from multiple angles. These imaging methods depend on radiotracers labeled with radioactive atoms whose decay produces a single gamma photon directly. Another imaging method, called *positron emission tomography* (PET), is based on radiotracers labeled with radioactive atoms whose decay produces a positron that is subsequently annihilated, producing two gamma photons. PET is implemented only as a tomographic imaging method and does not have a corresponding projection imaging mode. Together, SPECT and PET are referred to as *emission computed tomography*; they are described in the next chapter. Here, we focus on planar (2-D projection) scintigraphy.

8.2 Instrumentation

The three basic imaging modalities in nuclear medicine—planar imaging, SPECT, and PET—can be logically grouped in two ways. Planar imaging and SPECT use radiotracers that are gamma emitters, while PET uses radiotracers that emit positrons. On the other hand, SPECT and PET require tomographic reconstruction techniques (like CT), while planar imaging forms images by projection (as in projection radiography). In this section, we describe the instrumentation in planar scintigraphy, much of which is also used in SPECT, which is described in Chapter 9. Because of the peculiar nature of positron decay, PET scanners have significantly different instrumentation, which we also describe in Chapter 9.

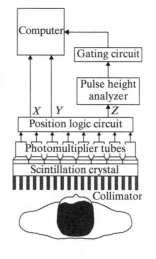

Figure 8.1
Components of an Anger scintillation camera.

As described in the Overview to Part III, nuclear medicine images are based on the distribution of activity. Unlike projection radiography and CT, we are thus interested not in total intensity but in the detected decay rate of the source, typically expressed as *counts* per time. Nuclear medicine images are thus built event-by-event.

The *Anger scintillation camera*, or *gamma camera*, was invented in the late 1950s by Hal Anger of the Donner Laboratory at the University of California, Berkeley. It is the most commonly used imaging instrument in nuclear medicine today. The complete camera system consists of a multihole lead collimator, a 10- to 25-inch circular, square, or rectangular sodium iodide scintillation crystal, an array of photomultiplier tubes on the crystal, a positioning logic network, a pulse height analyzer, a gating circuit, and a computer, as shown in Figure 8.1. We now describe each of these components in detail.

8.2.1 Collimators

The *collimator* is a 1- to 2-inch thick slab of lead the same dimensions as the scintillation crystal, with a geometric array of holes in it. The lead that separates each hole is called a *septum*; collectively the lead represent *septa*. The collimator provides an interface between the patient and the scintillation crystal, by allowing only those photons traveling in an appropriate direction (i.e., those that can pass through the holes without being absorbed in the lead) to interact with the crystal. Thus, the collimator discriminates against photons based on their direction of travel, and restricts the field of view of the crystal. There are several types of collimators used with Anger cameras: parallel-hole, converging, diverging, and pinhole (Figure 8.2).

The most commonly used collimator is the *parallel-hole collimator*, which consists of an array of parallel holes perpendicular to the crystal face, as shown in Figure 8.2(a). With parallel hole collimators, the image on the crystal face is the same size as the object. Such collimators are either cast or fabricated from corrugated lead sheets. Originally, collimator holes were circular in cross section. This meant that the lead septa were thicker in some areas than in others. Today, collimators

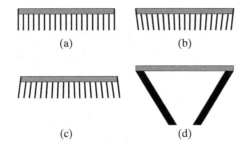

Figure 8.2
Different types of
collimators: (a) parallel-hole,
(b) converging, (c) diverging,
and (d) pinhole.

have square, hexagonal, or triangular holes, and the septa are of uniform thickness around each hole.

Converging and diverging collimators are available as well, but less frequently used in practice. A *converging collimator*, shown in Figure 8.2(b), has an array of tapered holes that aim at a point—the *focal point*—some distance in front of the collimator. The image that is presented to the crystal is a magnified version of the real object. A *diverging collimator*, shown in Figure 8.2(c), is essentially an upside-down converging collimator. Diverging collimators have an array of tapered holes that diverge from a hypothetical focal point behind the crystal. In this case, the image presented to the crystal face is smaller than the real object. Since converging and diverging collimators are simply flipped versions of each other, some collimator carriages actually have an insert that can be flipped either way, in effect producing two collimators in one.

EXAMPLE 8.1
When a converging or diverging collimator is used in an Anger camera, the image on the crystal face is not the same size as the object.

Question Using a diverging collimator with a focal length of z_f, what is the ratio between the size of the image and the size of the actual object when a planar object is placed parallel to the crystal face at a distance z away from the collimator face?

Answer For a diverging collimator, only the photons traveling towards the focal point can be detected. (Others will be absorbed by the collimator walls.) Using geometry and ignoring the height of the collimator, we have the ratio between the size of the image and the size of the actual object is

$$\frac{S_i}{S_o} = \frac{z_f}{z + z_f}.$$

∎

Pinhole collimators comprise a different class of collimators. As shown in Figure 8.2(d), these are thick conical collimators with a single 2 to 5 mm hole in the bottom center. As the source object is moved away from the pinhole, its image on the scintillation crystal gets smaller. In fact, the camera image is magnified (that is, larger than real size) from the collimator face to a distance equal to the length of the collimator and is then progressively smaller at larger distances.

The magnification produced by converging and pinhole collimators is useful in those cases when the object being imaged is significantly smaller than the field of view of the camera, and when the object's size or the amount of detail within the object challenges the intrinsic resolution capabilities of the system. With magnification, the image "presented" to the camera face is larger than life-size. This means that detail that is smaller (in the "real world") than the resolving capability of the camera may possibly still be resolved, as long as the actual "separation" or "spread" presented to the camera face is larger than the camera's intrinsic resolution. This same concept is what allows us to see small details with a magnifying glass that are invisible to the unaided eye.

8.2.2 Scintillation Crystal

The *scintillation detector* is the most commonly used detector in nuclear medicine, because it is more sensitive to electromagnetic radiation than is a gas-filled detector. This type of detector is based on the property of certain crystals to emit light photons (*scintillate*) after deposition of energy in the crystal by ionizing radiation. The most commonly used scintillation crystal in nuclear instrumentation is sodium iodide with "thallium doping," which is written NaI[Tl]. Since NaI[Tl] crystals absorb moisture from the air, the crystals are hermetically sealed in aluminum "cans." Because the aluminum absorbs alphas and betas, NaI[Tl] detectors are generally used only for detection of x-rays and gamma rays.

The scintillation crystals in a gamma camera are typically 10 to 25 inches in diameter and 1/4 to 1 inch thick. The thicker crystals are used for high-energy gamma rays, while the thinner crystals are used for low-energy gamma rays. Of course, any camera only has one crystal, so the user purchases a system with a crystal thickness matched to his or her needs. Like the screens that are used in projection radiography, thicker crystals stop more radiation than thinner crystals, but they also have poorer resolution—so there is a tradeoff between efficiency and resolution just as there is in projection radiography. We will have more to say about this later in the chapter.

8.2.3 Photomultiplier Tubes

Each gamma photon that interacts in the scintillation crystal (by a photoelectric or Compton scattering process) produces a burst of light in the crystal, comprising thousands of light or scintillation photons. This light is reflected and channeled out the back of the crystal, through a glass plate, and is incident upon an array of *photomultiplier tubes*. Each photomultiplier tube, shown in Figure 8.3, is a vacuum tube with two important components: a sensitive front surface, called the *photocathode*, and a series of electrodes, called *dynodes*. The photomultiplier tube serves two important functions: it converts a light signal into an electrical signal, and it amplifies this electrical signal.

The front surface of the photomultiplier tube has a transparent window through which light travels in order to hit a photosensitive surface—the *photocathode*, an extremely thin layer of an alloy such as cesium and antimony. For every 4 to 5 light

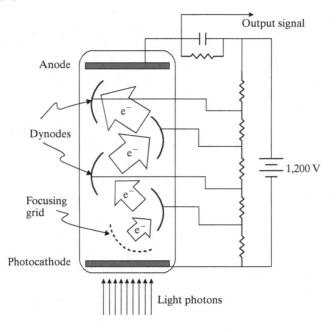

Figure 8.3
A schematic of a photomultiplier tube.

photons incident upon the photocathode, one electron is released or ejected from the photocathode by the photoelectric effect.

These photoelectrons are accelerated to the first *dynode*, an electrode that is positively charged and positioned a short distance from the photocathode (see Figure 8.3). For each electron reaching the first dynode, 3 to 4 electrons are released. The second dynode has a higher voltage than the first; thus, the electrons liberated from the first dynode are accelerated to the second dynode. Each of these electrons in turn liberates 3 to 4 electrons from this second dynode. This process is repeated at 10 to 14 successive dynodes in the photomultiplier tube, and 10^6 to 10^8 electrons reach the tube's anode (i.e., the final, most positively charged electrode) for each electron liberated from the photocathode.

These electrons comprise the output of the photomultiplier tube, and they form a current pulse. The photomultiplier tube outputs such a pulse each time a gamma photon interacts and deposits energy in the NaI[Tl] crystal, triggering the sequence of events in the tube as just described. This current pulse is directed into a preamplifier circuit, which amplifies the signal to provide a voltage pulse with a peak voltage—i.e., its *pulse height*—that ranges from a few millivolts to a few volts.

As stated above, an array of photomultiplier tubes covers the back of the crystal. This array means that light produced in the scintillation crystal is detected no matter where it is produced. An array is also necessary to produce the signals used in event positioning. The original Anger camera had seven photomultiplier tubes, but modern gamma cameras have 37, 61, 75, or 91 tubes, arranged in a hexagonal pattern, as shown in Figure 8.4. Generally, more photomultiplier tubes means better spatial resolution and image uniformity, but this comes with a higher initial cost, more difficult calibration procedures, and more costly maintenance.

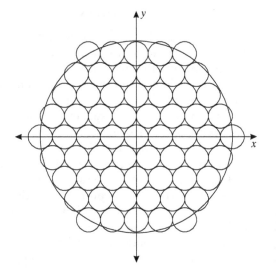

Figure 8.4
Arrangement of 61
photomultiplier tubes on the
face of an Anger camera.

The number of tubes is determined by the size and shape of both the crystal and each individual photomultiplier tube. The tubes must be smaller, of course, when there are more of them arrayed over the scintillation crystal. Early photomultiplier tubes had round cross-sections, whereas current photomultiplier tubes often have hexagonal cross sections to better "tile" the scintillation crystal without gaps between tubes.

8.2.4 Positioning Logic

When a gamma photon interacts with the crystal, thousands of scintillation photons are produced, and every photomultiplier tube produces an output pulse. The goal of the Anger camera's *positioning logic* circuitry is to determine both where the event occurred on the face of the crystal (see Figure 8.4) and the combined output of all the tubes, which represents the light output of the crystal (which in turn represents the energy deposited by the gamma photon). These output signals are denoted X and Y for the estimated two-dimensional position of the event, and Z for the total light output. The amplitude of a given tube's output is directly proportional to the amount of light (number of scintillation photons) its photocathode receives. The tubes closest to the scintillation event will have the largest output pulses, while those farther away will have smaller output pulses. By analyzing the spatial distribution of pulse heights, the location of a single scintillation event (X, Y) can be determined quite accurately—to within a fraction of the diameter of a photomultiplier tube, in fact. We will discuss this process later, but for now it is important to remember that the *rate* of (X, Y, Z) signals or pulses coming from the camera is proportional to total radioactivity, and the *size* or *height* of each Z-pulse is proportional to the energy deposited in the crystal by the gamma photon for the event that (X, Y, Z) represents.

8.2.5 Pulse Height Analyzer

The desired goal of the Anger camera is to create an image that portrays the distribution of radioactivity (i.e., sites and numbers of radioactive atoms) within the patient. Because the collimator only allows those photons traveling in predetermined directions to interact in the crystal, a line drawn from the scintillation event in the crystal through the nearest collimator hole is presumed to intersect the site of origin of the photon (i.e., the radioactive atom it came from) in the patient. If the photon has been Compton scattered in the patient, a line drawn through its direction of flight will not intersect its site of origin, only the site of the Compton interaction. Thus, photons scattered into the field of view could be falsely attributed to activity at the sites of Compton interactions in the patient. It is clearly not desirable to have these scattered photons contribute to the final image, as they significantly degrade resolution and contrast. It is important to note that a large percentage of photons striking the crystal have been scattered in the patient.

Fortunately, photons that have been Compton scattered can be distinguished from those that arrive directly by analyzing the energy deposited in the crystal via the Z-pulse, whose height is proportional to the total energy deposited in the crystal. In practice, we know the energy of gamma photons emitted from the radioactive source. If this amount of energy is deposited in the crystal, the gamma photon could not have Compton scattered in the patient, since it would have lost some of its energy in the process [as determined by (4.8)]. Of course, a gamma photon could emerge from the patient with all its energy (i.e., without having interacted in the patient), and only deposit part of its energy in the crystal (via a Compton scattering interaction). We cannot distinguish between this event and one in which the gamma photon Compton scattered in the patient and then deposited all its remaining energy in the crystal via a subsequent photoelectric event. In practice, we can only be certain that the gamma photon has not scattered if it deposits the full amount of energy with which it started in the crystal, and these are the only events we allow to contribute to image formation.

Scintillation spectrometry, or *pulse height analysis*, refers to the use of a scintillation counting system to obtain an energy spectrum from a radioactive source. This energy spectrum is a plot of the number of pulses with a given pulse height (within a pulse height "bin," actually) as a function of the pulse height, as illustrated in Figure 8.5(a). The measured spectrum is a function of the energies of gamma rays emitted by the source, and the interactions of these photons in both the body and the crystal. It should be noted that the pulse heights on the x-axis generally have arbitrary units (AU), although they could be scaled to units of energy (keV) if the camera were properly calibrated.

A pulse height spectrum, as shown in Figure 8.5(a), has two main features: a broad plateau called the *Compton plateau*, and a peak at the highest pulse heights, which is called the *photopeak*. The broad plateau represents Compton scatter interactions in the patient and/or crystal, broadly distributed in energy due to the random nature of the Compton scatter angle. The rightmost limit of this plateau, called the *Compton edge*, represents Compton interactions in which the incoming (unscattered) gamma photon is backscattered 180° in the crystal, thus depositing the maximum energy possible in a single Compton interaction. The photopeak

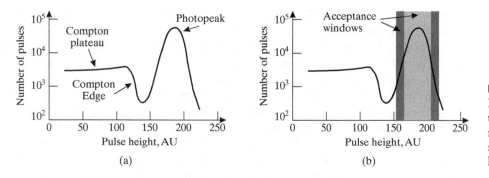

Figure 8.5
Pulse height spectrum for
technetium-99m, plotting
number of pulses or count
rate as a function of pulse
height or energy.

represents gamma photons that come directly from the source, and deposit all of
their energy in either a single photoelectric interaction or one or more Compton
interactions followed by a photoelectric interaction. Because a gamma photon cannot
lose all its energy in a single Compton scatter event, there is a separation between
the Compton plateau and the photopeak.

EXAMPLE 8.2
The pulse height analyzer is used to reject photons that have been Compton scattered.

Question What is the maximum angle through which a 140 keV photon can be scattered
and still be accepted within a 20 percent pulse height window?

Answer Note that 20 percent pulse height window is 10 percent on either side.

$$140 \text{ keV} \times 0.1 = 14 \text{ keV}.$$

$$140 \text{ keV} - 14 \text{ keV} = 126 \text{ keV}.$$

Since

$$h\nu' = \frac{h\nu}{1 + \frac{h\nu}{m_0 c^2}(1 - \cos\theta)},$$

we have

$$126 \text{ keV} = \frac{140 \text{ keV}}{1 + \frac{140 \text{ keV}}{511 \text{ keV}}(1 - \cos\theta)}.$$

Solving for θ yields $\theta = 53.54°$. ∎

8.2.6 Gating Circuit

The Z-pulse is proportional to the total energy deposited in the crystal by the
gamma ray photon and is used by the pulse height analyzer to discriminate against
Compton scattered photons. The pulse height analyzer is used to set an acceptance
window around the photopeak (the dominant energy for the particular radiotracer

being used), as shown in Figure 8.5(b). Its lower threshold is set to discriminate against Compton events, which have lower energy than a non-Compton event. Its upper threshold is set to discriminate against multiple (simultaneous, summed) events, which have more energy than a single event (and for which "position" is a meaningless concept). Because the window has a finite width, some scattered photons may still be accepted (those which scatter through a small angle, and thus retain most of their energy). For example, 140 keV photons can scatter by as much as 50° and still be accepted by the often-used 20 percent window (i.e., a window of ±10 percent around the photopeak energy). In practice, proper window setting is vital, as a window that is not centered around the photopeak (an *offset window*) can degrade field uniformity of response for many cameras; this is typically a result of the slightly better light collection efficiency directly under each photomultiplier tube.

Current cameras, with microprocessor-based correction circuitry (which will be described later), generally maintain good uniformity even with offset pulse height windows. Such cameras may be purposely peaked to the high side of the photopeak to further reduce scatter, by eliminating any Compton scattered photons that show up in the lower half of the photopeak (due to every camera's less than perfect energy resolution). Some cameras have two or three separate pulse height windows to simultaneously image the multiple emissions of some radionuclides (e.g., those from gallium-67). In this way, counts are acquired in a shorter amount of time as the multiple energy emissions are utilized.

8.2.7 Image Capture

Together, the positioning logic and pulse height analyzer (with its gating circuit) yield an estimated position (X, Y) of each scintillation event that falls within the energy discrimination window. Earlier Anger cameras were analog, and relied on photographic image formation. In these systems, the X and Y pulses associated with each Z-pulse that passed through the pulse height analyzer were used to position a finely focused dot of light on the cathode ray tube face. A collection of these light dots over time would produce an image. Since it took a period of time for a complete picture to be obtained, perhaps up to several minutes to obtain several hundred thousand counts, an integrating medium was used to record the image. The most frequently used media were various types of photographic film. A photographic camera was mounted on the cathode ray tube, and the shutter was left open during the entire image acquisition period. The film was developed and an image of the distribution of radioactivity was obtained.

In the past 20 years or so, it has become the norm for the camera to be interfaced to a computer. During the transition period from analog image capture, as described above, to digital capture via computer, the (X, Y, Z) signals were digitized (typically with 16-bit quantization) and used by software to create images. In today's cameras (see Figure 8.1), the output of each photomultiplier tube is digitized by its own analog-to-digital converter, and the positioning logic and image formation described below are implemented in a fully digital fashion. As a result, the computer and its software become an integral and necessary part of the planar scintigraphy image formation process, as they do in computed tomography. We describe the computer assisted image formation process in the following section.

8.3 Image Formation

The primary mechanism for creating images in planar scintigraphy is to detect and estimate the position of individual scintillation events on the face of an Anger camera. We now describe how the locations of these events are estimated, how these locations are combined to produce images, and what these images represent mathematically in terms of the distribution of radioactivity in the body.

8.3.1 Event Position Estimation

A single scintillation event produces a flash of light that yields a response from many, if not all, photomultiplier tubes attached to the Anger camera. The amplitude or height of each tube's response is proportional to its distance from the scintillation event, and this is how the position (X, Y) of the scintillation event is encoded. We start our discussion of event position estimation with a general mathematical framework, and then specialize the presentation to both analog and digital cameras.

Consider a coordinate system with the origin $(0, 0)$ at the center of the crystal. Let the photomultiplier tubes be indexed by $k = 1, \ldots, K$, and their positions be given by (x_k, y_k). Suppose the amplitudes of their response to a scintillation event are given by a_k, $k = 1, \ldots, K$. These heights are viewed as representing samples of a 2-D distribution of light, and the location of the maximum of this distribution represents the position (X, Y) of the event.

We could set (X, Y) equal to the location of the photomultiplier tube having the largest amplitude, but that would provide only a very crude estimate of the location, yielding a resolution on the order of the size of a tube. Instead, gamma cameras implement a center of mass calculation to estimate (X, Y). To calculate the center of mass of the tube responses, we first form

$$Z = \sum_{k=1}^{K} a_k, \tag{8.1}$$

which represents the *mass* of the light distribution. This signal is also proportional to the total light output from the scintillation event (it is the so-called *Z-pulse*), and is used in the pulse height analysis circuit, as previously discussed. Given the total mass Z, the components of the center of mass (X, Y) are calculated as

$$X = \frac{1}{Z} \sum_{k=1}^{K} x_k a_k, \tag{8.2}$$

$$Y = \frac{1}{Z} \sum_{k=1}^{K} y_k a_k. \tag{8.3}$$

These three values (X, Y, Z) represent the "heart" of an Anger camera. Together, they provide the location (X, Y) of the scintillation event and the total light output Z—or, equivalently, the total energy—of the event.

It is worth making a couple of comments about this process. First, it should be understood that the a_k's are actually short time waveforms, pulses arising from the scintillation event causing the cascading electrons in each photomultiplier tube. The individual numbers X, Y, and Z represent values obtained at the peak of such pulses or by integrating over a short time frame. Second, small signals occurring in photomultiplier tubes that are distant from the event can degrade the position estimates. Therefore, a threshold circuit is added to permit only signals above a certain level to contribute the estimates of X and Y. This so-called *discriminator circuit* does not apply to the estimate of Z, which still uses the output of all photomultiplier tubes. Do not confuse this discriminator circuit with the pulse height gating circuit, which determines whether the Z signal is in a range sufficient to qualify the event as both a non-Compton and nonmultiple event.

In a modern Anger camera, the event position estimation is implemented digitally. In most camera designs, the output of each photomultiplier tube goes directly to a separate analog-to-digital converter, typically with at least 16-bit precision. Then, software embedded in programmable read-only memory (PROM) chips determines the event position utilizing the center-of-mass calculation described above. A potential advantage of digital position estimation systems is that alternative algorithms could be easily implemented, although this advantage is rarely exploited.

In older, analog Anger cameras, (X, Y) position estimates are actually made by using four resistive circuits and two gain-controlled difference amplifiers. In particular, a separate resistive circuit is used to weight and sum all the signals for tubes whose x-position is positive, whose x-position is negative, whose y-position is positive, and whose y-position is negative. This notion of positive and negative position values arises because the conventional Anger camera coordinate system has the origin $(0, 0)$ at the center, as shown in Figure 8.6. This network of circuits creates four signals: X^+, X^-, Y^+, and Y^-, and the estimated event position is given by

$$X = \frac{X^+ - X^-}{Z}, \tag{8.4}$$

$$Y = \frac{Y^+ - Y^-}{Z}. \tag{8.5}$$

The values of the resistors in this network are determined by the tube positions, and are chosen so that the above calculation yields the same answer as the center of mass calculation as in (8.2) and (8.3).

EXAMPLE 8.3

Consider an Anger camera with 9 square photomultiplier tubes arranged in a 3 × 3 array, as shown in Figure 8.6. Each tube is numbered, and the recorded pulse heights of the tubes are: 5, 10, 15, 10, 40, 30, 0, 5, 10, in the order of the tube numbers.

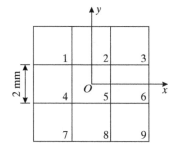

Figure 8.6
Arrangement of 9 square photomultiplier tubes of an Anger camera.

Question: What is the value of the Z-pulse? What is the position of the event?

Answer: The value of the Z-pulse is the summation of the pulse height recorded by each tube:

$$Z = 5 + 10 + 15 + 10 + 40 + 30 + 0 + 5 + 10 = 125 \, .$$

The centers of the tubes are

$$c_1 = (-2, 2), \quad c_2 = (0, 2), \quad c_3 = (2, 2)$$

$$c_4 = (-2, 0), \quad c_5 = (0, 0), \quad c_6 = (2, 0)$$

$$c_7 = (-2, -2), \, c_8 = (0, -2), \, c_9 = (2, -2)$$

The estimated position of the event is

$$X = \frac{1}{Z} \sum_{k=1}^{K} x_k a_k = 80/125 = 0.64 \text{ mm} \qquad \text{and}$$

$$Y = \frac{1}{Z} \sum_{k=1}^{K} y_k a_k = 30/125 = 0.24 \text{ mm} \, .$$

∎

8.3.2 Acquisition Modes

For each scintillation event whose pulse height falls within the photopeak energy window, an Anger camera provides the estimated position (X, Y) of the event, the time of the event, and the peak height of the Z-pulse. In *list mode acquisition*, the (X, Y) signals are transferred directly to computer memory in the form of a list of (X, Y) coordinates, as shown in Figure 8.7(a). In addition to the (X, Y) signals, typically the Z-pulse value is also recorded, and time markers are inserted every 1 to 10 msec. Physiological trigger marks, such as the occurrence of the R-wave from an electrocardiogram (ECG) monitor of the patient's heart, can also be inserted. In list mode acquisition, no matrices or images are formed within computer memory during acquisition. Although no images are produced immediately, list mode acquisition is useful because the (X, Y) signals are permanently recorded in computer memory, allowing flexible control over subsequent formatting into digital matrices. The precision of the event position is limited only by the number of bits used to store the

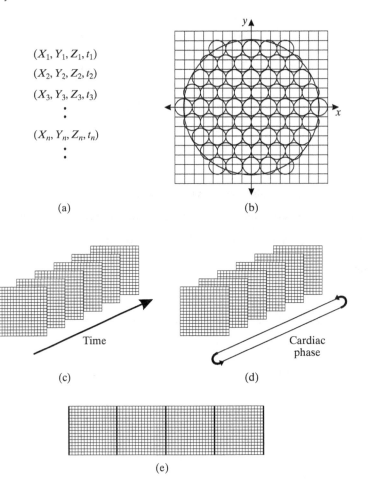

(X_1, Y_1, Z_1, t_1)

(X_2, Y_2, Z_2, t_2)

(X_3, Y_3, Z_3, t_3)

$\vdots$

(X_n, Y_n, Z_n, t_n)

$\vdots$

(a)   (b)

Time   Cardiac
                        phase

(c)   (d)

(e)

Figure 8.7

Anger camera acquisition modes: (a) list mode; (b) static frame mode; (c) dynamic frame mode; (d) multiple-gated acquisition; and (e) whole body.

X and Y positions. The downside is that a huge amount of storage space is required even for a single study.

Static frame mode acquisition, as shown in Figure 8.7(b), represents the face of the Anger camera as a computer matrix—i.e., a finite collection of discrete bins, or *pixels*, as introduced in Section 2.2. Accordingly, each pixel corresponds to a certain area of the camera face and is designated by a specific range of (X, Y) signal values. An image is created by starting with a zero image at the beginning of the scan and then incrementing the pixel containing (X, Y) with each (accepted) scintillation event. Finally, if there are 400,000 accepted events, then the sum of all the pixel values in the final image will be 400,000—i.e., the so-called *total count* for the scan.

Dynamic frame mode acquisition, as shown in Figure 8.7(c) is a temporal succession of frame mode images. After the first image is created and stored, its buffer (matrix) is zeroed and another image is accumulated in identical fashion. This mode allows the study of transient physiological processes such as radiotracer uptake, washout, or redistribution. Movies of the temporal progression of a dynamic process can be created by appending the frames and replaying them after acquisition is complete. Generally, the number of pixels in dynamic frame mode images is smaller in order to yield higher counts (and correspondingly lower noise) in each pixel. [Increasing pixel size to reduce noise has already been discussed in the context of projection radiography; see (5.37).]

A further extension of frame mode acquisition is *multiple-gated acquisition*, as depicted in Figure 8.7(d). In this mode, the data from the camera are distributed to a series of matrices in computer memory. A trigger signal (usually a physiological trigger such as the R wave of the electrocardiogram) controls the distribution of data among the matrices. Immediately after the trigger, data from the camera are placed in the first matrix for a fixed time interval. When the time interval has elapsed, data are then placed in the second matrix for the same time interval. This process continues until the occurrence of a new trigger signal, at which time data distribution restarts at the first frame, or until all the assigned matrices in computer memory are used, in which case no data are acquired until the occurrence of a new trigger signal.

Multiple-gated acquisition is used to study a repetitive (cyclic) dynamic process. For example, in a cardiac gated blood pool study, in which the circulating blood is labeled with radioactivity and the beating chambers of the heart are examined, the data from the corresponding phases of many heartbeats are superimposed during acquisition, resulting in a series of images representing one "average" cardiac cycle. Typically, the cardiac cycle is divided into 16 to 64 frames, with each frame representing 1/16 to 1/64 of the cycle.

Whole body acquisition is yet another variation of static frame mode acquisition. In this mode, the body is divided into a matrix of pixels, as depicted in Figure 8.7(e) and a series of static frames are acquired to cover the body, in a "step and shoot" sequence. As an alternative, the camera or bed can continuously move. The position of the patient relative to the camera is maintained so that the camera can be scanned over the whole patient, thereby creating a whole-body image.

The most common (2-D) matrix sizes used in nuclear medicine are 64×64, 128×128, and 256×256, although matrices up to 1024×1024 are available. The larger the matrix size, the better the digital spatial resolution in the image, up to the limit imposed by the intrinsic capabilities of the camera. The digital sampling requirements necessary to preserve the spatial resolution in the analog image are given by the Nyquist theorem in Section 2.8. This theorem states that, to accurately portray a signal, the sampling frequency must be twice the highest frequency present in the signal. Thus, pixel dimensions should be smaller than one-half the spatial resolution of the Anger camera. In practice, pixel dimensions range from 2 to 6 mm.

EXAMPLE 8.4

The multiple-gated acquisition mode can be used to study beating hearts. In this mode, each cycle of the acquisition is triggered by the R wave of the ECG, as shown in Figure 8.8.

Question Suppose the heart rate is 50 bpm (beats per minute). We want each frame to last for 75 ms. What is the total number of frames?

Figure 8.8
One cycle of heartbeat is divided into frames. The R wave of the ECG signal is used to trigger acquisition in multiple-gated frame mode.

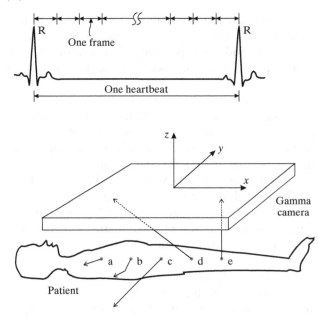

Figure 8.9
Imaging geometry showing the paths of five photons (a, b, c, d, and e) and a camera-centric coordinate system.

Answer The heart rate is 50 bpm, so each heartbeat lasts for $60/50 = 1.2$ sec. Each frame lasts for 75 ms. Therefore, a heartbeat is divided into

$$N = \frac{1.2 \text{ sec}}{75 \text{ ms}} = 16 \text{ frames.}$$

∎

8.3.3 Anger Camera Imaging Equation

Consider what happens to a photon emitted from within the body. Statistically, the photon has equal probability to propagate in any direction. As shown in Figure 8.9, sometimes the photon will get absorbed (a, b), sometimes it will be scattered (b), and sometimes it will leave the body (with or without being scattered) (c, d, e). If it leaves the body, sometimes it will hit the Anger camera (d, e); more often it will not (c). When a gamma ray hits the camera, it will usually be absorbed by the lead in the collimator because it will either be traveling in an improper direction or will miss a collimator hole.

We will ignore Compton scattering in our development of an imaging equation; hence, photons are assumed to travel in straight lines. Assume that the radioactivity in the body is given by $A = A(x, y, z)$ and that the energy of each photon is given by E—i.e., a monoenergetic beam. From (4.16), we see that the energy fluence rate, or intensity, on the detector due to photons emanating from (x, y, z), assuming there is no attenuation, is given by[1]

$$I_d = \frac{AE}{4\pi r^2}, \tag{8.6}$$

---

[1]The symbol A here—radioactivity—should not be confused with the symbol A in (4.16), which is a measure of area.

where r is the distance from (x, y, z) to a detector position $(x_d, y_d, 0)$, i.e.,

$$r = \sqrt{(x - x_d)^2 + (y - y_d)^2 + z^2}\,.$$

However, recall that we are not interested in intensity, but activity. In nuclear medicine, we estimate activity as the photon fluence rate coming from the patient. In (4.14), we defined the photon fluence rate, ϕ, and in (4.17) multiplied ϕ by E to obtain intensity. So, we will switch from the use of I_d to ϕ_d in subsequent equations.

Incorporating attenuation along the line between (x, y, z) and $(x_d, y_d, 0)$ yields [see (4.25)]

$$\phi_d = \frac{A}{4\pi r^2} \exp\left\{-\int_0^r \mu(s; E)\, ds\right\}, \tag{8.7}$$

where it is assumed that the origin $(s = 0)$ of the line is at the point (x, y, z).

Now we restrict our attention to a parallel-hole collimator, and consider the response through a single hole. In this case, the line of integration is parallel to the z-axis. We assume that the collimator's lead is sufficient to absorb all gamma rays that are not traveling in the z-direction. Under these assumptions, (8.7) becomes

$$\phi_d = \frac{A}{4\pi z^2} \exp\left\{-\int_z^0 \mu(x, y, z'; E)\, dz'\right\}, \tag{8.8}$$

where the detector position under consideration is directly "underneath" the point (x, y, z)—i.e., $(x_d, y_d, 0) = (x, y, 0)$; see Figure 8.9. This development, so far, has only considered radioactivity arising from the point (x, y, z). In fact, the photon fluence rate observed at the detector point $(x, y, 0)$ is the integral of all sources that are in line with the collimation. Because of the geometry (see Figure 8.9) and because we are considering a parallel hole collimator, the only sources that can contribute to this integrated photon fluence rate are those lying on the $-z$ axis. Therefore, the imaging equation becomes

$$\phi(x, y) = \int_{-\infty}^0 \frac{A(x, y, z)}{4\pi z^2} \exp\left\{-\int_z^0 \mu(x, y, z'; E)\, dz'\right\} dz\,. \tag{8.9}$$

We see from (8.9) that the photon fluence rate recorded on the face of a gamma camera is a type of projection, like projection radiography [e.g., (5.3)], but with important differences. On the one hand, this equation is simpler because there is (typically) only one energy to consider; therefore, there is no integral over an energy spectrum as there is in radiography and CT. On the other hand, (8.9) is not a simple integral over $A(x, y, z)$, like the ideal model (involving an effective energy) used in CT reconstruction. There are, in fact, two sources of depth-dependent signal loss, one arising from the inverse square law and the other from object-dependent attenuation between the radiation source and the detector. Because of these effects, we can expect that activity close to the camera will contribute a larger photon fluence rate than an equivalent activity far away from the camera (i.e., a larger contribution to the total counts at that position within the 2-D projection image).

Another consequence of depth-dependent attenuation is that we see dramatic differences in a source's observed photon fluence rate when the camera is located on the same side of the body as the source versus when it is located on the opposite side. This is especially true for low-energy radiotracers since the body's attenuation is larger. Like the reading of x-ray films, where magnification distorts sizes of objects that are near versus far, we can get accustomed to reading planar images in nuclear medicine, knowing about object-dependent attenuation and the anatomy of the body. It will turn out this object-dependent attenuation causes problems for SPECT, and we will require a simpler imaging model for computational tractability, much like the effective energy model employed in CT.

Planar Sources We saw above that the image formed on an Anger camera is a projection of the radioactivity within the body with depth-dependent effects. As in projection radiography, it is useful to consider a planar source in order to see what the image would be like without depth-dependent effects. This concept is particularly useful when we begin incorporating other effects such as blurring and noise in the next section.

The source or object in nuclear medicine imaging is the distribution of radioactivity, $A(x, y, z)$. Let us define a planar source $A_{z_0}(x, y)$ which has radioactivity restricted to the plane defined by $z = z_0$; in other words,

$$A(x, y, z) = A_{z_0}(x, y)\delta(z - z_0). \tag{8.10}$$

Of course, $z_0 < 0$ in order to conform to the geometry of Figure 8.9. Plugging (8.10) into (8.9) and applying the sifting property of the impulse function [see (2.6)] yields

$$\phi(x, y) = A_{z_0}(x, y)\frac{1}{4\pi z_0^2}\exp\left\{-\int_{z_0}^{0}\mu(x, y, z'; E)\,dz'\right\}. \tag{8.11}$$

Equation (8.11) shows that the measured photon fluence rate of a planar source represents the radioactivity within the source plane scaled by two factors. The first factor $1/(4\pi z_0^2)$ represents a signal loss due to the inverse square law. The signal loss, while dependent on z_0, is uniform—i.e., independent of x and y—so it does not affect the relative amplitudes of the radioactive sources as they appear on the detector plane. The second factor is the effect of the integrated attenuation of photons as they travel from the plane $z = z_0$ to the plane $z = 0$. Unless $\mu(x, y, z'; E)$ is itself independent of x and y—not likely in realistic scenarios—this term is not uniform, and it causes a variation of the relative source amplitudes across the image plane. This emphasizes the inherent difficulty, even for ideal planar sources, in comparing relative intensities between sources located at different points within a planar scintigraphic image.

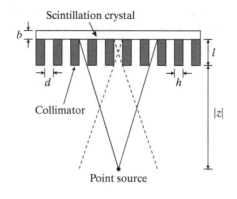

Figure 8.10
Collimator geometry demonstrating depth-dependent resolution.

8.4 Image Quality

There are many factors that affect the performance of Anger cameras, including spatial resolution, sensitivity, and field uniformity. Modern Anger cameras contain correction circuitry to improve performance in these areas as much as possible. We now explore these and other factors affecting image quality in planar scintigraphy.

8.4.1 Resolution

The ability of an Anger camera to localize a photon interaction is characterized by its *resolution*, defined by a full width at half maximum (FWHM) value, as discussed in Section 3.3. Two factors affecting resolution are most important: *collimator resolution* and *intrinsic resolution.* These two factors combine to yield the system resolution of the Anger camera. Since scintigraphic scans can be lengthy, patient motion enters as another factor affecting resolution, but we will not develop a model for patient motion here.

Collimator Resolution   Since a parallel hole collimator is most commonly used, we focus on its performance here. With reference to the geometry shown in Figure 8.10, the collimator resolution is given by (see Problem 8.12)

$$R_C = \frac{d}{l}(l + b + |z|),\qquad(8.12)$$

where d is the collimator hole diameter, l is the collimator hole length, b is the (effective) scintillator depth, and $|z|$ is the collimator surface-to-patient distance. The most important feature of this equation is that the FWHM is dependent on target range $|z|$—i.e., $R_C = R_C(|z|)$. In particular, collimator resolution degrades as $|z|$ increases; therefore, targets farther away are blurred more.

EXAMPLE 8.5
Making the collimator hole longer will reject more Compton scattered photons. This also has an impact on the resolution of the imaging system.

Question   If we double the hole length to reject more Compton scattered photons, what is the collimator resolution? Is there a limit on the collimator resolution?

Answer   Equation (8.12) gives

$$R_C = \frac{d}{l}(l + b + r) = d + \frac{b+r}{l}d.$$

When the collimator hole length is doubled, the collimator resolution becomes

$$R_C' = \frac{d}{2l}(2l + b + r) = d + \frac{b+r}{2l}d < R_C.$$

So by doubling the hole length, the collimator resolution is also improved. If we make the hole length go to infinity, the collimator resolution tends to the hole diameter, d, which is the limit to the resolution. We will see later, by using longer holes, we degrade the collimator sensitivity.   ∎

We can factor collimator resolution into our imaging equation for planar sources quite readily. Equation (8.12) does not give an explicit formula for the collimator PSF, but it is common to assume that the PSF is Gaussian with FWHM equal to R_C. For a Gaussian function (see Examples 3.4 and 3.5),

$$\text{FWHM} = 2\sigma\sqrt{2\ln 2}.$$

Therefore, since $\text{FWHM} = R_C$, a common approximation for the collimator PSF is

$$h_C(x, y; |z|) = \exp\left\{-4(x^2 + y^2)\ln 2/R_C^2(|z|)\right\}, \qquad (8.13)$$

where we have made the dependence on range $|z|$ explicit.

Consider our planar source $A(x, y, z) = A_{z_0}(x, y)\delta(z - z_0)$ from the previous section. By design, all the sources are at range $r = |z_0|$ from the collimator. Therefore, putting (8.13) together with (8.11) yields

$$\phi(x, y) = A_{z_0}(x, y)\frac{1}{4\pi z_0^2}\exp\left\{-\int_{z_0}^{0}\mu(x, y, z'; E)\,dz'\right\} * h_C(x, y; |z_0|). \qquad (8.14)$$

The PSF blurring is a convolution in this case, because the collimator is spatially the same in the x-y plane. We could not write a 3-D convolution for a 3-D source, however, since there is depth dependency in both the basic imaging equation and the collimator PSF.

Equation (8.12) is based on the assumption that photons are always absorbed in the septa. Some fraction of photons will penetrate the septa, however, which causes R_C to be somewhat larger than what is predicted above. A simple way to account for septal penetration (see Problem 8.5) is to assume in (8.12) that the hole length is slightly shorter than reality. The *effective hole length* (l_e) is defined as

$$l_e = l - 2\mu^{-1}, \qquad (8.15)$$

where μ is the linear attenuation coefficient of the collimator material, typically lead, at the photon energy under consideration. The effective hole length is shorter than the actual length and is shortest at the highest photon energies because these photons are more penetrating (recall that μ is smallest at the highest photon energies). Using this definition, the adjusted collimator resolution becomes

$$R_C = \frac{d}{l_e}(l_e + b + r). \tag{8.16}$$

This definition more accurately reflects the fact that a given collimator has a different (effective) resolution depending on the energy of the radiotracer used. In particular, a collimator designed for low-energy radiotracers will have poor resolution when used for high-energy radiotracers (because photons can penetrate the septa without being absorbed), and is generally an ill-advised practice for that reason. It should be noted that b, the effective scintillation crystal thickness, is also energy-dependent. Higher-energy photons will penetrate the crystal farther, on average, before being absorbed, which contributes an additional degradation of resolution (R_C becomes larger) with high-energy photons.

Intrinsic Resolution   The concept of collimator resolution incorporates both the geometry of the collimator and the energy-dependent septal penetration of photons. Additional blurring takes place in the scintillator itself, however, and this process is characterized by the *intrinsic resolution* of the Anger camera.

The concept of *intrinsic resolution* is not quite the same as that of resolution in an intensifying screen in radiography. In both projection radiography and planar scintigraphy, there is a flash of light in the scintillator (or phosphor) when a photon is absorbed. In projection radiography, the spread of the light is permanently recorded on the (film or solid state) detector. The amount that the light spreads represents the blurring of the image. In planar scintigraphy, the spreading of the flash of light is used to estimate the position of the single absorption event. It is inaccuracy in the estimation of (X, Y) that accounts for intrinsic resolution in planar scintigraphy.

There are two primary reasons for inaccuracy in the estimation of (X, Y) in an Anger camera. The first reason is related to the path of the absorbed photon. As discussed in Chapter 4, a photon that is absorbed by the photoelectric effect may have experienced some Compton scattering events prior to the final absorbing event. Those Compton events occurring in the scintillator generate flashes of light, smaller in intensity than the final event, but spread out in space. These flashes are detected in the array of photomultiplier tubes (as though they came from a single event) and they contribute to the error in estimating (X, Y) by skewing the peak light distribution along the photon path.

The second reason for inaccuracy in the estimation of (X, Y) is noise, or statistical fluctuation. Consider a photon that goes straight through a collimator hole and is absorbed promptly in a single photoelectric event. This is a "dream" event, devoid of geometric or Compton scattering problems, and should be localized perfectly if we lived in an ideal world. In fact, the physics and electronics that follow are beset with inherent statistical fluctuations that render the result less than perfect.

First, both the number and the spatial distribution of scintillation light photons resulting from the photoelectric event are random. Although detailed analysis of the physics of the scintillator can provide good models for the mean numbers of these photons, the exact number cannot be predicted. Thus, the number and spatial distribution of light entering into any photomultiplier tube are also random. The situation gets progressively worse within a photomultiplier tube, where the electron cascade process is also random. All this leads to statistical variations in the peak pulse heights a_k, $k = 1, \ldots, K$ within the photomultiplier array. These variations lead to a statistical error variance associated with the estimation of (X, Y), which causes additional blurring in the localization of radioactive sources in space.

Intrinsic resolution of an Anger camera can be characterized by another FWHM, R_I, which encompasses both sources of variation in the scintillation process. It is common to model the PSF associated with this process by a Gaussian function. Accordingly [see (3.23)], we define

$$h_I(x, y) = \exp \left\{ -4(x^2 + y^2) \ln 2 / R_I^2 \right\}, \tag{8.17}$$

the intrinsic PSF of an Anger camera. Unlike collimator resolution, intrinsic resolution is not dependent on the target or its range. We can factor intrinsic resolution into our imaging equation for a planar source as follows:

$$\phi(x, y) = A_{z_0}(x, y) \frac{1}{4\pi z_0^2} \exp \left\{ -\int_{z_0}^{0} \mu(x, y, z'; E) \, dz' \right\} * h_C(x, y; |z_0|) * h_I(x, y). \tag{8.18}$$

Generally, intrinsic resolution is much better than collimator resolution (i.e., $R_I \ll R_C$) at typical imaging depths, so the collimator's geometric characteristics dominate effective resolution. The effective resolution is the result of the subsystem cascade of R_I and R_C, as discussed in Section 3.3.4.

8.4.2 Sensitivity

As noted above, not every gamma ray leaving the body will be directed at the camera. If possible, we would like to detect those that are directed at the camera in the right direction. There are two major factors that may prevent such detection. First, the photon may be absorbed in the collimator; second, the photon may pass through both the collimator and the scintillation crystal. A very high sensitivity (or efficiency) camera will detect most photons; a very low sensitivity (or efficiency) camera will reject or miss most photons. Most cameras are somewhere in-between; we now characterize sensitivity in more detail.

Collimator Sensitivity From (8.12), we can see that longer collimator holes yield better resolution (smaller R_C). When the holes are made longer, however, fewer gamma ray photons that impinge upon the collimator will actually pass through it and reach the scintillator crystal. This means that the collimator is less efficient, or less sensitive. The *collimator efficiency* ϵ, also called *collimator sensitivity*, is given by

$$\epsilon = \left(\frac{Kd^2}{l_e(d+h)} \right)^2 , \qquad (8.19)$$

where $K \approx 0.25$. The quantity ϵ reflects the fraction of photons (on average) that pass through the collimator for each emitted photon directed at the camera. All else being equal, we would like to choose collimator parameters that make ϵ as large as possible.

The resolution of a parallel hole collimator is best at the collimator surface and degrades with distance $|z|$, as shown in (8.12). On the other hand, (8.19) shows that the sensitivity is independent of the distance between the source and the collimator. While this seems at first blush to contradict the inverse-square law, it really does not. The field of view of each hole increases with increasing distance. This means that each hole "sees" a larger area at a greater distance from the collimator (see Figure 8.10). Looking at it another way, more holes see the same source if it is farther away from the collimator. As a radioactive source is moved away from the face of the collimator, the count rate through the central hole decreases due to the inverse-square law. However, more and more holes see the source, and the total count rate remains constant. Since more holes see the source, its image is spread over a larger area of the crystal face (i.e., the image is progressively smeared out). Thus, the resolution gets worse with increasing distance.

Equation (8.19) presents the overall sensitivity of a collimator—that is, the total counts received by the detector. It may seem odd that it does not include a distance term (i.e., there is no inverse square law effect) given the development in Section 8.3.3. However, in that section, we focused on the detected photon fluence rate at one point on the detector, which is affected by the inverse square law. The reason why the overall sensitivity is not affected in the same way is because the counts from the source are not lost, but instead distributed to other parts of the camera.

Consider a fixed source range $|z|$ satisfying $|z| \gg l_e + b$. Then,

$$R_C \approx \frac{d}{l_e}|z| , \qquad (8.20)$$

and by substitution,

$$\epsilon \approx \left(\frac{R_C Kd}{|z|(d+h)} \right)^2 . \qquad (8.21)$$

This relationship shows that there is a tradeoff between resolution and sensitivity. Thus, while we already understand the need for separate high-energy, medium-energy, and low-energy collimators to account for differences in septal penetration, there is also the need to consider different collimators for the same energy range that trade off resolution and sensitivity. Table 8.1, for example, shows several low energy (i.e., for 140 keV) collimators and their resolutions and sensitivities.

TABLE 8.1

Resolution and Sensitivity for Several Collimators					
Collimator	d (mm)	l (mm)	h (mm)	Resolution @ 10 cm (mm)	Relative sensitivity
LEUHR	1.5	38	0.20	5.4	12.1
LEHR	1.9	38	0.20	6.9	20.5
LEAP	1.9	32	0.20	7.8	28.9
LEHS	2.3	32	0.20	9.5	43.7

LEUHR = low energy ultra-high resolution

LEHR = low energy high resolution

LEAP = low energy all purpose

LEHS = low energy high sensitivity

EXAMPLE 8.6

We saw in Example 8.5 that use of a collimator with long holes can help to reject Compton scattered photons and improve collimator resolution.

Question What is the change in sensitivity if we double the hole length and keep other parameters fixed?

Answer For simplicity, assume $l_e = l$, the sensitivity of a collimator is given by (8.19):

$$\epsilon = \left(\frac{Kd^2}{l_e(d+h)} \right)^2.$$

If the collimator hole length is doubled, we have

$$\epsilon' = \left(\frac{Kd^2}{2l_e(d+h)} \right)^2 = \epsilon/4.$$

So by doubling the collimator hole length, we reduce the sensitivity by 75 percent. ∎

Detector Efficiency Not every gamma ray that passes through the detector crystal will deposit energy in the detector material. If no energy is deposited, obviously no pulse will be generated. Scintillation detectors are between 10 and 50 percent efficient for electromagnetic radiation. The wide difference in efficiencies is mostly the result of different designs trading off efficiency, resolution, and target gamma ray energy.

The crystals used in Anger cameras vary from 10 to 25 inches in diameter, and from 1/4 to 1 inch thick. The thicker the crystal, the higher the probability that an incoming photon will interact, deposit its energy, and be detected;—thus, the higher the sensitivity of the camera. The thicker the crystal, however, the poorer the spatial resolution, due to the complex interaction between the crystal, the photomultiplier tubes, and the light pipe that is generally used to optically couple the two. One-quarter inch thick crystals have about 1 mm better intrinsic resolution than one-half inch thick crystals. When counting low-energy radionuclides such as thallium-201, there is no difference in sensitivity. However, when counting technetium-99m, one-quarter inch thick crystals have 15 percent less sensitivity than one-half inch thick crystals. At higher energies, the difference in sensitivity is even more significant.

8.4.3 Uniformity

Field uniformity is the ability of the camera to depict a uniform distribution of activity as uniform. At one time, it was thought that nonuniform response arose from changes in sensitivity across the crystal. To correct the nonuniformity, a uniform *flood* or sheet source of radioactivity was imaged and recorded, and used as a reference. Clinical images were corrected during acquisition by either adding counts to the image in areas where the flood had too few counts relative to the other areas, or by subtracting (i.e., purposely not recording) counts in areas with too many counts.

It is now well-known that the majority of the nonuniformity in a camera occurs as a result of spatial distortion; that is, the mispositioning of events (errors in determining the (X, Y) location). To correct this distortion, reference images are acquired, and digital correction maps generated and stored. Each map contains values that represent (X, Y) correction shifts. Sophisticated microprocessor circuitry is used to reposition each count in real time during acquisition using these shifts. With many current cameras, it is best to acquire correction maps with the same radionuclide that is used for patient imaging. In some cameras, several sets of corrections maps are stored on the computer, representing all the radionuclides used in the nuclear medicine department, and the appropriate set chosen for a given patient study.

Variation in the position of a pulse from different areas of the camera within the pulse height window can also produce nonuniformities. This spatially dependent energy variation may also be corrected by microprocessor circuitry. The combination of energy variation and spatial distortion is responsible for loss of spatial resolution and imperfect linearity and uniformity. In today's cameras, explicit uniformity correction is typically carried out with multiplicative factors, only after spatial distortion (and spatially dependent energy response) corrections.

8.4.4 Energy Resolution

Pulse height analysis is critical for rejection of scattered photons, whose inclusion in the image would reduce contrast. Thus, the performance of the pulse height analyzer, and especially its energy resolution, is critical. In an ideal detector, the photopeak would be a single vertical line at that pulse height representing the energy of the emitted x-ray or gamma ray. In reality, the statistical nature of the light emission and the finite energy resolution of a pulse height analyzer smears out this line, producing a bell-shaped photopeak. In essence, the observed pulse height spectrum can be treated as the convolution of the ideal spectrum with the energy impulse response function of the system.

The worse the energy resolution of a pulse height analyzer, the broader the photopeak. Energy resolution can be quantified as the FWHM of the photopeak in a fashion similar to the use of FWHM to characterize spatial resolution. This is measured by first determining the counts at the peak of the photopeak and locating the points on either side of the peak where the counts are half of the peak counts. The width of the photopeak in pulse height units is obtained by subtracting the lower pulse height from the upper. Finally, this width is divided by the pulse height at the apex of the photopeak to yield a percentage of energy resolution measurement. The smaller the number, the better the energy resolution; typical scintillation systems average 8 to 12 percent FWHM. It is important to note that the time resolution of the scintillation system also plays a role in energy resolution, in that two scintillation events occurring within the time resolution will produce a single, sum pulse.

In order to calibrate a pulse height analyzer to establish the quantitative relationship between energy deposited and pulse height, either the applied voltage across the photomultiplier tube or the amplification in the electronics is adjusted until the photopeak from a known energy source falls at the desired pulse height. Changing the applied voltage across the photomultiplier tube changes the size of pulses coming out of the detector by changing the amplification in the tube. If the voltage across each dynode is increased, electrons liberated from the previous dynode gain more kinetic energy. Upon striking the next dynode, a larger number of electrons are liberated. This leads to a larger signal from the photomultiplier tube. Changing the amplification in the electronics directly changes the size of each pulse.

8.4.5 Noise

The entire imaging process, from gamma ray emission to electron cascade in a PMT is governed by the Poisson probability law (as introduced in Section 3.4.3). In particular, the number of detected photons (per unit area) is a Poisson random variable (by way of analogy to the initial discussion in Section 6.4.2). In a Poisson process, the variance is equal to the mean, which we have used to simplify the analysis of noise in projection radiography. We now study the implication of this fact for planar scintigraphy.

Signal-to-noise Ratio   Because both projection radiography and planar scintigraphy are ultimately limited by the *total* number of photons per unit area contributing to image formation, the calculation of SNR is essentially the same for the two modalities. In Chapter 5, we related intensity to the total number of photons [see (5.35)] in projection radiography; this same relation appears in (4.16), where we used the term *energy fluence rate* instead of *intensity*. By way of analogy to Section 5.4.1, we will focus on $\overline{N}$, the mean total number of acquired photons.

The intrinsic SNR of the camera (for this acquisition) is

$$\text{SNR} = \frac{\overline{N}}{\sqrt{\overline{N}}} = \sqrt{\overline{N}}. \tag{8.22}$$

Therefore, the intrinsic SNR of the image is made larger by increasing the number of detected photons. This situation is identical to that of projection radiography.

We know that in frame mode, the face of the Anger camera is divided into a rectilinear grid of pixels. Suppose there are $J \times J$ pixels arranged in a square grid covering the camera. Then, without any prior knowledge about the source, we would expect to see the $\overline{N}$ detected photons spread evenly over all J^2 pixels. Accordingly, the intrinsic SNR per pixel is given by

$$\text{SNR}_p = \sqrt{\frac{\overline{N}}{J^2}} = \frac{\sqrt{\overline{N}}}{J}. \tag{8.23}$$

An obvious, and important, implication of this equation is that the per-pixel SNR is smaller as the pixel size becomes smaller. This should help to explain why it is not always advantageous to use a 512×512 matrix for a planar scintigraphic image. Aside from the fact that the camera may not have sufficient resolution to justify the implied pixel resolution, the SNR may drop precipitously, yielding a severely degraded image.

It is often very useful to characterize our ability to discern a hot spot or cold spot from the background. This requires the concept of contrast C, as introduced in Section 3.2. The contrast here is the fractional change in counts in a pixel or set of pixels from the background. If $\overline{N}_b$ is the mean background count and $\overline{N}_t$ is the mean target count, then the contrast is defined [in a fashion analogous to (3.12)] as

$$C = \frac{\overline{N}_t - \overline{N}_b}{\overline{N}_b}. \tag{8.24}$$

The local SNR is defined as

$$\text{SNR}_l = \frac{\overline{N}_t - \overline{N}_b}{\sqrt{\overline{N}_b}}, \tag{8.25}$$

which is readily simplified [see (5.38)] to yield

$$\text{SNR}_l = C\sqrt{\overline{N}_b}. \tag{8.26}$$

As a practical matter, the mean count is not known, anywhere, since it is a parameter of an unknown probability mass function. Instead, averages are used to approximate means. For example, if t_n, $n = 1, \ldots, N$ represents N pixel counts within a target and b_m, $m = 1, \ldots, M$ represents M pixel counts within the background (e.g., surrounding the target), then the means are estimated as

$$\hat{\overline{N}}_t = \frac{1}{N} \sum_{n=1}^{N} t_n, \tag{8.27}$$

$$\hat{\overline{N}}_b = \frac{1}{M} \sum_{m=1}^{M} b_m. \tag{8.28}$$

These estimates are then used in (8.26) to provide an estimate of the local SNR. This kind of calculation is used in practice to evaluate the detectability of objects, providing guidance about the choices of collimators, imaging times, pixel dimensions, etc.

8.4.6 Factors Affecting Count Rate

It is clear from the previous section that increasing the number of detected counts will improve the performance of the Anger camera. This situation is analogous to that in projection radiography, and some of the solutions are apparently the same—for example, improve detector efficiency, increase dose. But there are some practical considerations that must be addressed in planar scintigraphy that are largely missing in projection radiography. We know, for example, that it is straightforward to improve collimator efficiency, but that is at the expense of resolution. So, for a given desired resolution, perhaps geared toward a particular clinical goal, we cannot improve collimator efficiency beyond a certain value.

We know also that we could improve detector efficiency in order to increase the number of detected photons. If that efficiency is improved by changes to the detector crystal material, providing higher gamma ray stopping power with the same geometry, then that is good. A great deal of research has gone into this endeavor, leading to very efficient intrinsic materials today, and it is not clear how much farther this can be taken. The obvious solution is to increase the crystal's thickness, but we know that this will degrade resolution as well.

As in projection radiography, we can increase the dose, in this case by injecting a larger concentration of radiotracer into the patient. Aside from the obvious, undesirable properties associated with increasing the dose of ionizing radiation to the patient, planar scintigraphy (unlike projection radiography) has a problem handling the larger numbers of photons incident upon the camera. In particular, after the deposition of energy in the detector, it takes a certain amount of time for the scintillation photons to be given off, and for the electronics to register a "count." During this time, called the *dead time* or *resolving time* τ of the Anger camera, the detector is only partially responsive to the deposition of additional energy. The inverse of the resolving time is the maximum counting rate of the camera:

$$\text{Maximum counting rate} = \frac{1}{\tau}. \tag{8.29}$$

In clinical situations, resolving times are typically between 10 and 15 μs, and this leads to maximum counting rates of between 60K and 100K counts per second.

There are actually a large number of factors that must be considered in trying to push the count rate near to the maximum without exceeding it. Patient dose is one factor, of course. Orientation of the patient will affect the count rate as well. If the radiotracer settles in an organ nearer to the camera, the count rate will be higher. The collimator can have a major impact on count rate. More sensitive collimators will yield higher counts (while also producing poorer resolution).

Imaging scenarios that exceed the maximum counting rate will cause *pulse pileup*, a situation that occurs when photons are absorbed in the crystal too close together to be counted as an event. The photons may be both rejected due to energy window considerations or only the second may be rejected to electronic dead time required to count the first. Either case represents a loss of signal and an excess, unnecessary dose to the patient.

8.5 Summary and Key Concepts

Nuclear medicine produces images of function (body physiology and biochemistry). Planar scintigraphy is the nuclear medicine technique that produces 2-D projection images of radiotracer distribution, utilizing an Anger scintillation camera. In this chapter, we presented the following key concepts that you should now understand:

1. *Planar scintigraphy* is the nuclear medicine analog of projection radiography.
2. Planar scintigraphy makes use of an *Anger scintillation camera*, which consists of a *collimator*, a *scintillation crystal*, *photomultiplier tubes*, *positioning logic*, a *pulse height analyzer*, and an *image capture device* (today, a computer).
3. Typical *collimators* include *parallel-hole*, *converging*, *diverging*, and *pinhole*.
4. *Event positioning* is based on a *center-of-mass calculation*; unlike radiographic image formation, this takes place on a photon-by-photon basis.
5. The *basic imaging equation* includes terms for both *activity* (the desired parameter) and *attenuation* (an undesired, but extremely important, additional factor); these two terms are not separable.
6. *Image quality* depends on *resolution* (governed by both the collimator and the intrinsic resolution of the camera), *noise* (governed by the sensitivity of the system, the injected activity, and the acquisition time), and *count rate factors*.

Bibliography

Chandra, R. *Nuclear Medicine Physics*, 5th ed. Baltimore, MD: Williams and Williams, 1998.

Goris, M. L., and Briandet, P. A. *A Clinical and Mathematical Introduction to Computer Processing of Scintigraphic Images*. New York: Raven Press, 1983.

Rao, D. V., Chandra, R., and Graham, M. C. *Physics of Nuclear Medicine*. New York: American Institute of Physics, 1984.

Rollo, F. D. *Nuclear Medicine Physics: Instrumentation and Agents*. St. Louis, MO: C. V. Mosby, 1977.

Sorenson, J. A., and Phelps, M. E. *Physics in Nuclear Medicine*, 2nd ed. Philadelphia, PA: W. B. Saunders, 1987.

Swenbwrg, C. E., and Conklin, J. J. *Imaging Techniques in Biology and Medicine*. San Diego, CA: Academic Press, 1988.

Problems

Instrumentation

8.1 (a) Explain with diagrams the design and operation of an Anger gamma camera.

(b) What are some of the physical considerations when choosing which radionuclide to use for imaging?

8.2 Suppose you are designing a collimator to be used in an Anger camera. Assume the radiotracer used emits 140 keV gamma rays and the collimator surface-to-patient distance r is 10 cm.

(a) What is the minimum septal thickness h required for the septal penetration to be less than 60 percent for gamma-rays incident at 45°?

(b) Suppose the resolution and sensitivity of the collimator are given by

$$\text{Resolution} = d\left(1 + \frac{r}{l}\right), \qquad \text{Sensitivity} = \frac{d^4}{l^2(d+h)^2},$$

where d is the collimator hole diameter and l is the collimator hole length. Assume $h = 0.2$ mm. What values of l and d should be chosen for the resolution and sensitivity of the collimator to be 7.8 mm and 28.9×10^{-4}, respectively?

(c) Describe the advantage(s) and disadvantage(s) of making the holes longer (with no other changes to the design).

(d) Can you compensate for the longer holes by changing the scintillator in any way?

8.3 A photon that has undergone Compton scattering has both changed its direction of flight and lost some energy. Pulse height analysis is used to reject these scattered photons during nuclear imaging. A pulse height window is symmetrically set around the photopeak of the pulse height spectrum. The full width of this window is usually expressed as a percentage of the photopeak energy.

(a) Calculate the maximum angle through which a 140 KeV photon can be scattered and still be accepted within a 20 percent offset window centered at 150 KeV.

(b) What are the maximum acceptable scattering angles for 140 keV and 364 KeV photons if 20 percent windows centered at photopeak are used.

(c) Draw a conclusion about the "directional selectivity" of an Anger camera as a function of frequency.

8.4 Consider an Anger camera with only one parallel collimator hole. The measured intensity is the energy deposited on the camera per unit time per unit area. Suppose the hole diameter is d, and a point source with radioactivity of A is at a distance of r from the camera, directly below the hole.

(a) What is the measured intensity?

(b) What is the measured intensity if we double the hole diameter?

(c) What is the measured intensity if we double the source-camera distance?

8.5 Photons can penetrate (i.e., go through) collimator's septa, which means that a photon entering one collimator hole can sometimes be detected under another hole. Here we consider the effect of septal penetration on collimator design and collimator resolution. Let the hole diameter be d, the septal thickness be h, and the hole length be l. Ignore Compton scattering—i.e., photons go straight—and ignore the scintillator thickness.

(a) What is the minimum distance w through a septum that the photon can pass under the condition that it enters one hole and is detected in the adjacent one? Simplify your expression by assuming that $l \gg 2d + h$.

(b) Suppose we desire septal thickness to be large enough to stop 95 percent of the photons that pass through the minimum distance w. What inequality should h satisfy? [Use the simplified expression from Part (a)].

(c) How does septal penetration influence collimator resolution?

Image Formation

8.6 (a) Derive an imaging equation for an Anger camera having a converging collimator.

(b) Derive an imaging equation for an Anger camera having a diverging collimator.

8.7 Consider nine 1×1 cm photo multiplier tubes arranged in 3×3 array (Figure P8.1). The output of each photo multiplier tube to a scintillation event is modeled as

$$a_i = 20 \exp\left(-\frac{(x - x_i)^2 + (y - y_i)^2}{5}\right) \qquad \text{(P8.1)}$$

where (x, y) is the location of the scintillation event, and (x_i, y_i) is the center of the i-th PMT.

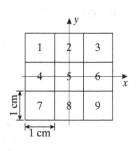

Figure P8.1
Problem 8.7.

(a) Find the output of each PMT for a scintillation event that occurs at $(-0.5, 0.5)$ cm.

(b) What is the estimated position of the scintillation event?

(c) Is the estimated position the same as its true position? If not, explain why.

8.8 A series of scintillation events produced responses from the photomultiplier tubes. The recorded pulses have a photopeak centered at pulse height 180. The pulse height is proportional to the energy deposited into the crystal, which is shown in Figure P8.2 (a). (This is not a pulse height spectrum.)

(a) We want to set an acceptance window around the photopeak to reject those photons that have undergone Compton scattering bigger than 50

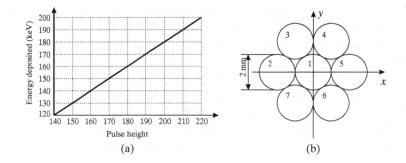

degrees. What is the range of photon energies that will be accepted by the acceptance window?

Consider seven photomultiplier tubes located around the origin of the x-y coordinates on the face of the Anger camera shown in Figure P8.2 (b). The diameter of the tubes is 2 mm and each tube has been assigned a number. A single scintillation event yields a response from the photomultiplier tubes. All other tubes recorded zero height except these seven tubes, which recorded height 40, 5, 15, 15, 20, 45, 30 in the order of tube numbers.

(b) Compute the Z-pulse. Will this pulse be accepted by the acceptance window?

(c) Estimate the position of the event (X, Y).

(d) Explain why we do not set (X, Y) equal to the location of the photomultiplier tube having the largest amplitude.

Image Quality

8.9 In practice the spatial resolution in a clinical nuclear medicine image is determined by three factors: the intrinsic resolution of the camera (I), the collimator resolution (C), and patient effects (P). Suppose the impulse response could be modeled as arising from the cascade of three linear shift-invariant systems, one for each factor.

(a) Suppose each system has an impulse response given by a rect function, and suppose their widths are r_I, r_C, and r_P, respectively. Find the FWHM of the cascade system.

(b) Suppose each system has a Gaussian impulse response function, and suppose their standard deviations are σ_I, σ_C, and σ_P, respectively. Find the FWHM of the cascade system.

8.10 We are conducting a study that requires us to acquire a number of images over a 2-hour period of time. Each acquisition lasts for 10 minutes. In the study, technetium-99m is used as the radiotracer. The images are recorded in frame mode with a matrix size of 128 × 128. In order to get satisfactory image quality, we need at least 2 million counts for each image.

(a) In order for the last image to have enough counts, what is the total count during the first minute of the study?

(b) What is the SNR per pixel as a function of the image number?

(c) Suppose there is a tumor with 10 percent contrast. What is the approximate count for the first image so that the tumor has a 5 dB local signal to noise ratio in the last image?

8.11 Looking again at the cardiac study described in Example 8.4, assume that the image for each frame is a 64 × 64 matrix. We need 1K counts for each pixel on average when the study is done. We can detect at most 128K photons per second on the entire camera.

(a) How many heartbeats are needed?

(b) How long does it take to complete the study?

(c) What is the SNR for each pixel?

(d) If we want double the SNR by prolonging the study, how long will it take to complete the study?

8.12 Using geometric arguments, derive the collimator resolution formula given in (8.12).

Applications

8.13 Suppose you replaced the standard collimator on an Anger camera with one that had only one hole with diameter d positioned at the center of the camera—i.e., at $x = 0$ and $y = 0$. Let the object be a small radioactive "point" source that is placed 20 cm away from the camera face in direct alignment with the center of the camera.

(a) When a gamma ray photon goes through the collimator hole, the camera produces a response. Explain in words what happens to this photon and what sequence of events takes place in order to produce the X and Y signals and the Z pulse.

(b) Suppose the X signal is held constant until the next photon hits the camera (so it is a piecewise constant signal $X(t)$ in time). Draw $X(t)$ when the hole has a small diameter [call this $X_1(t)$] and then draw $X(t)$ when the hole has a larger diameter (call this $X_2(t)$). Account for all effects that change when you increase the hole diameter and draw the signals on the same vertical and horizontal scales with clear labels.

(c) Would you expect the sequence of Z-pulses to change when the hole diameter (of this single hole) is increased? Explain your answer.

(d) If you double the hole diameter, what must you do to the hole length in order for the camera to have the same sensitivity?

8.14 A planar scintigraphy calibration experiment is being performed using a "uniform flood" phantom. With this phantom, each pixel out of 64 by 64 pixels on an Anger camera is struck by 4 gamma photons per second on average.

(a) If a total count of 2 million across the whole camera is considered to be a complete experiment, how long will it take (on average) to complete the calibration scan?

The burst of light from a scintillator—the Z-pulse—lasts for a brief moment every time a gamma ray hits it. Let us model this burst as a triangle with peak height A (volts, for example), dropping linearly to zero in 250 μs, as shown in Figure P8.3. The pulse height analyzer detects peaks and records their peak voltage. For the pulse shown in Figure P8.3, therefore, the pulse height analyzer records the value A. Assume that the response of two or more gamma rays is the sum of their Z-pulses, delayed in time according to their time-of-arrival.

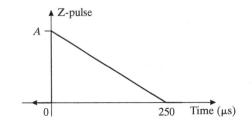

Figure P8.3
Problem 8.14.

(b) Draw the combined Z-pulse arising from two photons, one striking the camera at $t = 0$ and the other at $t = 100$ μs. What will the output of the pulse height analyzer be? (Hint: It should be a sequence of heights at different times.) Assume the photopeak is set at A and a ± 20 percent discriminator window is used.

(c) Because of the discriminator, the second photon may not be detected if it arrives too soon after the first. Explain, and determine the time separation required in order for the second photon to be detected as a separate event.

(d) Is it theoretically possible to complete the experiment in the time you computed in part (a)? Explain.

(e) Under what condition(s) will a photon be rejected because its Z-pulse height is too small? Explain.

(f) Assume that the experiment is concluded when 2 million counts are accepted no matter how long it takes. What is the *intrinsic SNR* in a single pixel?

8.15 A small pellet is located at the origin and filled with a radiotracer (left in Figure P8.4). Assume a two-dimensional scenario in which the pellet emits gamma rays only in the x-y plane and the radioactivity is equal to $A = 0.027$ mCi.

(a) What is the average rate of photons hitting the detector (1-D Anger camera)? Assume the detector is made of NaI(Tl) ($\mu = 0.644$ cm^{-1}) and its thickness is $b = 2.5$ cm.

(b) What is the detector efficiency at the center of the detector?

(c) If the Anger camera is rotated stepwise around the origin and requires 2×10^5 detected counts at each orientation, how many orientations can be captured in 10 sec? (Ignore rotation time.) A parallel hole collimator with septal thickness $h = 5$ mm, septal height $l = 12$ cm, and hole diameter $d = 5$ mm, is mounted on the camera, as shown in Figure P8.4 (b).

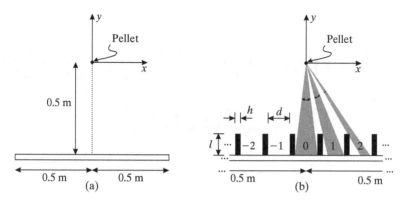

Figure P8.4
Problem 8.15.

(d) What is the collimator resolution R_c at the range of the pellet?

(e) Assume that there is no Compton scattering and that the lead septa always absorb gamma rays incident upon them. Find the rate of photons hitting the detector in the three central holes. Plot this as a function of hole number $n = -1, 0, +1$.

(f) What is the (approximate) resolution $\hat{R}_c$ that you would deduce from your result in part (e)? Compare with R_c and speculate on any difference you may observe.

8.16 An Anger camera collimator has the following dimensions: septal thickness $h = 6$ mm, septal height $l = 10$ cm, and hole diameter $d = 3$ mm. The detector thickness is $b = 2.5$ cm. The range $z = 0.5$ m.

(a) Find the collimator resolution.

(b) Suppose the intrinsic resolution of the Anger camera is 0.2 mm. Write an expression for the intrinsic PSF assuming it is Gaussian.

(c) Find the overall resolution of the Anger camera. The final image (frame mode) on the Anger camera is shown in Figure P8.5.

0	0	0	0	0	0	0	0	0
0	0	0	0	0	0	0	0	0
0	0	3	3	3	3	3	0	0
0	0	3	8	8	8	3	0	0
0	0	3	8	8	8	3	0	0
0	0	3	8	8	8	3	0	0
0	0	3	3	3	3	3	0	0
0	0	0	0	0	0	0	0	0
0	0	0	0	0	0	0	0	0

$\times 10^6$

Figure P8.5
Problem 8.16.

(d) Find the total count of the scan.

(e) Find the local contrast and SNR of the object at the center of the camera.

Emission Computed Tomography

Planar scintigraphy is the only direct imaging approach in nuclear medicine (i.e., an image exists directly at the end of frame mode acquisition). The other two imaging methods in nuclear medicine—SPECT and PET—require computed tomography image reconstruction methods. Like CT, these modalities require the acquisition of projection data, which are then reconstructed into transverse (or transaxial) slice images. Single photon emission computed tomography (SPECT) systems can be designed around a customized array of detectors, but are usually based on one or more high-quality Anger cameras, which rotate around the body to acquire a series of projection images with different angular orientations. Multiple cross-sections of the body can be reconstructed—in parallel, in principle—from these data. It is reasonable to say that SPECT is to planar scintigraphy as multislice CT is to projection radiography.

Positron emission tomography (PET) is in a class of its own, however, having no suitable analogy in other imaging modalities. PET measures the location of the line on which a positron is annihilated, an event that produces two simultaneous 511 keV gamma photons (or *annihilation photons*). After accumulating many such *coincidence lines*, line integrals of activity are produced and computed tomography methods are directly applied. The excitement about PET is due to both the chemistry and physics inherent in positron tomography. The most commonly used PET radionuclides, carbon-11, nitrogen-13, oxygen-15, and fluorine-18, are isotopes of elements that occur naturally in organic molecules. (Fluorine usually does not, but it is a bioisosteric substitute for hydrogen.) Thus, radiopharmaceutical synthesis is simplified, and the tracer principle (which mandates as small a change in the molecule to be traced as possible) is better satisfied. Indeed, useful PET radiopharmaceuticals are now available to measure *in vivo* such important physiologic and biochemical processes as blood flow, oxygen, glucose, and free fatty acid metabolism, amino acid transport, and neuroreceptor density. The short half-lives of the radionuclides (carbon-11, 20 min; nitrogen-13, 10 min; oxygen-15, 2 min; fluorine-18, 110 min) permit the acquisition of serial studies on the same day without background activity from prior injections interfering with the measurements. The physics of

PET permits greater quantitative accuracy and precision. The use of small, high-density crystals improves spatial resolution (about 4 mm in the best commercial PET scanners). The lack of collimation to determine photon direction dramatically increases sensitivity. Finally, coincidence detection allows mathematically accurate attenuation correction, as we shall see.

We now explore the instrumentation, reconstruction methods, and image quality of both SPECT and PET, which together comprise *emission computed tomography* (ECT) systems.

9.1 Instrumentation

9.1.1 SPECT Instrumentation

Single photon emission computed tomography (SPECT) refers to true transaxial tomography with standard nuclear medicine radiopharmaceuticals (i.e., those that emit a single photon upon decay). SPECT is performed with either specialized ring detector systems or rotating Anger cameras. Ring systems consist of an array of individual detectors (usually sodium iodide crystals) that surround the patient. These systems, while producing excellent tomograms, tend to be very expensive and relatively rare, and we will not discuss them any further. By far the most popular method of doing SPECT is with a rotating Anger camera mounted on a special gantry that allows 360-degree rotation around the patient. A typical rotating Anger camera system is shown in Figure 9.1.

In rotating Anger camera SPECT, the camera is rotated either 180° or 360° (assuming a single-head system). At each angle, a projection image (a standard planar image) is acquired so that, effectively, multiple 1-D projections from different (contiguous) cross-sections of the body are acquired simultaneously. After complete

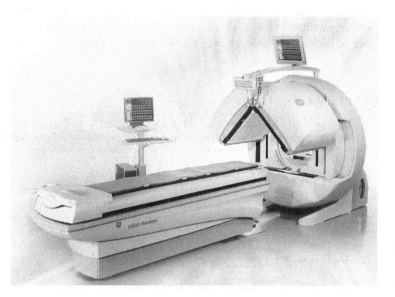

Figure 9.1
A dual-head SPECT system.
(Used with permission of GE
Healthcare.)

rotation of the camera, each cross-section is reconstructed separately, typically using convolution backprojection, yielding a collection of transverse images, stacked to form a reconstructed volume. We will have much more to say about this in Section 9.2.

Two types of collimators are used in SPECT (see Section 8.2.1). Usually, a parallel-hole collimator is used; this geometry corresponds to that of parallel-beam geometry in CT. Here, the geometry is parallel both in-plane and in the axial direction. In some cases, a *fan-beam* collimator is used. This collimator converges in-plane, like a converging collimator, but has parallel-hole geometry in the axial direction. With fan-beam collimation in each contiguous 1-D projection, the "stack" has fan-beam geometry, but the slice-to-slice geometry is parallel. In rare cases, a collimator with more complicated geometry is used, requiring more sophisticated reconstruction algorithms. Such systems are not considered here.

Perhaps the biggest recent advance in SPECT systems has been the incorporation of multiple Anger camera detectors (so-called *heads*)—e.g., a dual-head system is shown in Figure 9.1. As shown in Table 9.1, adding camera heads increases sensitivity. Dual-head variable-angle systems that are capable of orienting the heads at both 90° and 180° are particularly popular, as this optimizes the acquisition geometry for the two most commonly performed nuclear medicine procedures: cardiac SPECT and whole-body bone scans. These two types of studies account for at least 70 percent of all nuclear medicine imaging procedures. Cardiac SPECT is often performed with 180° acquisition (i.e., only a 180° arc of projection data around the heart is acquired), whereas many other SPECT studies use 360° of projection data.

The increased sensitivity provided by multihead systems can be used primarily in three ways. First, it can be used to decrease noise while using the same acquisition time as with a single-head system. Second, it can be used to decrease acquisition time to get the same counts as a single-head system; this decreased acquisition time increases patient throughput, decreases motion (and associated blur), and decreases tracer washout effects. Third, it may be "traded" for higher resolution through the use of higher resolution/lower sensitivity collimators.

TABLE 9.1

Comparison of Acquisition Times and Relative Sensitivities for Single- and Multihead Systems with Identical Camera Heads and Collimation				
	360°		180°	
	Acquisition Times	Relative Sensitivities	Acquisition Times	Relative Sensitivities
Single	30	1	30	1
Double (heads@180°)	15	2	30	1
Double (heads@90°)	15	2	15	2
Triple	10	3	20	1.5

EXAMPLE 9.1

Suppose a single-head system requires N counts in a 30 min scan using an all-purpose collimator. Let us imagine performing the same study with a two-head system instead, and let the two new collimators have higher resolution with just 75 percent of the sensitivity of the all-purpose collimator on the single-head system.

Question How long will it take to achieve the same counts as the single-head system, and will the image quality be equivalent?

Answer In 30 min, each (high-resolution) head collects $0.75 \times N$ counts (on average, of course). This corresponds to $2 \times 0.75 \times N$ counts in 30 min. The time t to collect just N counts satisfies

$$\frac{t}{30 \text{ min}} = \frac{N}{2 \times 0.75 \times N},$$

so

$$t = \frac{30 \text{ min}}{1.5} = 20 \text{ min}.$$

The image obtained in just 20 min will be better because, although the total counts are the same in the two scans, the resolution is better in each of the dual-head collimators. ∎

Multiple-head systems have permitted a wide variety of typical SPECT scans to be performed in under 30 min. This is an important time milestone because beyond 30 min patient motion becomes a very significant factor in image quality. Three-head systems exist and it would seem that N-head systems (where $N > 3$) would be in the offing. However, simple geometric constraints (packing that many heads in close to the patient), quality control issues (how do you keep all those detectors in calibration?), and cost (N heads means N collimators) make it infeasible to go much further in this design direction.

Studies have shown that Anger cameras must have significantly better performance for SPECT than for planar scintigraphy. For example, nonuniformity must be reduced to less than 1 percent in order to avoid reconstruction artifacts. This requires acquisition of a 30 to 120 million count reference image of a uniform field for subsequent computer correction of nonuniformities. (There should be around 10,000 counts per pixel in order for the relative Poisson variation or noise in the uniformity calibration to be < 1 percent ($\sqrt{10,000}/10,000$), so a longer reference scan is required for 128×128 acquisition versus 64×64 acquisition.) The camera image must also be mechanically properly aligned within the computer matrix, or an *axis-of-rotation correction* made. In addition, there are several other factors that must be accounted for to make SPECT truly quantitative. There is disagreement about the need for attenuation, scatter, and spatial resolution corrections for subjective visual interpretation of SPECT images.

9.1.2 PET Instrumentation

One of the most exciting tomographic techniques is positron emission tomography (PET). In PET, positron emitting radionuclides are used, which makes it a functional

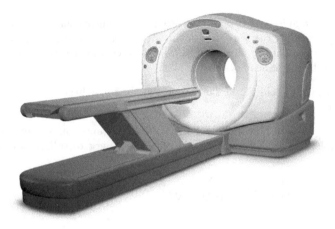

Figure 9.2
A PET system. (Used with permission of GE Healthcare.)

imaging technique and in the family of nuclear medicine modalities. A typical PET system is shown in Figure 9.2. From the outside we see the plastic housing, and a PET system looks just like a CT or MRI system; but the hardware is quite different in PET.

Consider a positron-emitting radiopharmaceutical distributed in a patient, and refer to Figure 9.3 for the following discussion. When a positron is emitted, it travels up to several millimeters in the tissue, depositing its kinetic energy. It then meets a free electron in the tissue, and mutual annihilation occurs (recall that a positron is an anti-electron). From conservation of energy—511 keV is the energy equivalent to the rest mass of an electron or positron—two 511 keV annihilation (gamma ray) photons appear. From conservation of momentum, the two 511 keV photons

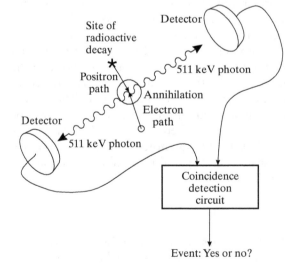

Figure 9.3
Coincidence detection due to positron decay and annihilation.

are emitted back-to-back; i.e., they propagate outward from the site of annihilation (almost exactly) 180° apart.

It is possible to use a single-head Anger camera to individually detect just one of the two 511 keV gamma ray photons that are emitted by positron-electron annihilation. Images could then be created using planar scintigraphy or SPECT reconstruction (if the camera rotates). With a multihead system, both photons might be detected if two of the Anger cameras were 180° apart and the photons made it through the opposing collimators. In this case, scintigraphy or SPECT techniques could be used to create an image as well.

It is more advantageous (and is what sharply distinguishes PET from SPECT), however, to exploit the fact that a positron annihilation produces two gamma rays at the same time. PET uses electronic circuits, as shown in Figure 9.3, to detect a pair of 511 keV photons at the same time, in what is called *annihilation coincidence detection* (ACD). With this approach, only simultaneous gamma rays are declared as events, and all other events are rejected. Whether two detected photons are declared as "simultaneous" depends on a user set time interval, called a *time window*, which is typically 2 to 20 ns for modern scanners.

Of major importance, if two opposing detectors simultaneously detect the 511 keV gamma rays, the annihilation event must have occurred somewhere along the line joining the two detectors (the so-called *coincidence line*). Said another way, ACD intrinsically provides information about the direction of travel of the photons. Detector collimation is not required with this approach, and is in fact undesirable because it reduces the sensitivity of the detectors and may cause undesirable directional dependencies if the detectors remain in fixed positions.

Commercial dual-head SPECT scanners can be used (and some are sold this way), together with ACD circuitry, for PET. In this case, the two heads are placed 180° apart and their collimators removed (or replaced with special collimators that isolate axial planes only). The heads must be rotated in order to pick up emissions from all angles around the patient, as tomographic reconstruction methods are used to reconstruct cross-sectional images (see Section 9.2). In order to efficiently stop 511 keV photons, it is generally necessary to make the NaI(Tl) crystals (in the Anger cameras) thicker. Because of this, positioning uncertainty increases when the same camera is used for lower energy SPECT (say with Tc-99m or Tl-201). Because of this and because of the inefficient angular acquisition, dual-head SPECT scanners are not an ideal solution for PET, and their use has fallen into disfavor.

Full-fledged PET systems surround the patient with multiple rings of detectors, each having a diameter of about 100 cm, as shown in Figure 9.4. Each detector is electronically coupled to every other detector in a comprehensive ACD circuit. Older systems used a single scintillation crystal coupled to a PMT as the basic detector. The requirement for better resolution, however, has led to the development of so-called *detector blocks* that are based on the Anger camera principle. As an example, each detector block might comprise a single scintillation crystal backed by four PMTs, which are arranged in a 2×2 matrix, as illustrated in Figure 9.5. The scintillation crystal is cut into an array of smaller crystals, typically 8×4, 8×7, or 8×8 arrays, which partially isolate the light burst from each scintillation event to one of these subcrystals. Relative PMT signal heights are used to identify

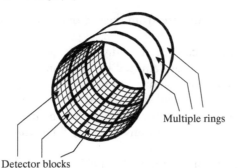

Figure 9.4
Geometry of a multiple-ring PET system.

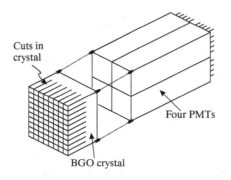

Figure 9.5
Geometry of a PET detector block.

which of the subcrystals contained the event in a crude approximation of the center of mass calculation, as in a conventional Anger camera (see Problem P9.1). A typical modern PET scanner has three primary rings of detectors with 48 detector blocks in each ring for a total of 144 detector blocks (3 × 48). Each detector block has 64 subcrystals backed by four PMTs, which means that we can also think of this PET scanner as having 24 rings (3 rings × 8 subcrystals) and 384 detectors per ring (48 detector blocks × 8 subcrystals).

PET scanners often have lead or tungsten *septa* between the detector rings. The purpose of the septa is to collimate the photons into a set of 2-D slices. In this mode, PET reconstruction methods reduce to 2-D reconstruction applied to each slice or cross-section separately. This geometry reduces the contribution of scattered photons between slices and is very similar to that of SPECT. When the septa are removed, the scanner operates in 3-D mode, and coincidence events between any crystal pair are accepted. This fully 3-D ("septa-less") acquisition and reconstruction is one of the biggest advances of the past several years. The use of a scanner with retractable septa for 3-D operation provides about a factor of 5 to 10 increase in sensitivity. This increase in sensitivity comes with a cost—significantly higher acquired scatter events (*scatter coincidences*, in which at least one of the two annihilation photons scatters before detection)—but accurate postacquisition software algorithms are now being developed for scatter correction in 3-D PET. Operating in "pseudo" 2-D mode is common. In this mode, septa between detector rings are used, but ACDs between adjacent rings are received and counted. In this way, slices between detectors are also reconstructed; so, if there are N detector rings, $2N - 1$ slices will be reconstructed. In this chapter, we will only consider the PET scanner operating in a 2-D tomographic mode.

The original commercial PET scanners utilized sodium iodide NaI(Tl) as the detector scintillation crystal, the same material used in planar scintigraphy and SPECT. Since 511 keV photons are harder to stop than the lower energy photons used in planar scintigraphy and SPECT, most current PET systems use bismuth germanate (BGO) instead of NaI(Tl) as the scintillation crystal. The linear attenuation coefficient of BGO at 511 keV is 0.964 cm^{-1} as opposed to 0.343 cm^{-1} for NaI(Tl). Therefore, BGO detectors can be about one-third the thickness of what would be required for NaI(Tl) in order to yield the same stopping power. A typical BGO crystal is about 3 cm deep, for example. Each crystal is typically "cut" into smaller subcrystals that are on the order of 6 × 6 mm. The downside of BGO as compared to NaI(Tl) is that it is only 12 to 14 percent as efficient in light conversion.

Both NaI(Tl) and BGO are relatively slow scintillators, having time constants of 230 ns and 300 ns, respectively. Faster scintillators would be advantageous because it would permit narrow time windows to be used in ACD circuits. Tighter time windows would in turn reject more *random coincidences*, which are the detection of two photons that are deemed to be from the same annihilation event but are actually from two different annihilations. There are a number of exciting new detector materials being vigorously investigated by industry and academia for PET. These include cerium-doped lutetium oxyorthosilicate (LSO) and gadolinium oxyorthosilicate (GSO), which are single crystal inorganic scintillators. The decay times of LSO and GSO are on the order of 50 ns, a very significant improvement over NaI(Tl) and BGO. The attenuation coefficients are comparable, as well, so these crystals can be made relatively small for good spatial resolution, and still retain good stopping power.

Just as in planar scintigraphy, the outputs of PET detectors are analyzed by energy discrimination circuits. An energy discrimination window, centered on the photopeak, is usually desirable. In this way, photons that have been detected after Compton scattering in the patient are rejected, and multiple simultaneous detection of photons is also rejected. Sometimes, it is desirable to lower the discrimination window to include part of the Compton plateau. The reason for this is that it is common for there to be a Compton interaction within the detector, depositing a fairly large amount of energy, but allowing the photon to ultimately escape the detector without being fully absorbed. In this case, the primary Compton event is actually the correct event to detect, as it represents the proper location of one of the two 511 keV photons created during positron annihilation. Setting the lower window threshold is therefore a matter of trading off the acceptance of two types of Compton events: those that occur in the patient, which we do not want to count, and those that occur in the crystal, which we do want to count. A common lower energy threshold is 300 to 350 keV.

Combined PET/CT Systems In discussing nuclear medicine, we always emphasize the "functional" nature of the image data. There is now ample evidence that clinical interpretation of nuclear medicine studies, especially in PET, can benefit from reference to anatomic images from CT or MRI. In the most common current method for combining PET and CT data, the physician reviews the PET images on a computer display and the CT films on a light box and then performs mental "image fusion." Alternatively, image registration and fusion may be accomplished through software. This process requires the images from both procedures to be on a common computer (typically via transfer of the CT exam to a nuclear medicine computer workstation); the PET images are interpolated and registered with the CT exam, and the fused images displayed. At least in theory, the best way of registering PET and CT is if both studies are acquired via a combined system with a common gantry and patient table. Several PET equipment manufacturers have developed PET/CT systems.

In a combined PET/CT system, both PET and CT exams may be performed without moving the patient relative to the table between the two exams. This provides functional and anatomical images that can be registered and fused with a higher degree of positional accuracy than is permitted by imaging on two separate scanners, especially because of differences in body shape and position introduced by the use of two different tables. The improved registration accuracy with a combined PET/CT system thus mainly results from the patient being in the same position on one table. However, physiologic motion such as breathing and abdominal motion cannot be avoided. These motions decrease the registration accuracy, and may produce artifacts in attenuation corrected PET images using the CT map (see below). Notice that the CT exam is typically taken during a breath hold, but the PET exam requires breathing. In addition to improved positional accuracy, there is easier availability of both the PET and CT data for the fusion software with a combined PET/CT system, since they are combined within one system (but note that the PET and CT data are not necessarily within the same computer immediately after acquisition).

In addition to their use for anatomic localization, the CT images may be used as attenuation maps for performing attenuation correction on the PET images. As noted above, one of the theoretical advantages of PET is that measured tissue attenuation values can be used to exactly correct for the effects of attenuation in the emission data. In practice, attenuation correction represents a significant limitation in PET imaging. Using radioactive sources to measure attenuation, as is usually done, means that attenuation maps frequently have low counts, and thus high noise. This noise propagates from the transmission data into the emission data when the attenuation correction is performed, resulting in degraded image quality. In whole-body FDG imaging, where transmission scans take from 1 to 3 minutes per axial field of view, attenuation correction typically accounts for 25 to 35 percent of the total data acquisition time. The attraction of using CT is that it provides high resolution, high contrast, and low noise attenuation maps, obtained quickly using a fast helical scan. Note that using x-ray CT images for attenuation correction requires translating attenuation values measured with a low-energy, polychromatic x-ray beam into 511 keV, monoenergetic attenuation values.

9.2 Image Formation

We focus on the reconstruction of a 2-D cross-sectional image from observations of line integrals within the cross-section. This approach is consistent with SPECT (with suitable approximation) and 2-D PET. It is completely analogous to our development of CT reconstruction methods in Chapter 6, upon which we draw heavily.

9.2.1 SPECT Image Formation

Coordinate Systems   Our first task is to set up a convenient (and conventional) coordinate system. Since the Anger camera rotates in SPECT, it is necessary to have both a laboratory (fixed) coordinate system and a rotating coordinate system. This differs from the "camera-centric" notation we were able to use for planar scintigraphy in Chapter 8. Here, we will use the notation $A(x, y, z)$ to denote the radioactivity within the (3-D) body, and want $A(x, y)$ to represent a reconstructed cross-section for a fixed z position. Accordingly, as shown in Figure 9.6, our fixed laboratory frame has the z-axis oriented along the patient's body so that when looking from the patient's feet (toes up), $A(x, y)$ will represent a so-called *axial cross-section*, which is a conventional tomographic view. We have chosen a right-handed frame here to correspond to traditional mathematics and physics notations. Please be aware that it is also common in medical imaging to use a left-handed system, so that the z-axis points from the patient's feet to head.

In principle, the projection imaging equation for SPECT (using an Anger camera) would be identical to the imaging equation for planar scintigraphy [(8.9)]. However, because of our choice of a laboratory coordinate system, it is necessary to use a coordinate system on the face of the Anger camera that is different from the one that is used in Chapter 8, and a somewhat different imaging equation will emerge below. As shown in Figure 9.6, we use the variable ℓ to coincide with the x-axis when the rotation angle θ is zero. The other coordinate on the camera is z, coinciding with the laboratory coordinate z regardless of orientation. One further point is that the Anger camera is assumed to remain at a fixed distance R from the origin. With this choice, the coordinate ℓ will represent the lateral position of a line integral, and the geometry used in the reconstruction of a cross-section in SPECT corresponds exactly to that in CT.

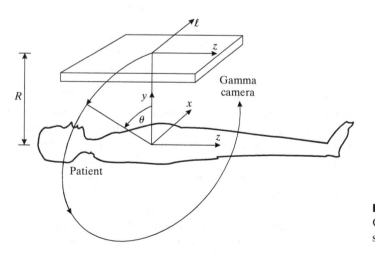

Figure 9.6
Geometry of a SPECT system.

Imaging Equation   We assume parallel-hole collimators, which implies parallel-ray geometry. When the camera is located in its "home" position $\theta = 0$, we can map the coordinate system used in planar scintigraphy (Chapter 8) into that of SPECT. Letting $x \to z$, $y \to \ell$, and $z \to y$ (compare Figures 8.9 and 9.6), we can rewrite (8.9) as

$$\phi(z, \ell) = \int_{-\infty}^{R} \frac{A(x, y, z)}{4\pi (y - R)^2} \exp\left\{-\int_{y}^{R} \mu(x, y', z; E) \, dy'\right\} dy. \tag{9.1}$$

Notice that there is a slight abuse of notation in keeping the arguments of A and μ the same as in the laboratory frame; the present functions refer to the radioactivity and linear attenuation coefficients defined on the laboratory frame in Figure 9.6.

The coordinate z is irrelevant in the selected geometry because the collimation confines (to good approximation) the integrals to within the x-y plane. In what follows, we drop the explicit dependence on z but do not forget that there is a separate reconstruction problem (which can be implemented on a parallel computer, in fact) for each z coordinate on the camera. When the camera is rotated, the lines of integration are no longer in the y direction. By design, the lines of integration are the same as those in CT [see (6.6)],

$$L(\ell, \theta) = \{(x, y) \mid x \cos \theta + y \sin \theta = \ell\}. \tag{9.2}$$

Because of the exponential term in (9.1), we cannot use the set form of the line integral (i.e., using the impulse function). Instead, we use the parameterization [introduced in (6.8) and (6.9)]

$$x(s) = \ell \cos \theta - s \sin \theta, \tag{9.3}$$

$$y(s) = \ell \sin \theta + s \cos \theta. \tag{9.4}$$

Examination of the geometry, as shown in Figure 9.7, and careful substitution yields

$$\phi(\ell, \theta) = \int_{-\infty}^{R} \frac{A(x(s), y(s))}{4\pi (s - R)^2} \exp\left\{-\int_{s}^{R} \mu(x(s'), y(s'); E) \, ds'\right\} ds. \tag{9.5}$$

Equation (9.5) is the starting point for a SPECT imaging equation. Unfortunately, there are two complications making (9.5) generally intractable as an imaging equation. The first complication has to do with the presence of a second unknown function, μ, the linear attenuation coefficient. There has been a considerable amount of research into this problem; even when μ can somehow be measured, inferred, or approximated, however, the presence of the additional term in the integrand leaves a relatively difficult reconstruction problem at hand. This is essentially one side of the second complication, which arises due to the position-dependent terms—the inverse-square law and the photon attenuation—in the integrand. Please be aware that the same collimator sensitivity considerations that were discussed in Section 8.4.2 also

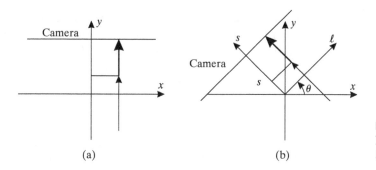

Figure 9.7
Integration geometry for SPECT imaging equation.

apply here. In this regard, the problems inherent in (9.5) are the same as those in (8.9).

There is no known closed-form reconstruction method when these terms are present. Iterative methods can be used, but only relatively recently are they entering into commercial systems. Even then, the methods seek only statistical optimality and do not address these additional complications.

Instead of trying to solve (9.5) directly, it is common practice to ignore the inverse square law and the attenuation terms in the integrand—a bold approximation that works reasonably well in practice. Ignoring scaling factors, (9.5) becomes

$$\phi(\ell, \theta) = \int_{-\infty}^{\infty} A(x(s), y(s)) \, ds, \tag{9.6}$$

where the upper limit has been set to ∞ assuming that $A = 0$ behind the collimator. Using the impulse function [refer to (2.3)], (9.6) can be rewritten as

$$\phi(\ell, \theta) = \int_{-\infty}^{\infty} \int_{-\infty}^{\infty} A(x, y)\delta(x \cos \theta + y \sin \theta - \ell) \, dx \, dy, \tag{9.7}$$

which is precisely the form of the CT imaging equation [see (6.10)]. In fact, we find that according to this equation, $\phi(\ell, \theta)$ is the 2-D Radon transform of $A(x, y)$. Using the notation from Chapter 6, we have the following identifications

$$f(x, y) = A(x, y) \qquad \text{and} \tag{9.8}$$

$$g(\ell, \theta) = \phi(\ell, \theta), \tag{9.9}$$

which should be compared to (6.11) and (6.12). One important difference between the CT and SPECT cases is that the SPECT imaging equation uses the actual intensities recorded by the detector (e.g., numbers of photons) in arriving at a Radon transform relation, while CT requires a logarithmic transformation of the data.

Reconstruction   Because of approximations made above, we can use the same reconstruction formula in SPECT as in CT. In particular, the convolution backprojection formula for SPECT is given by [see (6.26)]

$$A(x, y) = \int_0^\pi \int_{-\infty}^\infty \phi(\ell, \theta) \tilde{c}(x \cos \theta + y \sin \theta - \ell) \, d\ell d\theta \,. \qquad (9.10)$$

This inverse formula is commonly used in SPECT. Here, $\tilde{c}$ is an approximate ramp filter, defined in (6.27) and repeated here for convenience:

$$\tilde{c}(\ell) = \mathcal{F}_{1\text{D}}^{-1}\{|\varrho| W(\varrho)\} \,. \qquad (9.11)$$

In CT, the definition of the ramp windowing filter $W(\varrho)$ is not very critical. As long as its cutoff frequency ϱ_0 is high enough, there will be very little degradation in image quality. In SPECT, however, the windowing filter is very critical to image quality. In particular, if it is chosen to allow too many high frequencies through, then the SPECT images will be dominated by noise. In fact, noise amplification will occur due to the "differentiating nature" of the ramp filter. If, on the other hand, $W(\varrho)$ is chosen to pass only the very lowest spatial frequencies, then image detail will be absent. We will have more to say about the choice of W in Section 9.3.

9.2.2 PET Image Formation

Coordinate System   Here we consider only 2-D PET, the reconstruction of axial cross-sections from data collected within isolated axial detector rings. We need to consider only one detector ring, which can be taken to be in a fixed z plane. Geometry within this plane is considered relative to a conventional cross-sectional x-y coordinate system, as in CT and SPECT.

Lines of Response   As with planar scintigraphy and SPECT, PET counts the number of detected photons, but these are now based on *annihilation coincidence detections* (ACDs) rather than single gamma ray detections. As shown in Figures 9.8 and 9.9, the line joining any two opposing detectors for which an ACD could be identified is called a *line of response* (LOR). Over time, regions of high activity will generate more coincidence lines or coincidence events or counts along LORs passing through them than will regions of low activity. The total number of coincidence counts along any given LOR represents an integration of the total activity along that line. We note that it is not possible to know the position along an LOR where the annihilation took place.[1] Therefore, the measured number of counts on a particular line represents an integration of the activity on that line. Like SPECT, PET measures line integrals of activity.

---

[1] Faster scintillators, however, are allowing researchers to use so-called *time-of-flight PET*, which permits some degree of localization of events along the LOR.

Imaging Equation   Consider what happens at the site of a positron annihilation. Two gamma rays are generated, and they fly off in opposite directions. As in SPECT, the gamma rays will have potentially different experiences. A gamma ray can be absorbed in the body, be scattered, miss the detector ring, or hit a detector in the ring. To develop an imaging equation, we consider only those gamma rays that should hit the ring—i.e., those whose initial directions are in the x-y plane—and we ignore scattering. Therefore, attenuation is the only phenomenon we must consider.

Suppose the two gamma rays originate at position s_0 and travel in opposite directions on the line $L(\ell, \theta)$ [see (9.2)], as depicted in Figure 9.8. According to the line parameterization given in (9.3) and (9.4), one photon will travel in the $+s$ direction and other in the $-s$ direction. If not absorbed by the body, they will each strike a detector on the ring and be counted as a coincidence event (assuming perfect ACD behavior as well). However, because of the position of the positron annihilation in the body, one photon may experience more attenuation along its path to the detector than the other, and some fraction of the expected counts on this line may be missed. Like SPECT, we should expect an attenuation term in the imaging equation of PET.

It is easiest to understand the effect of attenuation by considering its statistical effects on a large number of photons, rather than just the effect on a single photon. Accordingly, suppose that over the course of the image acquisition period, there are N_0 positron annihilations at position s_0 whose gamma rays travel on line $L(\ell, \theta)$. This means that there will be N_0 gamma ray photons traveling in the $+s$ direction and N_0 gamma photons traveling in the $-s$ direction. We can predict how many photons will actually arrive at each detector—say, N^+ in the $+s$ direction and N^- in the $-s$ direction—if we had knowledge of the underlying attenuation $\mu(x, y; E)$ in the plane. With Figure 9.8 as reference, and referring to (4.29), we find

$$N^+ = N_0 \exp\left\{-\int_{s_0}^{R} \mu(x(s'), y(s'); E)\, ds'\right\} \tag{9.12}$$

$$N^- = N_0 \exp\left\{-\int_{-R}^{s_0} \mu(x(s'), y(s'); E)\, ds'\right\}. \tag{9.13}$$

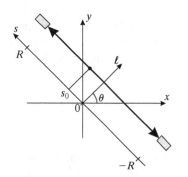

Figure 9.8
Integration geometry for PET imaging equation.

Notice that (9.13) appears to have its limits of integration backwards. However, attenuation is relative to the direction of photon propagation. If the limits were reversed, then ds would become $(-ds)$, and (9.13) would still result.

Unfortunately, knowledge of N^+ and N^- is not quite what we need. These numbers tell us how many individual photons made it from the site of annihilation to their respective detectors. However, a coincidence event will only register if both photons from the same annihilation arrive at their respective detectors. We are, in fact, concerned with a type of "pairwise" attenuation—this is something new. The key to understanding what is needed here is to realize that for both the $+s$ photon and the $-s$ photon to arrive, neither photon can be absorbed. Therefore, it is the *product* of their attenuation factors that matters. Accordingly, the number of coincidence events, $N_c(s_0)$, arising from positron annihilations at s_0 that will be detected (on average, assuming perfect ACD) is given by

$$N_c(s_0) = N_0 \exp\left\{-\int_{s_0}^{R} \mu(x(s'), y(s'); E)\, ds'\right\} \exp\left\{-\int_{-R}^{s_0} \mu(x(s'), y(s'); E)\, ds'\right\}. \tag{9.14}$$

It is quite satisfying to observe that the two exponential terms can be combined by summing their exponents, and that the sum of the integrals combine to yield just one integral over the whole line. This manipulation yields

$$N_c(s_0) = N_0 \exp\left\{-\int_{-R}^{R} \mu(x(s'), y(s'); E)\, ds'\right\}, \tag{9.15}$$

which shows that although there is attenuation dependence, it does not depend on the site of origin of the photons (i.e., the annihilation site), but only on the dimension of the whole line. This leads to an important simplification in the PET imaging equation over that of SPECT, as we shall see.

Equation (9.15) characterizes only those coincidence events arising from positron annihilations at position s_0 on line $L(\ell, \theta)$ (because we assumed there were N_0 annihilations at that location with that orientation over the life of the scan). To capture all events for this line, we need to integrate the activity over the line. To facilitate comparisons between SPECT and PET, we will use the same photon fluence rate notation, ϕ, with the understanding that in PET, ϕ should be interpreted as a *coincidence* fluence rate. Notice that our derivation is of a *rate*, while the considerations given above used total counts in an acquisition.

Accordingly, we have

$$\phi(\ell, \theta) = K \int_{-R}^{R} A(x(s), y(s)) \exp\left\{-\int_{-R}^{R} \mu(x(s'), y(s'); E)\, ds'\right\} ds, \tag{9.16}$$

where K is a single constant that includes all other (constant) factors, such as detector area and efficiency, that influence ϕ. The exponential factor in (9.16) does not depend on s, so we can remove it from the integrand. Rearranging terms yields our final imaging equation for PET:

$$\phi(\ell, \theta) = K \int_{-R}^{R} A(x(s), y(s)) ds \, \exp\left\{ -\int_{-R}^{R} \mu(x(s), y(s); E) \, ds \right\}. \qquad (9.17)$$

Equation (9.17) is similar to (9.5) for SPECT, but the differences in the limits of integration, which might seem subtle, have important consequences.

Image Reconstruction Coincidence counts can be organized into parallel-ray or fan-beam projections. Here, we consider only parallel-ray projections, as shown in Figure 9.9. From (9.17), we see that although we desire to reconstruct activity $A(x, y)$, attenuation $\mu(x, y)$ also appears in the PET imaging equation. Let us start by ignoring attenuation, as we did in SPECT. In this case, we have an approximate imaging equation given by

$$\phi(\ell, \theta) = K \int_{-R}^{R} A(x(s), y(s)) \, ds. \qquad (9.18)$$

The measured coincidence counts are directly proportional to the activity on the LOR. In other words, the PET scanner measures the Radon transform of the radioactivity, and we can use the reconstruction methods developed in Chapter 6.

As in CT and SPECT, it is helpful to identify in PET what corresponds to the measurement $g(\ell, \theta)$ and what is the function $f(x, y)$ to be reconstructed. From the previous discussion, we can identify the following in a fashion identical to (9.8)

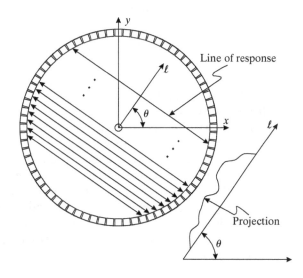

Figure 9.9
Lines of response organized into a parallel-ray projection.

and (9.9):

$$f(x, y) = A(x, y) \qquad \text{and} \tag{9.19}$$

$$g(\ell, \theta) = \phi(\ell, \theta)/K. \tag{9.20}$$

Therefore, convolution backprojection can be directly applied to the projection data acquired by the PET data:

$$A(x, y) = \int_0^\pi \int_{-\infty}^\infty \phi(\ell, \theta) \tilde{c}(x \cos \theta + y \sin \theta - \ell) \, d\ell d\theta. \tag{9.21}$$

This will reconstruct an approximation to the activity up to a (typically) unknown scaling constant K. As in SPECT, $\tilde{c}$ is an approximate ramp filter, defined using an appropriate window function as in (9.11). Although convolution backprojection ignores both the blurring caused by the physical size of the detectors (larger than x-ray CT) and the statistical noise made prominent by the relatively low photon counts, it has remained the routine method in PET reconstruction for many years.

Attenuation Correction Unlike SPECT, PET has a straightforward approach for handling attenuation. Suppose $\mu(x, y)$ were known. Then, using (9.17), we can form attenuation-corrected measurements as

$$\phi_c(\ell, \theta) = \frac{\phi(\ell, \theta)}{K \exp \left\{ - \int_{-R}^R \mu(x(s), y(s); 511 \text{ keV}) \, ds \right\}} = \int_{-R}^R A(x(s), y(s)) \, ds. \tag{9.22}$$

We see that $\phi_c(\ell, \theta)$ is now exactly the integral of activity along the line $L(\ell, \theta)$. Corrected coincidence counts can now be used in (9.21) to reconstruct the following more accurate image of activity:

$$A_c(x, y) = \int_0^\pi \int_{-\infty}^\infty \phi_c(\ell, \theta) \tilde{c}(x \cos \theta + y \sin \theta - \ell) \, d\ell d\theta. \tag{9.23}$$

In PET, it is standard practice to conduct a *transmission study*, which uses a small radioactive source outside the body to acquire projection data: integrals of $\mu(x, y)$. Alternatively, actual CT values can be used, especially with a combined PET/CT system, described above. We note that it is integrals of $\mu(x, y)$ that are required to correct the PET counts, so actual reconstruction of $\mu(x, y)$ need not be conducted. With attenuation correction in place, convolution backprojection would appear to be the correct approach to use, without any doubt. Indeed, it is the most commonly used reconstruction method in PET, with or without attenuation correction. But PET (and SPECT as well) is beset by noise, and new algorithms are emerging to handle noise as well as reconstruct from projections. We briefly explore this subject next.

9.2.3 Iterative Reconstruction

Filtered backprojection (FBP)—and, equivalently, convolution backprojection—is the most commonly used reconstruction algorithm in SPECT and PET. It can be efficiently implemented, and is relatively robust to those physical factors present in ECT imaging that corrupt the imaging process—e.g., limited detector resolution, Compton scatter, attenuation (at least in SPECT), and noise. Use of FBP is optimal, however, only if the imaging equation is an accurate depiction of reality, which we know from the preceding development is not the case. In both SPECT and PET, it is possible to model these physical factors to better reflect reality. Given these more accurate, and elaborate, imaging models, a number of algorithms have been developed to produce better reconstructions than FBP. Broadly speaking, these algorithms fall into a class of algorithms called *iterative reconstruction algorithms*.

In an iterative reconstruction algorithm, an initial "guess" of the distribution of radioactivity is made, quite frequently by using FBP. Let us denote this initial guess by $A^{(0)}(x, y)$. If this were the true distribution of radioactivity, then we should expect to see certain measurements as a result; generating an expected set of measurements, say $I^0(\ell, \theta)$, from this source object is called *solving the forward problem* or *forward projection*. This new sinogram data can now be compared to the measured sinogram. If the acquisition and reconstruction process is correctly modeled, and if the initial guess were correct, then these data would match. Otherwise, it is assumed that there are errors in the initial guess, and it should be modified. The iterative reconstruction algorithms use different methods to modify the estimated distribution in response to observation errors, and studying these differences is a long, involved subject. In the end, though, each algorithm produces a sequence of estimated cross-sections, $A^{(i)}(x, y)$, $i = 0, 1, 2, \ldots$, the last of which is considered to be the reconstruction.

An important feature of many of these algorithms is an attempt to "unify" compensation for attenuation, scatter, and depth-dependent blurring. These approaches include the general class of algorithms known as *maximum likelihood by expectation maximization* (ML-EM). This class of algorithms permits incorporation of models of blur, scatter, and attenuation effects into the projection/backprojection process. The ability to account for these three main physical effects has resulted in very accurate reconstructions.

The ML-EM algorithm seeks to maximize the logarithm of the Poisson likelihood objective function:

$$L(\mathbf{f}) = \sum_{j} \left\{ p_j \ln \left(\sum_{i} A_{ji} f_i \right) - \sum_{i} A_{ji} f_i \right\}. \tag{9.24}$$

The ML-EM algorithm iteratively updates voxel values by using the EM algorithm

$$f_i^{k+1} = \frac{f_i^k}{\sum\limits_{j=1}^{m} A_{ji}} \sum_{j=1}^{m} A_{ji} \frac{p_j}{\sum\limits_{i'=1}^{n} A_{ji'} f_{i'}^k}, \tag{9.25}$$

where f_k is the kth estimate of the image, p_j is the jth projection value, and A_{ji} is the system matrix that gives the probability of a photon emitted from voxel i being detected in projection bin j.

In the past, ML-EM approaches have suffered from increased reconstruction times and difficulty in converging to a useful solution, particularly in the presence of significant noise. In this regard, Hudson and Larkin made a major breakthrough when they developed an algorithm—*ordered subsets expectation maximization* (OSEM)—that utilizes "ordered subsets" to accelerate iterative reconstruction. In this approach, projection data are grouped into an ordered sequence of subsets or blocks. For example, if a full set of projections consists of 336 views and there are 14 subsets, each subset would consist of 24 views. Of major importance, these subsets represent mutually exclusive and exhaustive use of the projection data, and they are treated independently in parallel. One iteration of the algorithm is a single pass through all the subsets. Ordered subsets are an extension of iterative reconstruction. In Hudson and Larkin's original implementation, they used Shepp and Vardi's expectation maximization approach. Like any standard EM algorithm, this consists of a backprojection–projection–backprojection sequence. At each iteration, the reconstruction from the previous iteration is the starting point. In OSEM, the standard EM algorithm is applied to each of the subsets in turn. Such an approach can be applied to both SPECT and PET imaging. In SPECT, the subsets may correspond to natural groupings of projections—for example, each head in a triple-head system. In PET, projection data may be reorganized after acquisition to define blocks or subsets in a similar fashion. OSEM is rapidly becoming the preferred approach to reconstruction in emission tomography (including both SPECT and PET).

9.3   Image Quality in SPECT and PET

From a physics point of view, there are five major factors that affect image quality and the quantification of absolute radioactivity in emission tomography. These include finite spatial resolution (for example, the in-plane and axial spatial resolution of a SPECT or PET scanner); attenuation of photons by tissue; detection of scattered photons; accidental counting of "random" (nonpaired) photons in coincidence, applicable only to PET imaging; and "noise" resulting from the statistical nature of radioactive decay.

9.3.1   Spatial Resolution

As first introduced in Chapter 3, all imaging systems have a limited ability to resolve small objects. The spatial resolution of an emission tomography system can be thought of as that distance by which two small point sources of radioactivity must be separated to be distinguished as separate in the image (see again Figure 3.6). The spatial resolution of a SPECT system is mainly determined by the resolution in the projection data, which itself is determined by the factors discussed in Section 8.4.1, and the effects of reconstruction, especially the specific filter used, since the approximate ramp filter given in (9.11) usually includes a low-pass filter $W(\varrho)$ to reduce noise. The spatial resolution of a PET system is mainly determined by the physical cross-section dimensions of each detector element and the effects of reconstruction. In addition, because PET uses the annihilation site, rather than the emission site of the positron per se, *positron range* (i.e., the distance the positron travels before annihilation) also limits resolution.

In practice, we can model SPECT and PET resolution as a cascade of LSI systems [see (2.47)]. The development is identical to that for CT resolution, as fully derived in Section 6.4.1. The final expression for the relation between the reconstructed (but blurred) image and the object, identical to (6.56), is

$$\hat{f}(x, y) = f(x, y) * h(r),\qquad(9.26)$$

where $r = \sqrt{x^2 + y^2}$.

Here (as in CT), h(r) represents the (cascaded) PSF and h(r) for SPECT includes (1) the (1-D, in-plane) projection PSF [i.e., (8.13), with y set to 0, and $x = \ell$] and (2) w [identical to w in (6.48)]; h(r) for PET includes (1) a *positron range function*, (2) a rect function for detector width [identical to s in (6.48)], and (3) w.

Finite spatial resolution in SPECT and PET results in two important, related effects. First, the image is blurred, with the degree of blurring dependent on the spatial resolution. This blurring prevents the delineation of edges of larger structures and may not allow the visualization of smaller ones as distinct objects. Second, neighboring areas are smeared and averaged together, reducing the measured value in the areas with greater radioactivity, and increasing it in the areas with lesser radioactivity (i.e., reducing contrast). Said another way, referring back to (3.1), the reduction in f_{max} and the increase in f_{min} reduces modulation (or contrast). In this fashion, finite spatial resolution produces an underestimation of radioactivity in small structures, with progressive underestimation as the structures get smaller. The effect is not eliminated until the object size is approximately three times the resolution of the imaging system. These effects, which also apply to the axial resolution (sometimes mistakenly called "slice thickness") of a tomographic scanner, are sometimes referred to as "partial volume effects."

9.3.2 Attenuation and Scatter

The photons detected in nuclear medicine imaging are electromagnetic radiation. As such, they undergo two major types of interactions in tissue—photoelectric effect and Compton scattering—as discussed in Chapter 4. The photoelectric effect results in complete absorption of the photon, reducing observed count rate, while scattered photons are still detected. Since attenuation is a combination of absorption (by photoelectric effect) and scatter (by Compton scattering), correction schemes can either treat these processes independently or jointly. Event C in Figure 9.10 depicts a scatter coincidence in PET.

As first discussed in Section 5.4.3 in the context of projection radiography, large-angle scatter produces a low-level background "haze" in the image, which reduces contrast [see (5.41)]. This reduction in contrast arises from the addition of this scatter background, rather than a reduction in f_{max} or an increase in f_{min} per se, as discussed above in the context of resolution effects and in (3.1).

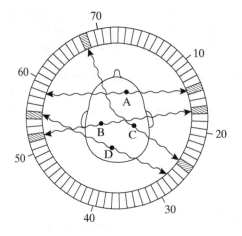

Figure 9.10
Coincidence counting in PET. B represents a true coincidence, A and D random or accidental coincidences, and C scattered coincidence.

Absorption produces a gradual, progressive underestimation of radioactivity from the edge to the center of the body, by about a factor of five to as much as 50, depending on body size, photon energy, and SPECT versus PET detection [since comparison of (9.5) and (9.17) indicates that, for SPECT, the photon only has to travel through part of the body, whereas for PET, the sum of the two annihilation photons' travel is always the complete body thickness]. The same physics and mathematics that make "perfect" attenuation correction at least theoretically achievable in PET, but not in SPECT, make the effects of attenuation itself more dramatic in PET than in SPECT. This is counterintuitive given the difference in photon energies of the commonly used radionuclides for PET and SPECT. In PET, the main photon energy and linear attenuation coefficient (for water) are 511 keV and 0.095 cm^{-1}, respectively, while for SPECT they are 140 keV and 0.145 cm^{-1}. The difference in energy is a factor of 3.7, but the difference in attenuation coefficients is only about 50 percent. The underlying basis for higher attenuation in PET is the fact that both photons must escape the patient and be detected in coincidence, meaning that the total attenuation path length is longer for coincidence imaging than for SPECT. This distinction increases exponentially with patient size. For example, with PET, a 30 cm diameter patient requires an average attenuation correction factor of 16, whereas a 40 cm diameter patient requires a correction factor of 50. For comparison, the attenuation correction factor in cardiac SPECT imaging, where the photons pass through 20 cm of tissue, is 20.

Since both attenuation and scatter influence the relative, as well as absolute, apparent distribution of activity within the image, it is important to try to correct for these effects even if only subjective visual interpretation—rather than absolute quantification—of the images occurs. In SPECT, the convolution backprojection formula given in (9.10) does not include a correction for attenuation. Of importance, inspection of (9.5) reveals that no analytical, closed-form correction is possible, because the attenuation term (the exp term) is not separable from the activity term. In practice, the nonattenuation-corrected reconstructed activity distribution from

(9.10) is corrected with an average attenuation correction factor, as a first-order correction, by

$$A_c(x, y) = \frac{1}{\mathrm{ACF}} A(x, y),$$
(9.27)

where

$$\mathrm{ACF} = \int_{-\infty}^{R} \exp\left\{-\int_{y}^{R} \mu(x, y', z; E) dy'\right\} dy.$$
(9.28)

In PET, an exact correction for attenuation, as presented in (9.22) and (9.23), can be made.

In both SPECT and PET, scatter is usually estimated at the projection level and subtracted prior to reconstruction.

9.3.3 Random Coincidences

In PET, random or accidental coincidences are possible. These occur when one annihilation photon from one electron-positron annihilation event, and another annihilation photon from a different electron-positron annihilation event are detected in coincidence. Events A and D in Figure 9.10 depict random or accidental coincidences.

The random coincidence count rate is given by the product of the coincidence time window and the "singles" (individual) count rates from the two opposing detectors in coincidence,

$$R = 2\tau S^+ S^-,$$
(9.29)

where 2τ is the coincidence time window width and S^+ and S^- are the individual count rates of opposing detectors.

As the activity in the field of view increases, the true coincidence rate increases linearly (ignoring dead-time effects), but the random coincidence rate increases as the square of the increase in activity. Sensitivity may be increased by increasing the width of the coincidence time window (up to the point when the full coincidence time spectrum has been included), but only at the expense of acquisition of more random coincidences. A great challenge in coincidence imaging is to balance sensitivity (true coincidences) with random coincidences in optimizing the time window setting. Some manufacturers specify the window in terms of the time resolution of the system; others do so in terms of twice the time resolution, which is usually the actual window width, and therefore the more traditional specification. The timing window, which typically ranges between 2 and 20 ns, is primarily determined by the "speed" of the detector material.

In practice, random or accidental coincidences may be subtracted from the observed projection data through use of (9.29).

9.3.4 **Contrast**

In the process of considering the effects of finite spatial resolution and scatter, we noted their influence on image contrast in SPECT and PET. In both cases, our consideration of contrast followed that first introduced in (3.1), and recast specifically for planar imaging in (8.24). In SPECT and PET, the definition of contrast remains the same, but f_{max} and f_{min} are from reconstructed images.

9.3.5 **Noise and Signal-to-Noise**

It is important to remember that these "deterministic" effects influence only the "signal" component of an image's signal-to-noise ratio; in other words, they influence only quantitative *accuracy*. Noise arises from the random statistical nature of radioactive decay, as well as potentially from the imaging hardware and certain image processing operations, and influences the "noise" component of the ratio; in other words, it influences *precision*. The noise in projection data in SPECT and PET is identical to the situation in planar scintigraphy and follows the development given in Section 8.4.5. However, the reconstruction process changes the noise magnitude and correlation in a fashion similar to the way that reconstruction changes the noise in CT compared with projection radiography.

In practice, estimating the noise in a reconstructed SPECT or PET image is difficult. In order to present a generic formulation, we make the same assumptions we made to derive an expression for variance in CT: (1) the measured line integrals are statistically independent, and (2) the object is sufficiently uniform that the line integrals are essentially equal. With these simplifying assumptions, the derivation of the variance in a reconstructed SPECT or PET image is identical to that presented in Section 6.4.2, is independent of (x, y) location, and is given by (6.72).

Ultimately, key performance parameters such as lesion detectability or quantitative accuracy and precision depend on the signal-to-noise ratio or total *error*. Since both contrast and noise are defined the same way in CT and SPECT/PET, the SNR formulation presented for CT in (6.73) applies, and the final SNR formulation is the same as (6.74).

9.4   Summary and Key Concepts

Cross-sectional—tomographic—imaging in nuclear medicine is accomplished with SPECT and PET. SPECT usually utilizes a rotating Anger camera; PET utilizes a dedicated imaging system. In this chapter, we presented the following key concepts that you should now understand:

1. *Emission computed tomography* includes *single photon emission computed tomography* (SPECT) and *positron emission tomography* (PET).
2. SPECT is based on an ensemble of projection images, each of which is a conventional planar scintigram.
3. PET has no projection analog and is based on *coincidence detection of paired gamma rays* ("annihilation photons") following positron-electron annihilation.

4. SPECT is typically performed with rotating Anger scintillation cameras (with multicamera systems being the most common); PET is typically performed with dedicated PET scanners.

5. The *basic SPECT imaging equation* is, in essence, identical to that in planar scintigraphy, except for a change in notation that facilitates consideration of reconstruction; reconstruction follows the approach outlined for CT.

6. The activity and attenuation terms in the basic SPECT imaging equation are not separable; thus, there is no closed-form solution for attenuation correction in SPECT.

7. The *basic PET imaging equation* is similar to that for SPECT, with one important difference: the use of coincidence detection leads to an attenuation term whose limits of integration span the entire body; thus, PET has a closed-form solution for attenuation correction.

8. *Iterative reconstruction* is a newer, more computer intensive approach that implicitly takes the random nature of decay into account and can incorporate models of attenuation, scatter, and blur.

9. *Image quality* in SPECT and PET is limited by *resolution*, *scatter*, and *noise*.

Bibliography

Bendriem, B. and Townsend, D. W. *The Theory and Practice of 3D PET*, Dordrecht, Netherlands: Kluwer Academic Publishers, 1998.

Bushberg, J. T., Seibert, J. A. Leidholdt, E. M. Jr., and Boone, J. M. *The Essential Physics of Medical Imaging*, Philadelphia, PA: Lippincott Williams and Wilkins, 2002.

Carlton, R. R. and Adler, A. M. *Principles of Radiographic Imaging: An Art and a Science*, 3rd ed. Albany, NY: Delmar, 2001.

Chandra, R. *Nuclear Medicine Physics, Fifth Edition*, Baltimore, MD: Williams & Williams, 1998.

Groch, M. W. and Erwin, W. D. Single-photon Emission Computed Tomography in the Year 2001: Instrumentation and Quality Control. *Journal of Nuclear Medicine Technology*, 29:9–15, 2001.

Hudson, H. M. and Larkin, R. S. Accelerated Image Reconstruction Using Ordered Subsets of Projection Data, *IEEE Transactions on Medical Imaging*, 13(4):601–609, 1994.

Rao, D. V., Chandra, R., and Graham, M. C. *Physics of Nuclear Medicine*. New York: American Institute of Physics, 1984.

Rollo, F. D. *Nuclear Medicine Physics: Instrumentation and Agents*. St. Louis, MO: C. V. Mosby, 1977.

Shepp, L. A. and Vardi, Y. Maximum Likelihood Reconstruction for Emission Tomography, *IEEE Transactions on Medical Imaging*, MI-1(2):113–122, 1982.

Sorenson, J. A. and Phelps, M. E. *Physics in Nuclear Medicine*, 2nd ed. Philadelphia, PA: W. B. Saunders, 1987.

Problems

9.1 Suppose a PET detector comprises four square PMTs (arranged as a 2 by 2 matrix) and a single BGO crystal with slits made in such a way that it is divided into an 8 by 8 matrix of individual detectors. Assume that the PMTs and the detectors cover the exact same square area, and that each PMT is 2 by 2 in. This

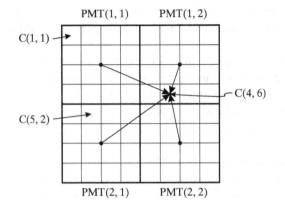

Figure P9.1
Problem 9.1.

geometry is shown in Figure P9.1. The response of a PMT to an event occurring in a particular subcrystal depends on the distance from the center of the PMT to the center of the subcrystal, r, as follows

$$\text{PMT Response} = \exp[-r/\tau],$$

where $\tau = 1$ inch.

(a) Find a general expression for the response in $\text{PMT}(i, j)$ to an event in crystal $C(k, l)$.

(b) Find the numerical responses in each PMT to an event in crystal $C(4, 5)$.

(c) Ignoring the possibility of noise, develop a scheme to uniquely identify the crystal in which an event occurred.

(d) Characterize a worst-case scenario in which the smallest possible additive noise in the signal of one PMT causes an error in event localization.

9.2 Both NaI(Tl) and BGO can be used as PET detectors. The linear attenuation coefficient of BGO at 511 keV is 0.964 cm^{-1} as opposed to 0.343 cm^{-1} for NaI(Tl). BGO does not convert gamma ray photons to light photons as efficiently as NaI(Tl), and is in fact roughly 13 percent as efficient.

In this problem, suppose two detectors are designed, one from NaI(Tl) and one from BGO. Assume both are designed to stop 75 percent of the 511 keV photons that strike the crystal.

(a) Find the detector thicknesses for NaI(Tl) and BGO.

(b) A light burst from NaI(Tl) has a higher intrinsic SNR than that from BGO. Find the ratio of intrinsic SNRs, NaI(Tl) to BGO.

9.3 You decide to build a PET scanner from a dual-headed SPECT camera you own. It has two 30 cm square Anger cameras mounted facing each other 1.5 meters apart.

(a) Would you use the low-energy collimators, the high-energy collimators, or no collimators? Explain.

(b) What significant piece of electronics would you have to add to the SPECT circuitry to make this work as a PET scanner?

(c) Explain how you would use the X and Y signals coming out of the opposing Anger cameras. Would you use the Z-pulses? Why or why not?

(d) You will have to rotate the Anger cameras at some point during the scan. What is the minimum number of angular positions of the two cameras that would be required in order to get full angular coverage for PET reconstruction?

(e) Consider the radiotracer concentration in the body. Would you expect to require a higher, lower, or about the same dose in this make-shift PET scanner versus a real PET scanner? Explain your rationale.

9.4 Consider the 2-D cross section shown in Figure P9.2 consisting of the three separate compartments R_1, R_2, and R_3.

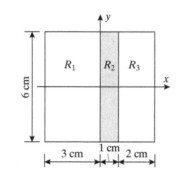

Figure P9.2
Problem 9.4.

(a) Suppose a solution containing a 511 keV gamma ray emitting radionuclide with concentration 0.3 mCi/cm^2 fills only R_2; R_1 and R_3 contain nonradioactive solutions. Let the linear attenuation coefficients (at 511 keV) in the three regions be $\mu_1 = 0.2$ cm^{-1}, $\mu_2 = 0.3$ cm^{-1}, and $\mu_3 = 0.1$ cm^{-1}, respectively. Suppose we image the radioactivity using a (2-D) SPECT scanner outside the object. Compute the projected radioactivities $g_{\text{SPECT}}(\ell, 90°)$ and $g_{\text{SPECT}}(\ell, 270°)$.

(b) Now assume the radionuclide in (a) is replaced by a positron emitting radionuclide with the same concentration. Assume the linear attenuation coefficients in the three regions are the same. This time the body is imaged using a (2-D) PET scanner. Compute $g_{\text{PET}}(\ell, 90°)$ and $g_{\text{PET}}(\ell, 270°)$.

(c) Explain how to compensate for attenuation in a PET scanner in order to reconstruct an accurate image of the radionuclide concentration. Can you do the same for a SPECT scanner?

9.5 Suppose a PET scanner has 1,000 detectors packed tightly around a circle with a diameter of 1.5 m.

(a) What is the approximate size (width) of each detector? Explain the tradeoff between using deep (long) and shallow (short) detectors.

(b) What is the purpose of coincidence detection in PET? What is a nominal time interval defining a coincidence "event window" for PET? Explain why it is undesirable to make the event window (1) smaller or (2) longer.

(c) Assume there is no "wobbling" or "dichotomic" motion of the PET gantry. What is the line integral sampling interval (which has the symbol T in CT) in the center of the scanner? Assuming the usual sampling rule-of-thumb from CT, how many pixels would be in a typical PET image? Explain why motion of the PET gantry is desirable.

(d) Explain why resolution of the PET scanner is typically worse away from the center of the scanner.

Ultrasound Imaging

Overview

In the preceding two parts of this book, we considered those imaging modalities that rely on ionizing radiation to create the signals. We saw that these modalities produce images of structure and function through a variety of mechanisms, all based on ionizing radiation. The modalities are of high clinical utility—that's why they're so commonly used—but the use of ionizing radiation carries some risk. At the low doses used in medical imaging, the main biological risk is cancer; the probability of inducing cancer from a diagnostic imaging examination is very, very low, but it is not zero. Sometimes, this is of concern; for example, when imaging *in utero*.

Here, we examine a modality considered to be fully noninvasive and risk free: ultrasound imaging. In projection radiography and computed tomography, we were interested in the *transmission* of a beam of radiation through the patient; the medical signal of interest was produced by the variable attenuation of this beam by body tissues. In contrast, with ultrasound, we are interested in the *reflection* of a beam of sound by the patient; the medical signal of interest here is produced by the variable reflectance of this sound beam by different body tissues.

In addition to depicting varying sound reflectance, ultrasound is capable of portraying the velocities of moving bodies, such as blood within vessels. This so-called *Doppler imaging* technique can provide a quantitative measure of blood flow. The combination of anatomical and functional information that ultrasound is able to provide—in a fully noninvasive manner—makes it a popular imaging modality.

A basic ultrasound imaging system consists of a transducer and associated electronics, including a display. The transducer itself is both a transmitter and receiver of ultrasound energy. In practice, the electronics "steer" the beam so that an arc is usually covered through the patient, producing a 2-D image on the display.

The most common ultrasound procedures involve imaging the developing fetus *in utero*, and *2-D echocardiography* for heart imaging. A representative echocardiogram is shown in Figure IV.1(a). Here, the transducer is positioned at the tip or

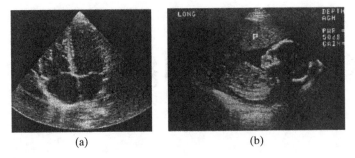

(a) (b)

Figure IV.1 Two common clinical ultrasound examinations: (a) an echocardiogram, showing the four chambers of the heart, and (b) fetal ultrasound, showing a normal fetus at the second trimester of gestation. (R. H. Daffner, Clinical Radiology, The Essentials, Lippincott, Williams, & Wilkins, 1993.)

apex of the heart, pointing in such a way as to cut through all four heart chambers. Cross-sectional images of the beating heart provide the physician with a lot of diagnostic and functional information about the health of the heart as a pump. You've probably all seen images of a fetus—looking quite babylike—produced by ultrasound; such an image was shown in Figure I.1(c). Figure IV.1(b) shows another such image, taken during the second trimester. If the angle of the view is just right, you can even tell the gender.

The fact that the method is noninvasive and readily performed in a doctor's office helps explain the popularity of 2-D echocardiography. Since the transducer is held by hand during most ultrasound procedures, the demand for well-trained, experienced ultrasonographers is quite high, as the details of acquisition, including the view angle, significantly influences both the field of view and the image quality.

The Physics of Ultrasound

10.1 Introduction

Ultrasound is sound with frequencies higher than the highest frequency that can be heard by human beings. This means that any sound above about 20 kHz is considered to be ultrasound. For several reasons we will explain fully in this chapter and in Chapter 11, medical ultrasound systems operate at much higher frequencies than this defining frequency, typically between 1 and 10 MHz, although there are experimental systems operating in ranges up to 70 MHz. The principles of ultrasound propagation, regardless of the exact frequency range, are the same as those of ordinary sound propagation and are defined by the theory of *acoustics*. In particular, ultrasound moves in a wavelike fashion by expansion and compression of the medium through which it is moving, and ultrasound waves travel at certain speeds, depending on the material through which they are traveling. In addition, ultrasound waves can be absorbed, refracted, focused, reflected, and scattered.

Ultrasound provides a noninvasive technique for imaging human anatomy. A *transducer*, which converts electrical signals to acoustic signals, generates pulses of ultrasound, which are sent through a patient's body. Organ boundaries and complex tissues produce echoes (by reflection or scattering) that return back and are detected by the transducer, which converts the acoustic signal to an electrical signal. The ultrasound imaging system then processes the echoes and presents a grayscale image of human anatomy on a display. Each point in the image corresponds to the anatomic location of an echo-generating structure, and its brightness corresponds to the echo strength. Figure 10.1 depicts several examples of ultrasound images arising from the dominant diagnostic areas in which ultrasound is used—i.e., obstetrics, oncology, cardiology, and gastroenterology.

The use of ultrasound in diagnostic imaging dates back to the mid 1950s. Rapid expansion of its use began in the early 1970s with the advent of two-dimensional real-time ultrasonic scanners. Additional milestones were the appearance of phased array systems in the early 1980s, color flow-imaging systems in the mid 1980s, and three-dimensional imaging systems in the 1990s. Manufacturers offer several types

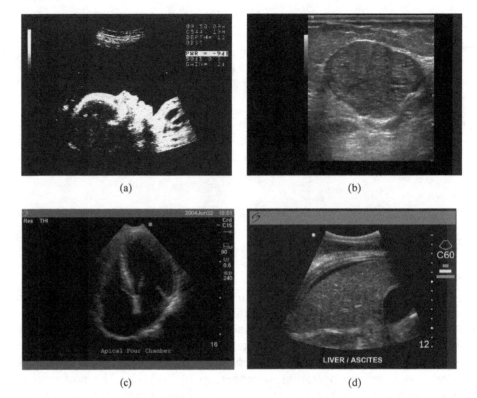

Figure 10.1
Examples of ultrasonic
images of human anatomy:
(a) fetal face (obstetrics),
(b) breast cyst (oncology),
(c) heart (cardiology), and
(d) liver (gastroenterology).
(b) (Used with permission of
GE Healthcare.), (c–d)
(Courtesy of SonoSite Inc.)

of ultrasonic medical imaging systems, each one having interchangeable transducers
and the ability to operate in a variety of modes. This is because different parts of the
human anatomy have different imaging requirements.

In this chapter, we study the physics of ultrasound and its interaction with
biological tissues. This background is necessary for the complete understanding of
medical ultrasound imaging systems, which will be discussed in the next chapter.

10.2   The Wave Equation

Acoustical waves are pressure waves that propagate through matter via compression
and expansion of the material. A wave can be generated by compressing a small
volume of tissue and then releasing it. The elastic properties of the material cause
the compressed volume to expand past its point of equilibrium, making neighboring
volumes compress. As this process continues through successive volumes of tissue, a
wave is generated.

As an acoustic wave propagates, small particles of the material move back and
forth in order to generate the compressions and expansions of the acoustic wave.
In medical ultrasound, where waves propagate only in soft tissues, these particles
move back and forth in the same direction that the acoustic wave is traveling. When
this happens, the acoustic wave is called a *longitudinal wave*. Harder materials also

support *shear waves*, in which the particles move at right angles to the direction of propagation of the acoustic wave, but these types of waves do not need to be considered in medical ultrasound.

10.2.1 Three-Dimensional Acoustic Waves

The *compressibility* κ and *density* ρ of a material, combined with the laws of conservation of mass and momentum, directly imply the existence of acoustic waves. ($B = 1/\kappa$ is called the *bulk modulus* of a material.) These waves travel at a *speed of sound c*, given by

$$c = \sqrt{\frac{1}{\kappa\rho}}.$$

(10.1)

Table 10.1 gives the speed of sound (as well as other information) for various materials and biological tissues. A good rule of thumb is that sound travels about 1540 m/s in tissue and 330 m/s in air.

TABLE 10.1

Acoustical Properties of Various Materials					
Material	Density, ρ [kg m^{-3}]	Speed, c [m s^{-1}]	Characteristic Impedance, Z [kg m^{-2} s^{-1}] ($\times 10^6$)	Absorption Coefficient, α [dB cm^{-1}] (at 1 MHz)	Approximate Frequency Dependence of α
Air at STP	1.2	330	0.0004	12	f^2
Aluminum	2700	6400	17	0.018	f
Brass	8500	4490	38	0.020	f
Castor oil	950	1500	1.4	0.95	f^2
Mercury	13,600	1450	20	0.00048	f^2
Polyethylene	920	2000	1.8	4.7	$f^{1.1}$
Polymethyl-methacrylate	1190	2680	3.2	2.0	f
Water	1000	1480	1.5	0.0022	f^2
Blood	1060	1570	1.62	[0.15]	
Bone	1380–1810	4080	3.75–7.38	[14.2–25.2]	
Brain	1030		1.55–1.66	[0.75]	
Fat	920	1450	1.35	[0.63]	
Kidney	1040	1560	1.62	—	
Liver	1060	1570	1.64–1.68	[1.2]	
Lung	400		0.26	[40]	
Muscle	1070		1.65–1.74	[0.96–1.4]	
Spleen	1060		1.65–1.67	—	
Water	1000	1484	1.52	[0.0022]	

Source: Data above the line are from P. N. T. Wells, *Biomedical Ultrasonics*, (New York: Academic Press, 1977). Data in square brackets below the line are taken from A. B. Wolbarst, *Physics of Radiology* (Norwark, CT: Appleton and Lange, 1993).

There are a variety of ways to mathematically characterize an acoustic wave. First, it is important to realize that an acoustic wave is inherently a 3-D phenomenon—i.e., it has spatial dependencies. For example, the sound generated in ordinary speech spreads out in all directions so that people can hear the speaker from many orientations. An acoustic wave also depends on time—i.e., it has temporal dependencies. In ordinary speech, for example, a sudden exclamation or shout dies away momentarily. So, whatever the physical quantity that is used to describe the wave, we realize that it must depend upon three spatial variables, x, y, z, and time, t.

In a traditional treatment of acoustics, one would begin by describing the *particle displacement* $\vec{u}(x, y, z, t)$ associated with the compression and expansion of the acoustic wave. The *particle velocity*, $\vec{v}(x, y, z, t)$, the temporal derivative of particle displacement, is another common choice for describing an acoustic wave. Since we are constraining our discussion to longitudinal waves, it is actually only necessary to describe the magnitudes of particle displacement and velocity—$u(x, y, z, t)$ and $v(x, y, z, t)$, respectively—since the orientation of these quantities is always identical to the direction of wave propagation. We are actually going to take this a step farther, however. Since the compression and expansion of a small volume are associated with a local change in the material's pressure, an acoustic wave can also be described in terms of a spatially dependent, time-varying pressure function $p(x, y, z, t)$, called the *acoustic pressure*. Acoustic pressure is zero in the absence of an acoustic wave, so it should be considered to be the variation of pressure caused by the acoustical disturbance around the ambient pressure.

In longitudinal waves, it is straightforward to relate the acoustic pressure to the underlying particle velocity. They are, in fact, linearly related by

$$p = Zv, \tag{10.2}$$

where

$$Z = \rho c \tag{10.3}$$

is called the *characteristic impedance*. The particle speed v and the speed of sound c are *not* the same quantity and are generally not equal. This is a common point of confusion and source of computational mistakes. The term *impedance* arises by analogy to electrical circuits or electromagnetic transmission lines. In particular, acoustical pressure is analogous to voltage, while particle speed is analogous to current. Since density has units kg/m^3 and speed has units m/s, the characteristic impedance Z has units kg/m^{2}s, which is given the name *rayls*, after Lord Rayleigh.

Throughout this chapter and the next, we use acoustic pressure to describe longitudinal acoustic waves. Where necessary (in deriving boundary conditions, for example), we refer to the velocity or displacement. From the physical properties of matter, it turns out that acoustic pressure p must satisfy the following *three-dimensional wave equation*:

$$\nabla^2 p = \frac{1}{c^2} \frac{\partial^2 p}{\partial t^2}, \tag{10.4}$$

where ∇^2 is the 3-D Laplacian operator given by

$$\nabla^2 = \frac{\partial^2}{\partial x^2} + \frac{\partial^2}{\partial y^2} + \frac{\partial^2}{\partial z^2}. \tag{10.5}$$

In general, a wave equation is a partial differential equation relating spatial partial derivatives of pressure to the temporal partial derivative of pressure. Solving this equation is difficult in general, but our needs will be satisfied if we consider two special cases: plane waves and spherical waves.

10.2.2 Plane Waves

If an acoustic wave varies in only one spatial direction and time, it is called a *plane wave*. For example, if $p(x, y, z, t)$ were constant for any choice of x and y given a fixed z and t, then the resulting pressure function $p(z, t) = p(x, y, z, t)$ describes a plane wave moving in the $+z$ or $-z$ direction. In this case, plugging $p(z, t)$ into (10.4) yields the *one-dimensional wave equation*:

$$\frac{\partial^2 p}{\partial z^2} = \frac{1}{c^2} \frac{\partial^2 p}{\partial t^2}. \tag{10.6}$$

Notice that a one-dimensional wave equation exists for any direction of propagation, not just the z direction. In the following, it will be convenient to think of the z direction in the sense of a rotated coordinate system—i.e., z is the direction of plane wave propagation, not a fixed "laboratory" coordinate direction.

The general solution to (10.6) can be written as

$$p(z, t) = \phi_f(t - c^{-1}z) + \phi_b(t + c^{-1}z), \tag{10.7}$$

which can be verified by direct substitution into (10.6). The function $\phi_f(t - c^{-1}z)$ is interpreted as a *forward-traveling wave*, since the basic waveform $\phi_f(c^{-1}z)$—i.e., the spatial pressure function at $t = 0$—is shifted in the positive z direction as t increases. The function $\phi_b(t + c^{-1}z)$ is interpreted as a backward-traveling wave for an analogous reason. The only requirement of the functions $\phi_f(\cdot)$ and $\phi_b(\cdot)$ is that they be twice differentiable, although even this assumption can be relaxed by using generalized derivatives. Since each function ϕ_f and ϕ_b satisfies the wave equation independently, one of them might be identically zero, so we can have only a forward traveling wave in a given medium, for example. In Chapter 11, we will use the plane wave to approximate the acoustic wave produced by certain types of ultrasound transducers.

An important example of a function that satisfies the one-dimensional wave equation is the *sinusoidal* function

$$p(z, t) = \cos k(z - ct). \tag{10.8}$$

Viewed as a function of t only (holding z fixed), we see that the pressure around a fixed particle varies sinusoidally with (radial) frequency $\omega = kc$. This corresponds to a (cyclic) frequency of

$$f = \frac{\omega}{2\pi} = \frac{kc}{2\pi}, \tag{10.9}$$

which has units of Hertz (Hz) or cycles per second. Viewed as a function of z only (holding t fixed), we see that the pressure at a particular time varies sinusoidally with (radial, spatial) frequency k, a quantity that is known as the *wave number*. The *wavelength* λ of this sinusoidal wave is the spacing between crests or troughs at any given time and is given by

$$\lambda = \frac{2\pi}{k}, \tag{10.10}$$

which has units of length. Solving (10.9) for k and substituting the resulting expression into (10.10) yields the important relationship between wavelength, speed of sound, and frequency

$$\lambda = \frac{c}{f}. \tag{10.11}$$

Sinusoidal waves are not realizable in practice because they need infinite space and infinite time. However, medical ultrasound systems use waveforms that are approximately sinusoidal when viewed locally over a short period of time. For example, it is common for ultrasound imaging systems to operate at 3.5 MHz, which corresponds to a wavelength of 0.44 mm when $c = 1540$ m/s. Although it cannot be made entirely clear at this point why this would be the case, the wavelength of the acoustic wave is approximately equal to the achievable resolution of the system (at least in the direction of the propagating wave). Therefore, we should understand that submillimeter resolution might be achievable in ultrasound imaging.

EXAMPLE 10.1
Suppose a steady-state sinusoidal wave with frequency 2 MHz is traveling in the $+z$ direction in the liver.

Question What is its wavelength?

Answer From Table 10.1, we find that $c = 1570$ m/s in liver. Therefore, the wavelength is $\lambda = 1570 \text{ ms}^{-1}/(2 \times 10^6 \text{ Hz}) = 0.785$ mm. ∎

Another important example of a function that satisfies the 1-D wave equation is

$$p(z, t) = \delta(z - ct), \tag{10.12}$$

where $\delta(\cdot)$ is the unit impulse (delta) function (see Section 2.2.1). Since $\delta(\cdot)$ is a generalized function, it satisfies the wave equation only through the use of generalized derivatives. Like sinusoidal waves, the impulse function cannot exist in practice. In this case, the impulse plane wave is supposed to have infinite extent orthogonal to the direction of propagation and an infinitesimally thin extent in the direction of propagation, neither of which can be achieved in practice. Nevertheless, the impulse plane wave is useful for approximating the short pulses that are used in medical ultrasound systems, in order to better understand the overall properties of these systems.

EXAMPLE 10.2

An ultrasound transducer is pointing down the $+z$ axis. Starting at time $t = 0$, it generates an acoustic pulse with form

$$\phi(t) = (1 - e^{-t/\tau_1})e^{-t/\tau_2}.$$

Question Assume the speed of the sound is $c = 1540$ m/s. What is the forward traveling wave down the $+z$ axis? At what time does the leading edge of the impulse hit the interface 10 cm away from the transducer?

Answer The forward traveling wave down the $+z$ axis is

$$\phi_f(z, t) = (1 - e^{-(t-z/c)/\tau_1})e^{-(t-z/c)/\tau_2}.$$

It will take $0.1 \, \mathrm{m}/(1540 \, \mathrm{ms}^{-1}) = 64.9 \mu s$ for the leading edge of the ultrasound impulse to travel 10 cm and hit the interface. ∎

10.2.3 Spherical Waves

In an isotropic material, a spherical wave can be generated by a small, local disturbance in the pressure. A spherical wave depends only on time t and the radius $r = \sqrt{x^2 + y^2 + z^2}$ from the source of the disturbance, which is assumed to be at $(0, 0, 0)$. Using (10.4), evaluating the Laplacian (viewing r as a function of x, y, and z), and manipulating the resulting equation (see Problem 10.4), the pressure $p(r, t)$ of a wave traveling in the radial direction can be shown to satisfy

$$\frac{1}{r}\frac{\partial^2}{\partial r^2}(rp) = \frac{1}{c^2}\frac{\partial^2 p}{\partial t^2}, \tag{10.13}$$

which is known as the *spherical wave equation*.

The general solution to the spherical wave equation is

$$p(r, t) = \frac{1}{r}\phi_o(t - c^{-1}r) + \frac{1}{r}\phi_i(t + c^{-1}r), \tag{10.14}$$

which can be verified by direct substitution. The wave $\phi_o(t - c^{-1}r)$ is an *outward-traveling wave*, and $\phi_i(t + c^{-1}r)$ is an *inward-traveling wave*. Generally, there is no source that will create an inward-traveling wave, so the most general solution we need to consider here is

$$p(r, t) = \frac{1}{r}\phi_o(t - c^{-1}r).$$ (10.15)

As in the 1-D plane wave solution, $\phi_o(\cdot)$ should be twice differentiable, but otherwise it is an arbitrary function. This solution looks very much like the forward-traveling wave in the solution to the 1-D wave equation (10.7), except for the factor of $1/r$, which causes the outward spherical wave to lose amplitude. This loss is simply due to the increasing overall surface area of the wave as it propagates radially outward from its source.

10.3   Wave Propagation

10.3.1 Acoustic Energy and Intensity

An acoustic wave carries energy with it. Particles in motion have kinetic energy, and those that are poised for motion have potential energy. To characterize these energies in a wave, we define the energy per unit volume, leading to the *kinetic energy density*

$$w_k = \frac{1}{2}\rho v^2,$$ (10.16)

and the *potential energy density*

$$w_p = \frac{1}{2}\kappa p^2.$$ (10.17)

The *acoustic energy density* is defined as the sum of these two:

$$w = w_k + w_p.$$ (10.18)

This quantity captures the idea of a change in energy at a point in space as a wave passes through it.

To capture the idea of energy that moves with the wave, we define the *acoustic intensity* as

$$I = pv,$$ (10.19)

which is also called the *acoustic energy flux*. Notice that by using the analogy to electric circuits, I is analogous to electric power, which gives a strong intuitive connection to the present application. In particular, by substituting (10.2) into (10.19), we find that

$$I = \frac{p^2}{Z},$$ (10.20)

which is analogous to the expression in electric circuits, V^2/R, that gives the power in a resistor. There is power in acoustic waves. Unlike the case of the resistor, which converts electrical power to thermal energy, the acoustic wave simply carries the power along with it as it propagates (unless there is acoustic attenuation, as described in Section 10.3.4 below). In fact, it can be shown that the acoustic energy density and the acoustic intensity are related by the equation of energy conservation

$$\frac{\partial I}{\partial x} + \frac{\partial w}{\partial t} = 0, \tag{10.21}$$

which describes the propagation of power in an acoustic wave.

10.3.2 Reflection and Refraction at Plane Interfaces

Figure 10.2 depicts a plane wave incident upon a plane interface with incidence angle θ_i (measured from the nearest surface normal). Assuming that the wavelength of the sound is small with respect to the spatial extent of the interface, the reflected and transmitted wave directions will obey the laws of geometric optics:

$$\theta_i = \theta_r \tag{10.22}$$

and

$$\frac{\sin \theta_i}{\sin \theta_t} = \frac{c_1}{c_2}, \tag{10.23}$$

where c_1 and c_2 are the speeds of sound in medium 1 and medium 2, respectively. Equation (10.23) is known as *Snell's law*.

EXAMPLE 10.3
Suppose medium 1 is fat and medium 2 is liver, and a plane wave is incident upon their interface with incidence angle $\theta_i = 45°$.

Question What are the reflection and transmission angles θ_r and θ_t, respectively?

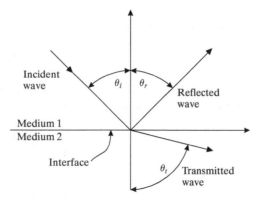

Figure 10.2
An incident plane wave reflects some energy and transmits the rest at a plane interface.

Answer $\theta_r = \theta_i = 45°$. Since $c_1 = 1450$ m/s and $c_2 = 1570$ m/s,

$$\sin \theta_t = \frac{1570 \sin 45°}{1450} = 0.7656 .$$

Solving this yields $\theta_t = 49.96°$. It makes intuitive sense that the transmission angle should be larger than the incidence or reflection angle, since $c_2 > c_1$. ∎

It may be that the quantity $c_2 \sin \theta_i / c_1$ is greater than 1.0. In this case, the inverse sine does not exist, and we conclude that all the energy is reflected. For a given pair of materials with speeds c_1 and c_2 in which $c_2 > c_1$, all incident angles above the so-called *critical angle* θ_c, given by

$$\theta_c = \sin^{-1}(c_1/c_2) \quad \text{for } c_2 > c_1, \tag{10.24}$$

will result in total reflection.

10.3.3 Transmission and Reflection Coefficients at Plane Interfaces

In the previous section, we examined the geometric properties of plane waves at plane intersections. Now we consider the energy characteristics in this same situation. Since all three waves—i.e., the incident, reflected, and transmitted waves—meet at the interface, the tangential particle motion caused by the incident wave must coincide with the sum of the tangential particle motions of transmitted and reflected waves. That is,

$$v_i \cos \theta_i = v_r \cos \theta_r + v_t \cos \theta_t .$$

Since $p = Zv$ [see (10.2)], we may replace the velocities in the above expression to get

$$\frac{\cos \theta_t}{Z_2} p_t + \frac{\cos \theta_r}{Z_1} p_r = \frac{\cos \theta_i}{Z_1} p_i . \tag{10.25}$$

Since pressure must be continuous across the interface, we also have

$$p_t - p_r = p_i . \tag{10.26}$$

Equations (10.25) and (10.26) are simultaneous linear equations with unknowns p_t and p_r, which when solved yield

$$R = \frac{p_r}{p_i} = \frac{Z_2 \cos \theta_i - Z_1 \cos \theta_t}{Z_2 \cos \theta_i + Z_1 \cos \theta_t}, \tag{10.27}$$

$$T = \frac{p_t}{p_i} = \frac{2 Z_2 \cos \theta_i}{Z_2 \cos \theta_i + Z_1 \cos \theta_t}, \tag{10.28}$$

where we have used the fact that $\theta_r = \theta_i$. The quantity R is called the *pressure reflectivity*, and the quantity T is called the *pressure transmittivity*.

Using the relationship $I = p^2/Z$ [see (10.20)], the *intensity reflectivity* is given by

$$R_I = \frac{I_r}{I_i} = \left(\frac{Z_2 \cos \theta_i - Z_1 \cos \theta_t}{Z_2 \cos \theta_i + Z_1 \cos \theta_t} \right)^2, \qquad (10.29)$$

and the *intensity transmittivity* is given by

$$T_I = \frac{I_t}{I_i} = \frac{4 Z_1 Z_2 \cos^2 \theta_i}{(Z_2 \cos \theta_i + Z_1 \cos \theta_t)^2}. \qquad (10.30)$$

In deriving the intensity transmittivity, the fact that p_t and p_i are in different media must be taken into account.

EXAMPLE 10.4

Consider an acoustic wave encountering a fat/liver interface at normal incidence.

Question What fraction of acoustic intensity is reflected back from a fat/liver interface at normal incidence? Does it matter from which direction the incident energy arrives?

Answer At normal incidence, we have

$$R_I = \left(\frac{Z_2 - Z_1}{Z_2 + Z_1} \right)^2.$$

The acoustic impedance of fat is 1.35×10^{-6} kg m^{-2}s^{-1}. The acoustic impedance of liver is 1.66×10^{-6} kg m^{-2}s^{-1} (nominal value). For propagation from fat to liver, we have

$$R_I = \left(\frac{1.66 - 1.35}{1.66 + 1.35} \right)^2 = 0.0106.$$

Only about 1 percent of the incident power is reflected back from the interface; about 99 percent is transmitted through. From the expression above, it does not matter from which direction the incident wave originates; the same fractions are calculated in either direction. (Is this also true for nonnormal incidence?) ∎

10.3.4 Attenuation

In practice, the amplitude of a real acoustic wave decreases as the wave propagates. *Attenuation* is the term used to account for loss of wave amplitude (or "signal") due to all mechanisms, including absorption, scattering, and mode conversion. *Absorption* is the process by which the wave energy is converted to thermal energy, which is then dissipated in the medium. *Scattering* is the process by which secondary spherical waves are generated as the wave propagates. *Mode conversion* is the process by which longitudinal waves are converted to transverse shear waves (and back again). In this section, we develop a phenomenological expression for attenuation, which includes all mechanisms for signal loss.

Consider a forward-traveling plane wave $p(z, t)$ in the $+z$-direction, where $p(0, t) = A_0 f(t)$. Then, under ideal circumstances $p(z, t) = A_0 f(t - c^{-1}z)$. Due to attenuation, however, we actually have

$$p(z, t) = A_z f(t - c^{-1}z), \qquad (10.31)$$

where A_z is the actual amplitude of the traveling wave and is dependent on the z-position of the wave. We model the amplitude decay as

$$A_z = A_0 e^{-\mu_a z}, \qquad (10.32)$$

where μ_a is called the *amplitude attenuation factor* and has units cm^{-1}. This model of attenuation is *phenomenological*, meaning it agrees well in practice but is not easily supported by theory. In particular, the pressure function given in (10.31) no longer satisfies the wave equation.

The unit given to the natural logarithm of an amplitude ratio is nepers (Np), pronounced "nay-pers," and the units of μ_a are sometimes quoted in nepers/cm. This is because

$$\mu_a = -\frac{1}{z} \ln \frac{A_z}{A_0}, \qquad (10.33)$$

Since $20 \log_{10} \frac{A_z}{A_0}$ is the amplitude gain in decibels (dB), it is useful to define the *attenuation coefficient* α as

$$\alpha = 20(\log_{10} e)\mu_a \approx 8.7\mu_a, \qquad (10.34)$$

which has the units dB/cm. Notice that 1 Np = 8.686 dB. Be careful when calculating the amplitude loss when given an attenuation coefficient α. You must first convert α to the amplitude attenuation factor μ_a [via (10.34)], and then use the amplitude loss equation (10.32).

When attenuation is due solely to the conversion of acoustic energy to thermal energy, the attenuation coefficient is called the *absorption coefficient*. Most measurements of attenuation in materials are conducted so that absorption is the dominant loss mechanism. Some typical absorption coefficients for biological and nonbiological materials are given in Table 10.1.

The absorption coefficient of a material is generally dependent on frequency f. A good model for this dependency is

$$\alpha = af^b, \qquad (10.35)$$

where b is just slightly greater than 1 in biological tissues. For example, in homogenized liver, $a = 0.56$ and $b = 1.12$. The rough approximation that $b = 1$ is often used, leading to a linear relationship between α and frequency f. The values of a in Table 10.2 are based on the assumption that $b = 1$.

TABLE 10.2

Frequency Dependence of Various Biological Tissues	
Material	$a = \alpha/f$ [dB cm^{-1} MHz^{-1}]
Fat	0.63
Skeletal muscle	
Along fibers	1.3
Across fibers	3.3
Cardiac muscle	1.8
Blood	0.18
Bone	20.0
Lung	41.0
Liver	0.94
Kidney	1.0
Brain	
White matter along fibers	2.5
White matter across fibers	1.2
Gray matter	0.5–1.0

EXAMPLE 10.5

Suppose a 5 MHz acoustic pulse travels from a transducer through 2 cm of fat, then encounters an interface with the liver at normal incidence.

Question   At what time interval after the transmitted pulse will the reflected pulse—i.e., the *echo*—arrive back at the transducer? Taking both attenuation and reflection losses into account, what will be the amplitude loss in decibels of the returning waveform?

Answer   Table 10.1 shows that the speed of sound in fat is 1450 ms^{-1}. The round-trip travel distance is 4 cm. Therefore, the echo will return after the interval

$$t = \frac{0.04 \text{ m}}{1450 \text{ m/s}} = 27.6 \ \mu\text{s}.$$

Table 10.2 shows that $a = 0.63$ dB cm^{-1} MHz^{-1}. Therefore the absorption coefficient at 5 MHz is

$$\alpha = 0.63 \text{dB cm}^{-1} \text{ MHz}^{-1} \times 5 \text{ MHz}$$

$$= 3.15 \text{ dB cm}^{-1}.$$

The amplitude attenuation factor is $\mu_a = 3.15$ dB cm^{-1}/(8.686 dB/Np) = 0.363 Np/cm. The roundtrip amplitude ratio (ignoring the reflection loss) is therefore

$$\frac{A_z}{A_0} = \exp\{-0.363 \text{ Np/cm} \times 4.0 \text{ cm}\} = 0.234.$$

From the previous example, we know that the intensity reflectivity is 0.0106. The amplitude reflectivity is therefore $\sqrt{0.0106} = 0.103$. Putting these facts together yields

$$\text{dB loss} = 20 \log_{10} \frac{A_z}{A_0} = 20 \log_{10}(0.234 \times 0.103) = -32.4 \text{ dB}.$$

■

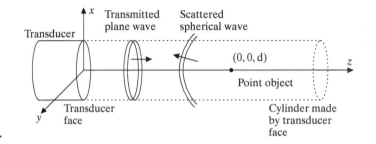

Figure 10.3
Basic pulse-echo experiment.

10.3.5 Scattering

Many targets within the body are much smaller than the acoustic wavelength. For these targets, the geometric optics equations for reflection and refraction do not hold. Instead, we assume that when these targets are excited by an incident acoustic plane wave, they vibrate as small spherical bodies, giving rise to spherical waves whose amplitude is some fraction of the incident wave amplitude.

Consider the plane (attenuated) wave

$$p(z, t) = A_0 e^{-\mu_a z} f(t - c^{-1}z)$$

incident upon a small pointlike target residing at $(0, 0, d)$, as shown in Figure 10.3. Notice that this wave is traveling in the $+z$ direction. The small target acts as a spherical wave source, converting a fraction R of the incident plane wave into a spherical wave centered at $(0, 0, d)$. Called the *reflection coefficient*, R is a property of the individual target and the embedding medium. Treating $(0, 0, d)$ as the origin [which means that r is the radius from $(0, 0, d)$], the resulting scattered wave is given by

$$p_s(r, t) = \frac{R e^{-\mu_a r}}{r} A_0 e^{-\mu_a d} f(t - c^{-1}d - c^{-1}r). \tag{10.36}$$

Here, we have included attenuation terms for both the incident plane wave and the scattered spherical wave, in addition to the reflection coefficient and the natural $1/r$ decay of the spherical wave. The wave is also delayed in time by $c^{-1}d$ due to the propagation delay from the true origin to $(0, 0, d)$. This equation is at the core of our ultrasound imaging equation, presented in the next chapter.

10.4   Doppler Effect

The *Doppler effect*, also known as the *Doppler shift*, is the change in frequency of sound due to the relative motion of the source and receiver. The Doppler effect we most commonly experience is that of the siren of an emergency vehicle. The pitch of the siren is higher as the vehicle approaches and becomes lower as the vehicle passes and moves away.

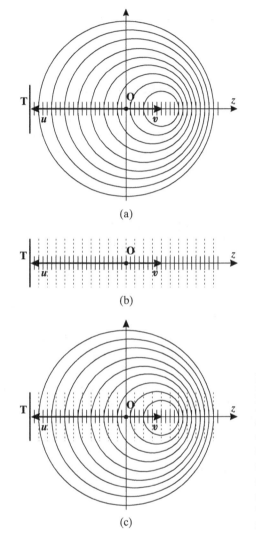

Figure 10.4
Basic ideas behind the
Doppler effect: (a) a moving
source O and stationary
observer **T**; (b) a stationary
source **T** and moving
observer O; and (c) a
stationary source/receiver **T**
and moving scatterer O,
which also acts as both a
receiver and a source.

The idea behind the Doppler effect for a moving source and stationary observer
is illustrated in Figure 10.4(a). Here, the object **O**, which is emitting sound and
is therefore the source, moves to the right, away from the transducer **T**, which is
receiving the sound and is therefore the observer. The frequency f_T of the sound
received by **T** can be derived by considering the separation of the wave crests that
are emanating from the moving source. If the source is producing a sinusoid with
frequency f_O, then the crests occur at the interval $1/f_O$—i.e., the period. In one
period, the crest will propagate a distance of c/f_O, where c is the speed of sound in
the medium. In this same interval the source moves a distance of v/f_O, where v is
the speed of the source, and at this moment another crest will be produced by the
source. So, the physical separation of crests—i.e., the wavelength—observed by **T**

is $(c + v)/f_O$. Therefore, **T** will receive (i.e., hear) an acoustic wave having frequency

$$f_T = \frac{c}{c+v} f_O . \tag{10.37}$$

This situation is generalized for an arbitrary observer velocity by recognizing that the frequency shift depends only on the component of source velocity in the direction of the observer (relative to where the source was when it generated the sound that is being received). Accordingly, the general expression for the sound frequency f_T observed by a stationary observer from a moving source with frequency f_O is

$$f_T = \frac{c}{c - v\cos\theta} f_O , \tag{10.38}$$

where θ is the angle between the vector pointing from the source to the receiver and the vector pointing from the source in its direction of its motion. Specifically, if x_T and x_O give the position of the transducer and object, respectively and if v is the velocity (vector) of **O**, then θ is the angle between vectors v and the vector

$$u = x_T - x_O , \tag{10.39}$$

as shown in Figure 10.4. In the situation depicted in Figure 10.4(a), $\cos\theta = -1$, and (10.38) reduces to (10.37).

The *Doppler frequency* f_D is the difference between the observed frequency and the source frequency. For the moving source situation in Figure 10.4(a), the Doppler frequency is given by

$$f_D = f_T - f_O . \tag{10.40}$$

Substituting (10.38) into (10.40) yields

$$f_D = \left(\frac{v\cos\theta}{c - v\cos\theta} \right) f_O , \tag{10.41}$$

after some algebra. Since v is ordinarily much smaller than c, this can be approximated as

$$f_D \approx \left(\frac{v\cos\theta}{c} \right) f_O . \tag{10.42}$$

The sign of the Doppler frequency indicates whether the source is moving toward or away from the observer. In the example shown in Figure 10.4(a), $\cos\theta = -1$, so $f_T < f_O$, and $f_D < 0$. Thus, negative Doppler frequencies mean that the source is moving away from the observer. This is consistent with our impression that the pitch of an emergency vehicle is lower as it moves away.

In medical ultrasound, the Doppler effect will be observed in pulse-echo mode, which requires that the above situation be extended. In pulse-echo mode, the transducer acts as both the source of the sound and the receiver of the Doppler-shifted echo returning from the object. Often this is done with a transducer that has two separate crystals located in the same housing, each performing one of these functions. The sound that returns to the transducer has been both *received* by the moving object and *retransmitted* by the moving object, thereby forming the echo that propagates back to the transducer. Thus, the object acts as both a moving receiver and a moving source, in order to create the echo. To fully understand what is happening, we therefore need to consider a second scenario in which there is a stationary source and a moving receiver; in this case, the transducer and object have switched their transmit and receive roles with respect to the previous scenario.

Following the preceding discussion, we now consider the transducer **T** as a stationary source, transmitting an acoustic wave with frequency f_S, and the object **O** as a moving receiver, as shown in Figure 10.4(b). The wavelength of the source in the medium is $\lambda = c/f_S$; but as far as the moving observer (the object) is concerned, the *acoustic speed* of the medium appears to be different than c because the object encounters wave crests and troughs at a lower or faster rate than a stationary observer would. In particular, the speed of sound appears to be lower if the object moves away from the transducer and higher if the object moves toward the transducer. Using the same definition of θ as before, it can be shown that the moving object observes a frequency

$$f_O = \frac{c + v\cos\theta}{c} f_S.$$ (10.43)

In Figure 10.4(b), $\cos\theta = -1$, so $f_O < f_S$, which is consistent with our intuition that the observed frequency should be less than the source frequency when the source and observer are moving apart.

Applying the fundamental definition of the Doppler frequency—the difference between the observed frequency and the source frequency—the Doppler frequency in the present scenario is given by

$$f_D = f_O - f_S.$$ (10.44)

Using (10.43), we find

$$f_D = \left(\frac{v\cos\theta}{c}\right) f_S.$$ (10.45)

The Doppler frequency is negative in the case depicted in Figure 10.4(b), but if the object is moving toward the transducer, the Doppler frequency will be positive.

In pulse-echo mode, the echo received by **T** will be shifted by *both* the effects of a moving receiver and a moving source; this will yield essentially twice the Doppler frequency than in either case alone. To see this, we start by assuming that the transducer **T** generates an acoustic wave having frequency f_S. The object **O**, moving with velocity v at an angle θ relative to u, receives a wave having frequency f_O, as in (10.43). The object reflects or scatters this received wave acting like a moving source with frequency f_O, and this yields a receive transducer frequency given by (10.38). Substituting (10.43) into (10.38) yields

$$f_T = \frac{c + v\cos\theta}{c - v\cos\theta} f_S, \tag{10.46a}$$

$$= \left(1 + \frac{2v\cos\theta}{c - v\cos\theta}\right) f_S, \tag{10.46b}$$

where the second equation follows after some algebra.

The Doppler frequency in pulse-echo mode is

$$f_D = f_T - f_S = \left(\frac{2v\cos\theta}{c - v\cos\theta} f_S\right). \tag{10.47}$$

Since ordinarily $c \gg v$, the Doppler frequency in this case is well-approximated by

$$f_D = \frac{2v\cos\theta}{c} f_S. \tag{10.48}$$

Again, for the example depicted in Figure 10.4(c), $\cos\theta = -1$ and therefore $f_D < 0$. To a good approximation, the Doppler frequency in the case of a pulse-echo scenario is twice that of either a moving source or moving receiver situation. In addition, a negative Doppler frequency means that the object is moving away from the transducer.

If the angle θ of movement of the scatterers were known and not equal to 90 degrees, then it is possible in principle to measure f_D and invert equation (10.48) in order to deduce the speed v of the scatterers. This is the principle of so-called Doppler-shift velocimeters. It is also possible to simply listen to $|f_D|$ as a function of time; this is the principle behind Doppler motion monitors, which are popular in fetal monitoring. Finally, it is also possible to measure f_D as a function of spatial position and to display its magnitude and sign in an image. This is the principle of Doppler imaging.

EXAMPLE 10.6

Suppose it is known that a 5 MHz transducer axis makes an angle of 30 degrees relative to the direction of motion of blood in a vessel.

Question   If the Doppler frequency is measured to be +500 Hz, what is the velocity of the blood? Is it moving toward or away from the transducer?

Answer   Since $f_D > 0$, the blood is moving toward the transducer. The velocity is given by

$$v = \frac{c f_D}{2 f_S \cos\theta}$$

$$= \frac{+500 \text{ Hz} \times 1540 \text{ m/s}}{2 \times 5 \times 10^6 \text{ Hz} \times \cos 30°}$$

$$= 0.0889 \text{ m/s}$$

10.5 Beam Pattern Formation and Focusing

In Section 10.3.5, we used a plane wave confined to a cylinder to study scattering. As it turns out, the acoustical energy of a real transducer is not confined to a cylinder but instead tends to spread out in a cone. The spatial distribution of the acoustic intensity of a transducer undergoing steady-state sinusoidal excitation is called the *field pattern* of the transducer.

In this section, we first study the field patterns created by the sinusoidal vibrations of a flat plate. This models flat transducers, which are elements of phased arrays (see Section 11.2.2), for example. We then study the field patterns created by focused transducers, which are composed of either curved plates or flat plates with lenses.

10.5.1 Simple Field Pattern Model

Treating the acoustic wave as if it were confined to the cylinder extended from the transducer's (flat) face is called the *geometric approximation*. It is only valid—and even then only approximately—very near to the face of the transducer. At a farther distance, the so-called *Fresnel approximation* holds, and beyond that the *Fraunhofer approximation* applies. Before developing a mathematical treatment of these approximations, it is useful to begin with a very simple model that captures many of the characteristics of field patterns from a geometric point of view.

Suppose a transducer has a "diameter" of D and is pointing down the z-axis, as shown in Figure 10.5. Then, the *geometric region* begins at the transducer face and extends in range out to $D^2/4\lambda$ (circular transducer) or $D^2/2\lambda$ (square transducer). The *Fresnel region* extends beyond this point out to range D^2/λ; the *Fraunhofer region* is beyond this range. In the Fraunhofer region, also called the *far field*, the beam is spreading. At range z, the beamwidth is approximately $w = \lambda z/D$. To summarize these observations, the beamwidth $w(z)$ of a flat transducer at range z is approximately given by

$$w(z) = \begin{cases} D, & z \le D^2/\lambda, \\ \lambda z/D, & z > D^2/\lambda \end{cases}. \tag{10.49}$$

This approximation ignores the "waist" occurring in the Fresnel region but is still very useful.

In the following section, we will develop the mathematics leading to the beamwidth formula in the far field or Fraunhofer region. As we will see, the width

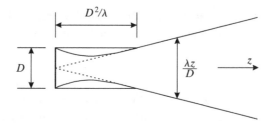

Figure 10.5
Simple field pattern
geometry.

$\lambda z/D$ is approximately where the signal strength drops off by a factor of two from the peak strength, which is on the z-axis. We will also see that as we go further off the z-axis, the acoustic energy continues to drop until it hits zero, and then it rises again, then drops, and so forth. The acoustic energy confined between the first zeros on either side of the z-axis is called the *main lobe*. The "spokes" of energy confined between the other zeros are called *side lobes*. Most of our discussion is directed at the main lobe. The side lobes are ignored in developing imaging concepts, but when they have significant energy, they can be a source of artifacts.

EXAMPLE 10.7

Consider a flat 1 cm transducer operating at 2 MHz in water.

Question   What is the approximate beamwidth at 5 cm range? What about at 20 cm range?

Answer   The wavelength in water at 2 MHz is $\lambda = c/f = 1484$ m/s$/2 \times 10^6$ s$^{-1} = 0.742$ mm. The range at which the beam transitions from geometric to far field is $z = D^2/\lambda = (10$ mm$)^2/0.742$ mm $= 134.8$ mm. Since 5 cm $= 50$ mm < 134.8 mm, the range of interest remains in the geometric region. Therefore, the beamwidth is approximately $w = D = 1$ cm. A range of 20 cm, however, exceeds the transition range. In this case, the beamwidth is approximately

$$w = \frac{\lambda z}{D}$$

$$= \frac{cz}{Df}$$

$$= \frac{1484 \text{ m/s} \times 0.2 \text{ m}}{2 \times 10^6 \text{ s}^{-1} \times 0.01 \text{ m}}$$

$$= 1.48 \text{ cm}$$ ∎

10.5.2 Diffraction Formulation

The *diffraction formulation* is used to derive a more accurate model for the field pattern of a transducer comprising a vibrating flat plate. This formulation requires a series of simplifying acoustical approximations, and also requires that we ignore the electronics, the electrical-mechanical interface to the transducer, and the mechanical-acoustical interface to the medium. Despite these simplifications, the diffraction formulation gives a reasonably accurate formula for the field pattern, especially in the far field, or Fraunhofer zone.

Narrowband Pulse   It is useful to model the pressure signal of a transmitted pulse as a narrowband pulse, which can be written as

$$n(t) = \text{Re}\{\tilde{n}(t)e^{-j2\pi f_0 t}\}, \tag{10.50}$$

where $\tilde{n}(t) = n_e(t)e^{j\phi}$ is the *complex envelope* and $n_e(t)$ is the *envelope*. The envelope of a narrowband pulse is the low-frequency (so-called baseband) signal that "rides"

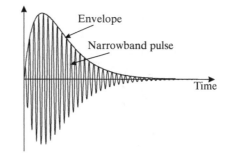

Figure 10.6
A narrowband pulse and its envelope.

the crests of the narrowband pulse. A narrowband pulse and its envelope are shown in Figure 10.6. We require that the envelope is long with respect to the period $1/f_0$ in order that $n(t)$ be considered narrowband. This is true for the narrowband pulse depicted in Figure 10.6, and is generally a good approximation in medical ultrasound imaging systems, although the number of high-frequency signal periods within the envelope is usually much fewer (on the order of three to five, typically).

Our analysis is made easier by presuming that the input to the acoustical system is the *complex signal*

$$n(t) = \tilde{n}(t)e^{-j2\pi f_0 t}. \tag{10.51}$$

With this model, the true input is given by

$$n(t) = \mathrm{Re}\{n(t)\}, \tag{10.52}$$

and the envelope of the input is given by

$$n_e(t) = |n(t)|, \tag{10.53}$$

where $|\cdot|$ gives the complex modulus. Since the acoustical system itself is a real physical system, the output must also be real. Therefore, the true output of the system driven by a complex input is just the real part of the complex output. Likewise, the envelope (baseband signal) of the output is just the complex modulus of the complex output. The primary reason that we use this complex formulation is that the output of an ultrasound system is typically the envelope, not the narrowband signal, and this formulation makes it easier to write the equations describing the envelope.

Received Signal with Field Pattern The geometry we consider is shown in Figure 10.7. The plate is assumed to be vibrating in the z direction, and each point $(x_0, y_0, 0)$ on the plate is assumed to produce an independent acoustic wave. The linear superposition of all these waves comprises the field pattern of the flat plate. Because the plate vibrates in the z direction, each point on the plate acts like an acoustic dipole rather than a monopole (spherical source). Accordingly, the acoustic wave produced by each point is not a spherical wave but is instead given by

$$p(x, y, z, t) = \frac{z}{r_0^2}n(t - c^{-1}r_0), \tag{10.54}$$

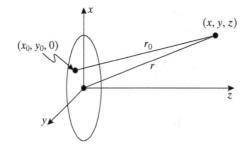

Figure 10.7
Geometry for field pattern
analysis.

where $r_0 = \sqrt{(x - x_0)^2 + (y - y_0)^2 + z^2}$ is the distance between the point on the plate and (x, y, z), an arbitrary point in space. A complex narrowband excitation $n(\cdot)$ is assumed; hence, the true signal is recovered by taking the real part, and the envelope is recovered by taking the modulus. On the z-axis, this pressure wave has exactly the form of a spherical wave propagation pattern, but at 90 degrees away from the source, the pressure is zero (because $z = 0$).

The total pressure at (x, y, z) is the superposition of all (dipole) sources on the transducer face

$$p(x, y, z; t) = \int_{-\infty}^{\infty}\int_{-\infty}^{\infty} s(x_0, y_0)\frac{z}{r_0^2}n(t - c^{-1}r_0)dx_0dy_0,$$

where $s(x, y)$ is the transducer face *indicator function*, given by

$$s(x, y) = \begin{cases} 1, & (x, y) \text{ in face} \\ 0, & \text{otherwise.} \end{cases} \tag{10.55}$$

Now suppose that a spherical wave scatterer of strength $R(x, y, z)$ is at (x, y, z). Then, a spherical wave will be sent back to the transducer, and the pressure at face point (x_0', y_0') will be

$$p_s(x_0', y_0'; t) = R(x, y, z)\frac{1}{r_0'}p(x, y, z; t - c^{-1}r_0'), \tag{10.56}$$

where r_0' is the distance from (x, y, z) to $(x_0', y_0', 0)$. Since the transducer is sensitive only to the dipole pattern, the response at each point on the face must be weighted by the dipole pattern. Accordingly, the received electrical (still modeled as complex) waveform due to a single scatterer at (x, y, z) is

$$r(x, y, z; t) = K \int_{-\infty}^{\infty}\int_{-\infty}^{\infty} s(x_0', y_0')\frac{z}{r_0'}p_s(x_0', y_0'; t)dx_0'dy_0', \tag{10.57}$$

where K is an arbitrary gain factor accounting for the transmit and receive sensitivity of the transducer and any preamplification hardware. Substituting all previous expressions gives

$$r(x, y, z; t) = KR(x, y, z) \tag{10.58}$$

$$\times \int_{-\infty}^{\infty}\int_{-\infty}^{\infty} s(x'_0, y'_0) \frac{z}{r'^2_0}$$

$$\times \int_{-\infty}^{\infty}\int_{-\infty}^{\infty} s(x_0, y_0) \frac{z}{r^2_0} n(t - c^{-1}r_0 - c^{-1}r'_0) dx_0 dy_0 dx'_0 dy'_0 .$$

Plane Wave Approximation   At this point, we use the first of a series of approximations. The *plane wave approximation* maintains that the envelope of the excitation pulse arrives at all points in a given z-plane simultaneously. This statement can be written mathematically as

$$n(t - c^{-1}r_0 - c^{-1}r'_0) \approx \tilde{n}(t - 2c^{-1}z)e^{-j2\pi f_0(t - c^{-1}r_0 - c^{-1}r'_0)} , \tag{10.59}$$

which can also be written as

$$n(t - c^{-1}r_0 - c^{-1}r'_0) \approx n(t - 2c^{-1}z)e^{jk(r_0 - z)}e^{jk(r'_0 - z)} . \tag{10.60}$$

Using this approximation, the receive equation (10.58) simplifies so that the quadruple integral separates into two identical double integrals. Defining the *field pattern* as

$$q(x, y, z) = \int_{-\infty}^{\infty}\int_{-\infty}^{\infty} s(x_0, y_0) \frac{z}{r^2_0} e^{jk(r_0 - z)} dx_0 dy_0 , \tag{10.61}$$

the complex received signal for a single scatterer located at (x, y, z) is

$$r(x, y, z; t) = KR(x, y, z)n(t - 2c^{-1}z)[q(x, y, z)]^2 . \tag{10.62}$$

Assuming superposition holds, the total response for a distribution of scatterers is

$$r(t) = \int_0^{\infty}\int_{-\infty}^{\infty}\int_{-\infty}^{\infty} r(x, y, z; t) dx \, dy \, dz \tag{10.63}$$

$$= \int_0^{\infty}\int_{-\infty}^{\infty}\int_{-\infty}^{\infty} KR(x, y, z)n(t - 2c^{-1}z)[q(x, y, z)]^2 dx \, dy \, dz . \tag{10.64}$$

For completeness, we include a round-trip attenuation factor, yielding the *basic pulse-echo signal equation*,

$$r(t) = K \int\int\int R(x, y, z)n(t - 2c^{-1}z)e^{-2\mu_a z}[q(x, y, z)]^2 dx \, dy \, dz . \tag{10.65}$$

The importance of this equation as the basis of an imaging equation will become clear in the next chapter. For now, it is important to continue examining the field pattern to see how it might be approximated in order to better understand it and to make this triple integral more tractable.

Paraxial Approximation   The *paraxial approximation* assumes that we are primarily interested in the pattern near the transducer axis, in which case $r_0 \approx z$. This approximation can be applied only to the amplitude terms, not the phase terms. Applying this approximation to (10.61) yields

$$q(x, y, z) \approx \frac{1}{z} \iint s(x_0, y_0) e^{jk(r_0 - z)} dx_0 dy_0 . \qquad (10.66)$$

Notice that this is the same result that we would get assuming that all the elements of the transducer acted as spherical wave generators and receivers instead of dipoles.

Fresnel Approximation   The *Fresnel approximation* simplifies the phase term in (10.66) by first noting that

$$r_0 = \sqrt{(x - x_0)^2 + (y - y_0)^2 + z^2}$$

$$= z \sqrt{1 + \frac{(x - x_0)^2}{z^2} + \frac{(y - y_0)^2}{z^2}} . \qquad (10.67)$$

Then, if z is large enough, two terms of the binomial expansion are sufficient to approximate the square root as

$$r_0 \approx z \left[1 + \frac{1}{2} \left(\frac{(x - x_0)^2}{z^2} + \frac{(y - y_0)^2}{z^2} \right) \right]$$

$$\approx z + \frac{(x - x_0)^2}{2z} + \frac{(y - y_0)^2}{2z} . \qquad (10.68)$$

Applying this to the phase term in (10.66) yields

$$q(x, y, z) \approx \frac{1}{z} \int_{-\infty}^{\infty} \int_{-\infty}^{\infty} s(x_0, y_0) e^{jk\left(\frac{(x-x_0)^2}{2z} + \frac{(y-y_0)^2}{2z} \right)} dx_0 dy_0 . \qquad (10.69)$$

This is in the form of a convolution in x and y; hence, we write the *Fresnel beam pattern* as

$$q(x, y, z) = \frac{1}{z} s(x, y) ** e^{jk(x^2 + y^2)/2z} , \qquad (10.70)$$

where $**$ denotes 2-D convolution [see (2.39) and (2.40)].

Substituting (10.70) into (10.65),

$$r(t) = K \int_0^{\infty} \int_{-\infty}^{\infty} \int_{-\infty}^{\infty} R(x, y, z) n(t - 2c^{-1}z) \frac{e^{-2\mu_a z}}{z^2} \left[s(x, y) ** e^{jk(x^2 + y^2)/2z} \right]^2 dx \, dy \, dz . \qquad (10.71)$$

If n(t) is sufficiently short, however, both the attenuation and $1/z^2$ factors may be moved outside the integral by setting $z = ct/2$. This gives a received (complex) signal

$$r(t) = K\frac{e^{-\mu_a ct}}{(ct)^2} \int_0^\infty \int_{-\infty}^\infty \int_{-\infty}^\infty R(x, y, z)n(t - 2c^{-1}z)\left[s(x, y) ** e^{jk(x^2+y^2)/2z}\right]^2 dx\,dy\,dz,$$
(10.72)

where the constant 4 has been absorbed by K.

EXAMPLE 10.8

An ultrasound transducer is placed in the x-y plane, pointing down in the z-axis. The transducer face is of dimension 1 mm × 10 mm, as shown in Figure 10.8. The transducer is working at 2 MHz, and the speed of sound is 1540 m/s.

Question   At range of 15 cm, where are the peaks of the first sidelobes along x- and y-axes?

Answer   The wavelength of the ultrasound wave is

$$\lambda = c/f = 1540\text{m/s}/2 \times 10^6\text{Hz} = 0.77\,\text{mm}.$$

For the ultrasound field in x-z plane, the transducer has width 1 mm, so the far field approximation holds for $z \geq D^2/\lambda = 1.2$ mm. For the ultrasound field in y-z plane, the transducer has width 10 mm, so the far field approximation holds for $z \geq D^2/\lambda = 129.8$ mm= 12.98 cm. So at range of 15 cm, the far field approximation holds for both directions. The widths of the main lobes along the x- and y-axes are

$$w_x = \lambda z/D_x = 11.55\,\text{cm}\quad\text{and}$$

$$w_y = \lambda z/D_y = 1.16\,\text{cm}.$$

So the peaks of the first sidelobes along the x- and y-axes are approximately at

$$p_x = \pm 1.5 w_x = 17.32\,\text{cm}$$

$$p_y = \pm 1.5 w_y = 1.73\,\text{cm}.\qquad\blacksquare$$

Fraunhofer Approximation   The Fresnel approximation for r_0 in (10.68) can be expanded as

$$r_0 \approx z - \frac{xx_0}{z} - \frac{yy_0}{z} + \frac{x^2 + y^2}{2z} + \frac{x_0^2 + y_0^2}{2z}.$$
(10.73)

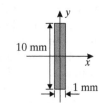

Figure 10.8
A transducer in the x-y plane, pointing in the z-axis.

Using this form in (10.66) yields the following alternate expression for the Fresnel field pattern,

$$q(x, y, z) = \frac{1}{z} e^{jk(x^2+y^2)/2z} \int_{-\infty}^{\infty} \int_{-\infty}^{\infty} s(x_0, y_0) e^{jk(x_0^2+y_0^2)/2z} \, e^{-jk\left(\frac{x_0 x}{z} + \frac{y_0 y}{z}\right)} dx_0 dy_0 \,.$$

(10.74)

To see how to further approximate $q(x, y, z)$, we define

$$D = 2\sqrt{\max_{x_0,y_0 \in \text{face}} (x_0^2 + y_0^2)} \,,$$

(10.75)

which is essentially the diameter (or maximum lateral dimension) of the transducer. If $z \geq D^2/\lambda$, then the term $\exp[jk(x_0^2 + y_0^2)/2z]$ is approximately 1. This approximation is called the *Fraunhofer approximation*, and only applies in the far field or Fraunhofer region. Under this approximation, the double integral in (10.74) is revealed as the Fourier transform of $s(x, y)$, evaluated at the spatial frequencies $u = x/\lambda z$ and $v = y/\lambda z$. Therefore, the Fraunhofer approximation to the field pattern is written as

$$q(x, y, z) \approx \frac{1}{z} e^{jk(x^2+y^2)/2z} S\left(\frac{x}{\lambda z}, \frac{y}{\lambda z}\right), \quad z \geq D^2/\lambda \,,$$

(10.76)

where

$$S(u, v) = \int_{-\infty}^{\infty} \int_{-\infty}^{\infty} s(x, y) e^{-j2\pi(ux+vy)} dx \, dy \,.$$

(10.77)

Using (10.76) in (10.65), and making similar approximations as was done in the Fresnel approximation, yields

$$r(t) = K \frac{e^{-\mu_a ct}}{(ct)^2} \int_0^\infty \int_{-\infty}^{\infty} \int_{-\infty}^{\infty} R(x, y, z) n(t - 2c^{-1}z) e^{jk(x^2+y^2)/z} S^2\left(\frac{x}{\lambda z}, \frac{y}{\lambda z}\right) dx \, dy \, dz \,.$$

(10.78)

The Fraunhofer approximation holds well for ranges greater than about D^2/λ, where D is the maximum lateral dimension of the transducer. For example, a 1 cm transducer operating at 2 MHz hits the far field at a range of about 13 cm. If the transducer is square, then the width of the beam is about $\lambda z/D$ in the far field. Therefore, at $z = D^2/\lambda$, the width of the field is D, equal to the width of the transducer. Doubling the range doubles the width of the beam, so that (in this example) at 26 cm, our beamwidth is 2 cm. Suppose we desire a lateral resolution of about 0.5 cm. Roughly, the geometric region holds until about D^2/λ—that is, up to the onset of the far field. Therefore, a transducer with $D = 0.5$ cm satisfies our criteria up to about 3 cm. Unfortunately, at 6 cm range, the beamwidth has increased to about 1 cm; at 12 cm range, the beamwidth has gone up to about 2 cm; and at 24 cm the beam width has gone to about 4 cm. Therefore, although the smaller transducer has better resolution up to its own far field than a larger transducer, the far field comes much sooner (i.e., at a distance much closer to the transducer), and the beam exceeds the dimensions of the larger transducer fairly quickly.

10.5.3 Focusing

Virtually all transducers used in medical ultrasound are focused to at least some degree. Here, *focusing* means to "shape" the beam into a narrower beam than is achieved using a flat vibrating plate. Focusing can be accomplished by manufacturing the transducer crystal in a curved shape, by applying a lens to a flat crystal, or by electronic focusing using multiple transducer crystals arranged either in a linear array or a set of concentric rings. Electronic focusing will be described in Chapter 11, as will the materials and construction of ultrasound transducers. In this section, we explore the physics of focusing using curved vibrators or lenses.

Consider the geometry shown in Figure 10.9. Here, the curved surface has radius d, and an acoustic pulse emitted simultaneously from the surface would arrive at the point $(0, 0, d)$ simultaneously. The focal depth is therefore designed to be at the range $z = d$. In order to analyze the field pattern for this geometry, it is convenient to assume that the actual pulse (or waveform) is generated from a flat plate at $z = 0$, and that we can control the timing of the pulses as a function of position (x_0, y_0). Our goal is to generate the transmit waveform first at the farthest point $(x_m, y_m, 0)$ (or set of points) and last at the origin in such a way that a round trip from the flat transducer to the focal point and back again takes the same amount of time across the entire face. This simulates the curved transducer as well as flat transducers with lenses, as we shall see.

It can be shown from the geometry (see Problem 10.16) that the time delay required for the pulse generated at point $(x_0, y_0, 0)$ is approximately

$$\tau \approx \frac{1}{dc}[r_m^2 - (x_0^2 + y_0^2)], \tag{10.79}$$

where $r_m^2 = x_m^2 + y_m^2$. Applying this time delay to the complex signal, and carrying out the plane wave and Fresnel approximations using analogous steps as above, yields the following Fresnel field pattern at the focal distance $z = d$:

$$q(x, y, d) = \frac{1}{d}e^{jk(x^2+y^2)/2d}S\left(\frac{x}{\lambda d}, \frac{y}{\lambda d}\right). \tag{10.80}$$

The field pattern of (10.80) is identical to that of the Fraunhofer field for a flat transducer (10.76), except that z is replaced by d. In practical terms, this means that

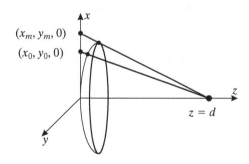

Figure 10.9
Geometry for the analysis of field patterns from focused transducers.

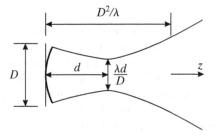

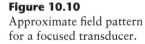

Figure 10.10
Approximate field pattern
for a focused transducer.

a relatively large transducer can produce a relatively narrow beam at the focal depth, as depicted in Figure 10.10. We will see in the next chapter that a narrower beam produces improved spatial resolution. The penalty for this improved resolution is that the beam diverges more rapidly than with an unfocused transducer. Therefore, although the resolution is better at or around the focal depth, it can be considerably worse at smaller and larger distances than with an unfocused transducer.

10.6 Summary and Key Concepts

Ultrasound imaging is based on the use of ultrasound, whose behavior is characterized by the 3-D wave equation. Ultrasound imaging is of interest because it is truly noninvasive, and the systems are typically quite portable. In this chapter, we presented the following key concepts that you should now understand:

1. *Ultrasound* is sound with frequencies above 20 kHz.
2. Ultrasound systems generate and detect ultrasound via a *transducer*, which converts electrical signals to acoustic signals and vice versa.
3. Ultrasound images are based on the *reflection* and *scattering* of ultrasound by body tissues; the returning *echoes* are detected and displayed.
4. *Acoustical waves* are pressure waves that propagate through materials via *compression* and *expansion* of the material; these waves can have different "geometries" or "patterns."
5. A *plane wave* is the most common wave pattern, which is represented by a simple *geometric approximation*, the *Fresnel approximation*, or the *Fraunhofer approximation*, depending on the distance from the transducer.
6. Plane waves are *reflected*, *refracted*, or *attenuated/transmitted* by/through *plane interfaces*.
7. The *Doppler effect* is the change in frequency of sound due to the relative motion of the source and receiver.

Bibliography

Kremkau, F. W. *Diagnostic Ultrasound: Principles and Instruments*, 5th ed. Philadelphia, PA: W. B. Saunders, 1998.

Macovski, A. *Medical Imaging Systems*. Englewood Cliffs, NJ: Prentice Hall, 1983.

Pierce, A. D. *Acoustics: An Introduction to Its Physical Principles and Applications*. New York: McGraw-Hill, 1981.

Wells, P. N. T., *Biomedical Ultrasonics*. NY: Academic Press, 1977.

Problems

The Wave Equation

10.1 Show that the function $w_1(z, t) = \xi(z - ct)$, where $\xi(\tau)$ is a twice differentiable function, is a solution of the 1-D wave equation (10.6). Moreover, show that the same is true for the function $w_2(z, t) = \xi(z - ct) + \xi(z + ct)$. What is the physical meaning of the wave function $w_2(z, t)$?

10.2 Show that the wavelength of a sinusoidal plane wave modeled by (10.8) is given by $\lambda = 2\pi/k$.

10.3 Consider the situation in Example 10.2 and let $\tau_2 = \tau_1 = 5$ μsec.

   (a) At what time does the peak of the wave with maximum pressure hit the interface?

   (b) After the wave hit the interface, a backward-traveling wave will be generated. Write down an expression for the backward-traveling wave;

   (c) When will the peak of the backward traveling wave arrive at the transducer face?

10.4 Derive the spherical wave equation (10.13) for a source at the origin. Start with the full 3-D wave equation (10.4) and assume that the amplitude of particle displacement depends only on the radius r and on time t.

10.5 Show that the function $w(r, t) = \xi(r - ct)/r$, where $\xi(\tau)$ is a twice differentiable function, is a solution of the spherical wave equation (10.13).

10.6 Prove the validity of the general solution to the spherical wave equation

$$p(r, t) = \frac{1}{r}f(t - c^{-1}r) + \frac{1}{r}g(t + c^{-1}r),$$

by direct substitution into (10.13).

Wave Propagation

10.7 Prove that for a plane wave in which $v(0, t) = \text{Re}\{Ve^{j\omega t}\}$ and $p(0, t) = \text{Re}\{Pe^{j\omega t}\}$, that the average power in the acoustic wave at $x = 0$ is $I_{av} = (1/2)\text{Re}\{VP^*\}$ where $*$ denotes complex conjugate.

10.8 (a) Solve Equations (10.25) and (10.26) to find the expressions for pressure reflectivity R and pressure transmittivity T given by Equations (10.27) and (10.28), respectively.

   (b) Derive the expressions for intensity transmittivity and reflectivity given in Equations (10.29) and (10.30), respectively.

10.9 An acoustic dipole can be modeled by two point sources (unipoles) residing close together but vibrating exactly out of phase. Suppose point source 1 is located at $(0, 0, -d)$ and point source 2 is located at $(0, 0, d)$, where d is very small. Further suppose that source 1 is generating the narrowband signal $f(t)$, leading to a spherical pressure wave given by $p(r_1, t) = f(t - c^{-1}r_1)/r_1$ and source 2 is generating the narrowband signal $-f(t)$, leading to a spherical pressure wave given by $p(r_2, t) = -f(t - c^{-1}r_2)/r_2$, where the two radii r_1 and r_2 are measured from their respective sources.

   (a) Show that the measured pressure in the far field is approximately $p(r, t) = zf(t - c^{-1}r)/r^2$.

   (b) Sketch isopressure lines in the x, z plane for the above field pattern.

10.10 Consider an outward propagating spherical wave.

   (a) Write down a general expression of an attenuated spherical wave whose source is located at the origin.

   (b) If there is a small target located at (x, y, z), derive an expression for the scattered wave.

   (c) Generalize the above result to a spherical wave generated at an arbitrary location (x_0, y_0, z_0).

10.11 (a) Derive the intensity reflectivity and transmittivity from the pressure reflectivity and transmittivity.

   (b) Show that $T - R = 1$.

   (c) Find a concise relationship between T_I and R_I.

Doppler Effect

10.12 Consider a stationary source that generates a plane wave with frequency f_0. Assuming that a receiver is moving toward the source with speed v, and that the speed of sound in the media is c, prove that the frequency observed is $f_R = (c - v)f_0/c$. What happens if the receiver moves away from the source?

10.13 Using the above derivation, discuss the Doppler effects in the following situations when the receiver moves both towards and away from the source:

   (a) $c > v$.

   (b) $c = v$.

   (c) $c < v$.

Ultrasound Field Pattern

10.14 Suppose the Fourier transform of a real signal $n_e(t)$ is $N_e(\omega)$. Find the Fourier transform of $\tilde{n}(t) = n_e(t)e^{j\phi}$ and $n(t) = \text{Re}\{\tilde{n}(t)e^{-j2\pi f_0 t}\}$.

10.15 An ultrasound imaging system is equipped with two square transducers. One operates at 5 MHz, and the other one operates at 12 MHz. The 5 MHz transducer is 2.0 cm by 2.0 cm and the 12 MHz one is 0.4 cm by 0.4 cm. The imaging system is tested in a medium having a speed of sound of 1560 m/s at both frequencies.

(a) What are the absorption coefficients $\alpha_{5\text{MHz}}$ and $\alpha_{12\text{MHz}}$ of the medium at the frequencies of the transducers?

(b) What ranges are considered as far field for each transducer?

10.16 A transducer can be focused by curving the crystal to conform to the shape of a sphere, as shown in Figure P10.1. To analyze the field pattern, however, it is convenient to assume that the transducer is flat but that we can independently control the time delay of each element on its face. Assume that the point $(x_m, y_m, 0)$ is the point farthest out on the flat face.

(a) Show that the time delay for a point $(x_0, y_0, 0)$ on the flat face is approximately

$$\tau \approx \frac{1}{dc}[r_m^2 - (x_0^2 + y_0^2)],$$

where $r_m^2 = x_m^2 + y_m^2$ and d is the desired focal depth.

(b) Use the steady-state approximation and show that at $z = d$ the Fresnel field pattern is

$$q(x, y, d) = \frac{1}{d}e^{jk(x^2+y^2)/2d}S\left(\frac{x}{\lambda d}, \frac{y}{\lambda d}\right),$$

where $S(u, v)$ is the 2-D Fourier transform of $s(x, y)$.

(c) Discuss the merits of focusing.

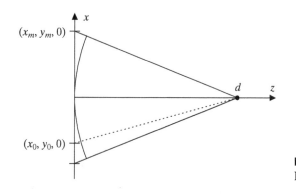

Figure P10.1
Problem 10.16.

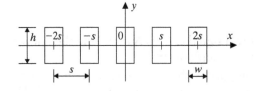

Figure P10.2
Problem 10.16.

10.17 A portion of a linear array is shown in Figure P10.2. Note that $h > w$. The array is composed of five transducers (called elements), whose faces are contained in the x, y-plane and vibrate in the z direction. Assume the far-field (Fraunhofer) approximation in all of the parts below.

(a) What is the far-field pattern $q_0(x, y, z)$ of the central element taken by itself?

(b) Find the range z_0 at which the first zeros of this pattern (in the x-direction) coincide with $x = -s$ and $x = +s$.

(c) What condition must the separation s satisfy in order that z_0 is in the far-field of the central element?

(d) Find the far-field pattern $q(x, y, z)$ of the five elements operating in unison.

(e) What is the beamwidth (out to the first zeros) of this pattern at range z_0 in both the x and y directions? (Think carefully about this one before you proceed blindly ahead; it's easier than it appears at first glance.)

Ultrasound Imaging Systems

11.1 Introduction

In this chapter, we study the principles of ultrasonic imaging systems. After plane film x-ray systems, these are the most widely used medical imaging systems in the world. In part, this is because at diagnostic intensities, ultrasound poses no known risk to the patient. Another factor is that these are among the least expensive medical imaging systems and are portable—easily moved from bedside to operating room.

11.2 Instrumentation

A block diagram of a typical ultrasound imaging system is shown in Figure 11.1. The vast majority of medical ultrasound imaging systems use the same transducer for both generation and reception of ultrasound; this is the so-called *pulse-echo mode* of operation. The transducer is coupled to the body using an "acoustic gel," and a brief pulse-like acoustic wave is generated. This wave propagates into the body, where it encounters reflecting surfaces and small scatterers. These objects reflect or scatter the sound, a part of which returns to the transducer. The transducer then converts the acoustic wave sensed at its face to an electrical signal that can be amplified, stored, and displayed. In this section, we will describe each of the components required to carry out this basic imaging paradigm. This sets the stage for the development of an imaging equation, the analysis of image quality, and an introduction to pulsed Doppler imaging systems in subsequent sections.

11.2.1 Ultrasound Transducer

Transducer Materials Medical transducers use *piezoelectric crystals* to both generate and receive ultrasound. These crystals have the property that an induced electric field produces a *strain* (mechanical displacement), which in turn causes an acoustic wave. They also satisfy the reciprocal property that a mechanical displacement

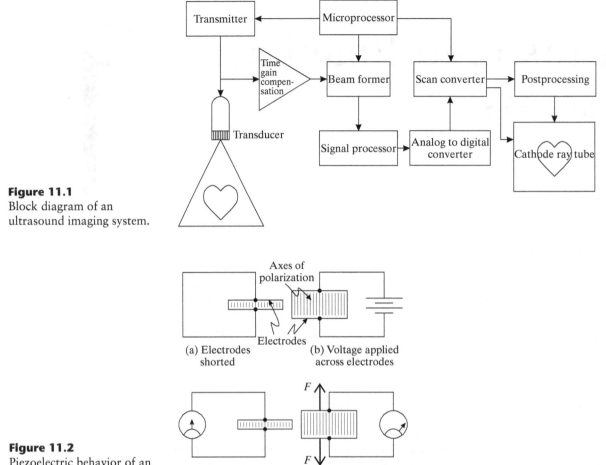

Figure 11.1
Block diagram of an
ultrasound imaging system.

Figure 11.2
Piezoelectric behavior of an
ultrasonic transducer crystal.

creates an electric potential, which means that they can also sense an acoustic wave. These concepts are illustrated in Figure 11.2.

Lead zirconate titanate, or PZT, is the piezoelectric material used in nearly all medical ultrasound transducers. It is a ceramic ferroelectric crystal exhibiting a strong piezoelectric effect. These crystals can be manufactured in nearly any shape, and their axes of polarization (see Figure 11.2) can be oriented in nearly any direction, as shown in Figure 11.3. The most common transducer shapes are the circle, for single crystal transducer assemblies, and the rectangle, for multiple transducer assemblies such as those found in linear and phased arrays, as described later.

The ideal transducer material, when used in pulse-echo mode, should be both an efficient producer of and sensitive receiver of ultrasonic waves. The *transmitting constant d* of a transducer is the strain produced by a unit electric field and has units of meters per volt. The *receiving constant g* is the potential produced by a unit stress and has units of volt-meters per Newton. For PZT, $d = 300 \times 10^{-12}$ m/V and $g =$

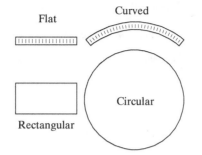

Figure 11.3
Shapes and axial orientations for PZT.

2.5×10^{-2} V/(N/m). By comparison, for quartz (a natural piezoelectric material), $d = 2.3 \times 10^{-12}$ m/V and $g = 5.8 \times 10^{-2}$ V/(N/m). This implies that quartz is two orders of magnitude less efficient in production of ultrasound as PZT, but comparable in its reception efficiency. Polyvinylidene fluoride (PVDF), a polymer film, also has piezoelectric properties and is commonly used as a probe when measuring the acoustic properties of other transducers. For PVDF, $d = 15 \times 10^{-12}$ m/V and $g = 14 \times 10^{-2}$ V/(N/m). Clearly, PVDF has poor efficiency as a transmitter of ultrasound but is much more efficient on reception than either PZT or quartz.

Resonance   Transducer crystals exhibit *resonance*—that is, they tend to vibrate sinusoidally after electrical excitation has ended. The frequency of this vibration is called the *fundamental resonant frequency f_T* of a transducer. In most systems, different transducers having resonant frequencies (potentially) in the range 1–20 MHz can be "plugged" into the ultrasound scanner. The difference in resonant frequency between the different transducers has a profound influence on image quality, as we shall see.

The resonant frequency of a transducer is largely determined by the thickness of its piezoelectric crystal. When the front face of the transducer is moved forward (by the electrically induced strain, for example), an acoustic wave is initiated forward into the medium (perhaps a human body) and backward into the transducer crystal itself. The backward-traveling wave will hit the back face of the transducer and reflect toward the front face again. Resonance is set up when the returning wave strikes the front face at the time when the wave reaches a "crest," thus reenforcing the wave. This condition occurs when

$$f_T = \frac{c_T}{2d_T}, \tag{11.1}$$

where c_T and d_T are the speed of sound in the transducer and the thickness of the transducer, respectively. Since $\lambda_T = c_T/f_T$, the resonance condition is equivalently given by

$$\lambda_T = 2d_T .$$

EXAMPLE 11.1
The thickness of the piezoelectric crystal determines the resonant frequency of a transducer.

Question   Consider a transducer made of a PZT crystal, which has a speed of sound of $C_T = 8000$ m/s. If we want the transducer to work at a frequency of 10 MHz, what should the thickness of the crystal be?

Answer   The relation between the thickness of the crystal and the resonant frequency is given by

$$f_T = \frac{c_T}{2d_T}.$$

where $f_T = 10$ MHz and $c_T = 8000$ m/s. So the thickness of the crystal should be

$$d_T = \frac{c_T}{2f_T} = 0.4 \, \text{mm}.$$

∎

Medical transducers are usually *shock excited*, meaning the electrical signal resembles an impulse—i.e., they have a very large peak voltage and a very short duration. Once excited, the transducer will resonate until the in-transducer wave loses energy, which causes a damping of the acoustic wave. In-transducer acoustic energy is lost in three ways: to the body, out the back of the transducer, and due to absorption within the crystal. The reflection coefficient between PZT and tissue is large, and the absorption coefficient of PZT is small. Therefore, to remove excessive vibrations, the transducer is backed with an epoxy material with impedance nearly matched to PZT, but with a high absorption coefficient. All factors are controlled so that the outgoing acoustic pulse has a well-defined center frequency (important in imaging, as we will see), but damps out after about three to five cycles. A typical transmit pulse is shown in Figure 11.4.

11.2.2 Ultrasound Probes

Single-Element Probes   The simplest transducer assembly, also called an *ultrasound probe*, is shown in Figure 11.5. This single-element assembly has a field pattern that is well-modeled using the vibrating plate model developed in Section 10.5.2. This

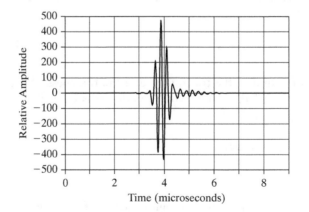

Figure 11.4
Typical transmit pulse from an ultrasonic transducer.

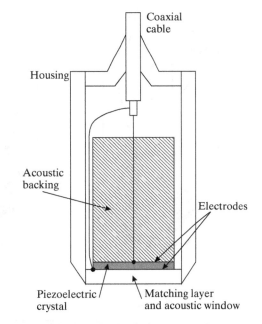

Coaxial
cable

Housing

Acoustic
backing

Electrodes

Piezoelectric
crystal

Matching layer
and acoustic window

Figure 11.5
A simple ultrasound probe.

basic design accommodates either a lens or a curved crystal for focusing. For imaging, the ultrasound beam must be *steered* (also called *scanned* or *swept*) within the body, typically within a plane. Early ultrasound systems used handheld single-element probes that were manually scanned within a body cross-section. Modern systems, however, use either mechanical or electrical means to scan the beam. This is what allows real-time imaging.

Mechanical Scanners   Mechanical scanning is accomplished by rocking or rotating a transducer crystal or set of crystals within the transducer assembly, as shown in Figure 11.6. In these designs, a transducer crystal is rapidly pulsed as it moves through the sector being imaged. Each acoustical pulse goes through an acoustical window and then propagates according to the speed of sound and the transducer's field pattern, as studied in the previous chapter. The echoes from any one scan line are received and processed before the transducer is pulsed again. In the *rocker* (or *wobbler*) design, shown in Figure 11.6(a), the transducer continues to travel through the same sector in a repeating fashion first clockwise, then counterclockwise, and so forth. In the *rotating* design shown in Figure 11.6(b), a new transducer is "switched in" as it enters the sector, so the scan sequence is always counterclockwise. Regardless of the specific design, the field of view for mechanical scanners is always shaped like a slice of pie.

Electronic Scanners   Transducers assemblies having multiple elements can be electronically scanned in order to sweep the field of view. The basic arrangement of the elements in these assemblies is linear, as shown in Figure 11.7. Each element is rectangular, and is focused in the longer dimension using a lens (or curved elements).

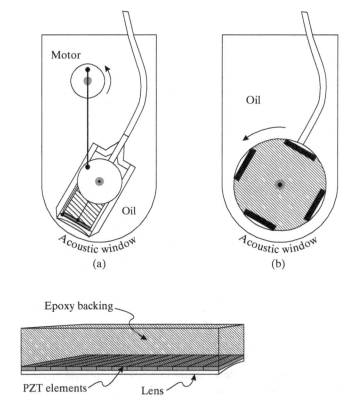

Figure 11.6
Mechanical scanner probe designs.

(a)   (b)

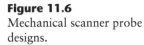

Figure 11.7
Basic arrangement of transducer elements in linear and phased arrays.

Epoxy backing

PZT elements   Lens

The width of each element and the attached electronics determine the manner in which the array is used to image the body. In particular, when the elements have widths on the order of a wavelength and are simply electronically grouped together making several elements appear as one, then the array is called a *linear array probe*. On the other hand, when the elements have widths on the order of a quarter wavelength and the timing of the firing of the elements are electronically controlled in order to steer and focus the beam, then the array is called a *phased array probe*. We now expand on each of these basic designs in more detail, describing the basic operation, the required electronics, and several variations on these basic designs.

EXAMPLE 11.2
Consider a mechanical scanner with rotating design. The angle of the field of view is 90° (see Figure 11.8).

Question Suppose N pulse-echo experiments are to be acquired for each transducer element over 90° window. If no echo is received from further than 15 cm in range, and the speed of sound is $c = 1540$ m/s, what is the maximum rate of revolution of the transducer?

Answer At a range of 15 cm, it takes

$$T = 2 \times \frac{15\,\text{cm}}{1540\,\text{m/s}} = 195\,\mu\text{s}$$

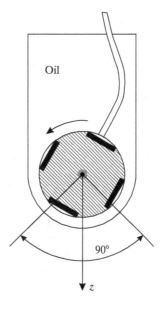

Figure 11.8
A mechanical scanner probe with rotating design.

for the transducer to receive the echo. During this period of time, the scanner probe cannot rotate more than 90°, otherwise the transducer leaves the window before it receives the echo. So the maximum rate of revolution is

$$r = \frac{1/4}{T} = 1283 \text{ revolutions per second.}$$ ∎

11.3 Pulse-Echo Imaging

In the following sections, we develop a mathematical framework for basic ultrasound imaging. We rely on simplifying assumptions throughout, generally making bold assumptions at the outset and then refining them as our understanding grows. In the following sections, we will explore two very important augmentations to basic imaging: phased arrays (and electronic focusing) and Doppler imaging.

An ideal ultrasound imaging system would reconstruct and display $R(x, y, z)$, the spatial distribution of reflectivity. This is not possible for two reasons. First, the transducer's impulse response function, defined in part by its field pattern, will blur out the true reflectivity. Second, the use of envelope detection will create a noiselike artifact called *speckle*; this accounts for the somewhat "blotchy" appearance of ultrasound images. There are other reasons, not the least of which is that some objects reflect, rather than scatter, ultrasound; thus, they do not even have the scattering property that we have called reflectivity.

11.3.1 The Pulse-Echo Equation

Analysis of ultrasound imaging systems begins with the pulse-echo received signal equation developed in Chapter 10. We derived (10.72) using the Fresnel approximation and (10.78) using the Fraunhofer approximation. Either equation can be written more generally as

$$r(t) = K \frac{e^{-\mu_a ct}}{(ct)^2} \int_0^\infty \int_{-\infty}^\infty \int_{-\infty}^\infty R(x, y, z) n(t - 2c^{-1}z) \tilde{q}^2(x, y, z) \, dx \, dy \, dz, \quad (11.2)$$

where

$$\tilde{q}(x, y, z) = zq(x, y, z). \quad (11.3)$$

Here, $q(x, y, z)$ is the transducer field pattern. Because the input waveform in (11.2) is complex, the actual received signal is the real part of $r(t)$. The envelope (which is usually desired instead of the high-frequency signal itself) is found by taking the complex modulus of $r(t)$.

The time-dependent terms outside the integral in (11.2) cause very severe signal loss if not compensated. All systems are equipped with circuitry that performs *time-gain compensation* (TGC), a time-varying amplification, as shown in Figure 11.1. Nominally, the default gain is set to cancel the gain terms appearing in (11.2), which requires a gain of

$$g(t) = \frac{(ct)^2 e^{\mu_a ct}}{K}. \quad (11.4)$$

In practice, most systems have additional (frequency dependent) slide potentiometers, which allow the gain to be determined interactively by the operator. This permits the user to manually adapt the system to special circumstances requiring either more or less gain so that subtle features can be seen in the images. In our development here, we assume the nominal gain compensation, which leads to the gain-compensated (complex) signal

$$r_c(t) = g(t)r(t) = \int_0^\infty \int_{-\infty}^\infty \int_{-\infty}^\infty R(x, y, z) n(t - 2c^{-1}z) \tilde{q}^2(x, y, z) \, dx \, dy \, dz. \quad (11.5)$$

The actual gain-compensated received signal is the real part of $r_c(t)$ and its envelope is the complex modulus of $r_c(t)$.

EXAMPLE 11.3

Suppose a system with nominal gain of 80 dB is working at a frequency of $f = 5$ MHz. The speed of sound is $c = 1540$ m/s, and the coefficients for attenuation are $a = 1 \text{dB} \cdot \text{cm}^{-1} \cdot \text{MHz}^{-1}$, and $b = 1$.

Question What is the nominal time-gain compensation?

Answer The attenuation coefficient is

$$\alpha = af^b = 5 \, \text{dB} \cdot \text{cm}^{-1}.$$

The amplitude attenuation factor μ_a is

$$\mu_a = \alpha/8.7 = 0.575 \, \text{Np} \cdot \text{cm}^{-1}.$$

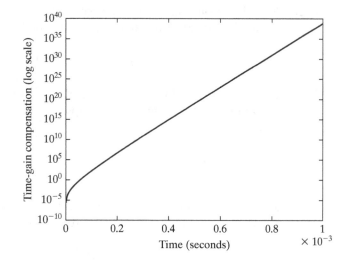

Figure 11.9
Time-gain compensation.

The nominal gain of the system is 80 dB. So the amplitude gain of the system is $K = 10000$. Based on this information, the required time-gain compensation is

$$g(t) = \frac{(ct)^2 e^{+\mu_a ct}}{K} = 2.37 \times 10^6 t^2 e^{88550t}.$$

The time-gain compensation plotted in log scale is shown in Figure 11.9.   ∎

Suppose our transducer has a resonant frequency of 5 MHz; then, the transmitted signal has most of its spectral energy concentrated around 5 MHz. For example, a typical system might have around a 60 percent fractional bandwidth, meaning that a nominal 5 MHz system uses frequencies in the range 3.5–6.5 MHz. Because the system (transducer, body, amplification electronics) is modeled as linear, the received signal will have its spectral energy concentrated around 5 MHz. For imaging purposes, however, it is the signal strength of the returning echoes that is important, and the strength of a narrowband signal is captured by its envelope. Therefore, the first step in an ultrasound imaging system after amplification is envelope detection or demodulation.

Inexpensive ultrasound systems use a simple envelope detection procedure given by the two-step procedure depicted in Figure 11.10. The absolute value of the received signal is accomplished using a rectifier. The low-pass filter can be accomplished using a simple R-C circuit. Electrical engineers will recognize this as a simple AM demodulation scheme. Mathematically, it is simpler to model

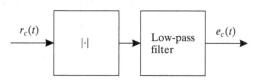

Figure 11.10
Envelope detection.

the envelope detection phase using the received complex signal. Since the received complex signal is a narrowband signal, it can be modeled as

$$r(t) = r_e(t)e^{j\phi}e^{-j2\pi f_0 t} .$$

Mathematically, the envelope of this signal is extracted by taking its modulus (complex magnitude). Cascading gain compensation and envelope detection yields the (real) signal

$$e_c(t) = \left| \int_0^\infty \int_{-\infty}^\infty \int_{-\infty}^\infty R(x,y,z)n(t - 2c^{-1}z)\tilde{q}^2(x,y,z)dxdydz \right| . \tag{11.6}$$

Equation (11.6) can be further simplified by recalling from Section 10.5.2 that n is a narrowband signal, which can be written as [see (10.50)]

$$n(t) = n_e(t)e^{j\phi}e^{-j2\pi f_0 t} . \tag{11.7}$$

Substitution of (11.7) into (11.6) yields

$$e_c(t) = \left| \int_0^\infty \int_{-\infty}^\infty \int_{-\infty}^\infty R(x,y,z)n_e(t - 2c^{-1}z) \right.$$

$$\left. \times e^{j\phi}e^{-j2\pi f_0(t - 2c^{-1}z)}\tilde{q}^2(x,y,z)\, dx\, dy\, dz \right| . \tag{11.8}$$

Phase terms that do not depend on x, y, or z disappear because of the modulus operator, leaving

$$e_c(t) = \left| \int_0^\infty \int_{-\infty}^\infty \int_{-\infty}^\infty R(x,y,z)n_e(t - 2c^{-1}z)e^{j2\pi f_0 2c^{-1}z}\tilde{q}^2(x,y,z)\, dx\, dy\, dz \right| . \tag{11.9}$$

We use the fact that $k = 2\pi f_0/c$ to get

$$e_c(t) = \left| \int_0^\infty \int_{-\infty}^\infty \int_{-\infty}^\infty R(x,y,z)n_e(t - 2c^{-1}z)e^{j2kz}\tilde{q}^2(x,y,z)\, dx\, dy\, dz \right| . \tag{11.10}$$

We will see in Section 11.5 that the signal $e_c(t)$, called the *A-mode signal*, is the fundamental signal in all of ultrasound imaging. It is used directly in A-mode, M-mode, and B-mode imaging.

11.4   Transducer Motion

So far in our analysis, our transducer has remained motionless, pointed down the z-axis. To acquire images, however, the transducer must move. In this section, we analyze the imaging equations that result when our transducer is allowed to move in the (x, y) plane.

Imagine moving the transducer in the $z = 0$ plane. The object does not move, but the transducer's field pattern does. Accordingly, with the transducer placed at $(x_0, y_0, 0)$ (still pointing parallel to the z-axis), the A-mode signal is

$$e_c(t; x_0, y_0) = \left| \int_0^\infty \int_{-\infty}^\infty \int_{-\infty}^\infty R(x, y, z) n_e(t - 2c^{-1}z) \right.$$

$$\left. \times e^{j2kz} \tilde{q}^2(x - x_0, y - y_0, z) \, dx \, dy \, dz \right| . \tag{11.11}$$

Since the envelope of the transmit pulse is fairly brief, it is nearly correct to say that a signal returning at time t must have originated at range

$$z_0 = ct/2, \tag{11.12}$$

where c is the speed of sound in the body. Notice that the "2" is needed to account for the round-trip delay (to get to range z and back). Equation (11.12) is known as the *range equation* and is fundamental in relating range to time in pulse-echo imaging.

Since the range equation relates z_0 and t, e_c can be thought of as a function of x_0, y_0, and z_0 instead of x_0, y_0, and t. Therefore, e_c can be thought of as an estimate of the reflectivity function as a function of spatial position. We formalize this notion by setting

$$\hat{R}(x_0, y_0, z_0) = e_c(2z_0/c; x_0, y_0) . \tag{11.13}$$

Combining (11.11) and (11.13) yields

$$\hat{R}(x_0, y_0, z_0) = \tag{11.14}$$

$$\left| \int_0^\infty \int_{-\infty}^\infty \int_{-\infty}^\infty R(x, y, z) e^{j2kz} n_e(2(z_0 - z)/c) \tilde{q}^2(x - x_0, y - y_0, z) \, dx \, dy \, dz \right| .$$

The Geometric Assumption   To get some insight into (11.14), let us make the drastic assumption that

$$\tilde{q}(x, y, z) = s(x, y) . \tag{11.15}$$

This is equivalent to assuming that the transducer energy travels down a cylinder having the same shape as its face. In this case, (11.14) reduces to

$$\hat{R}(x, y, z) = K \left| R(x, y, z) e^{j2kz} * \tilde{s}(x, y) n_e \left(\frac{z}{c/2} \right) \right| , \tag{11.16}$$

where $\tilde{s}(x, y) = s(-x, -y)$ and $*$ is 3-D convolution.

If we ignore the term e^{j2kz} in (11.16), then all terms are real, and the modulus bars are not needed. In this case, we see that the estimated reflectivity is the true reflectivity convolved with a "blurring" function, or impulse response, given by the product of the face shape and the envelope shape. This 3-D shape is known as the *resolution cell* of an ultrasonic imaging system. It might seem that better estimates of R could be obtained if the pulse were very short and the face were very small—i.e., if the resolution cell were small. Unfortunately, in reality, if the face gets small, the beam actually spreads (as discussed in Section 10.5.2); this is the problem with the geometric approximation. Also, if the pulse shortens, it is no longer narrowband, and problems with frequency-dependent attenuation arise. As always, there is a compromise between these various factors that lead to well-balanced ultrasound systems.

Now we turn to the factor e^{j2kz}, which we ignored above. It turns out that this term is the cause of *speckle*, the dominant artifact appearing in ultrasound images. To explain speckle, we note that the phase term e^{j2kz} multiplies the true reflectivity distribution $R(x, y, z)$. Since the resolution cell is far larger than π/k, e^{j2kz} can be expected to vary over many cycles within a resolution cell. At any given time, a scatterer within the resolution cell will receive a phase, which can be modeled as a random variable, uniform over $[0, 2\pi)$.

Integration over the entire resolution cell adds a large collection of complex numbers $R(x, y, z)e^{j2kz}$, where the angle of each is completely random. The modulus of this integral is therefore a Rayleigh random variable, whose mean is dependent on the underlying reflectivity distribution. Since resolution cells overlap in time [which is equivalent to range via (11.12)] and generally in azimuth (lateral position) in an image, the received signals are correlated; therefore, the Rayleigh random variables produce a spatially correlated random pattern, which multiplies the underlying desired image of reflectivity. This pattern, called a speckle pattern, appears as a splotchy granulation, oscillating from bright to dark across the image, with granule size that depends on the size of the resolution cell, the spatial distribution of scatterers, and the frequency of sound. It is an undesirable, but generally dominant, feature in all medical ultrasound images.

Fresnel and Fraunhofer Approximations   In order to further simplify (11.14), we can replace $\tilde{q}$ with an expression for the actual transducer field pattern. Here, we consider the Fresnel and Fraunhofer approximations, two approximations that were developed in Section 10.5.2.

The Fresnel approximation for $\tilde{q}$ is found from (10.70) and (11.3). Substituting the resulting expression for (10.70) into (11.14) yields

$$\hat{R}(x_0, y_0, z_0) = \int_0^\infty \int_{-\infty}^\infty \int_{-\infty}^\infty R(x, y, z)e^{j2kz}$$

$$\times\, n_e\left(\frac{z_0 - z}{c/2}\right)[s'(x - x_0, y - y_0, z)]^2 \, dx\, dy\, dz, \quad (11.17)$$

where

$$s'(x, y, z) = s(x, y) ** e^{jk(x^2 + y^2)/2z}. \quad (11.18)$$

Unfortunately, (11.17) cannot be simplified further, and it does not yield an intuitively satisfying or practically useful expression describing the relationship between R and $\hat{R}$.

The Fraunhofer approximation for $\tilde{q}$ is found from (10.76) and (11.3). Substituting the resulting expression for (10.70) into (11.14) yields

$$\hat{R}(x_0, y_0, z_0) = \int_0^\infty \int_{-\infty}^\infty \int_{-\infty}^\infty R(x, y, z) e^{j2kz} e^{jk(x^2+y^2)/z}$$

$$\times \left[S\left(\frac{x}{\lambda z}, \frac{y}{\lambda z} \right) \right]^2 n_e \left(\frac{z_0 - z}{c/2} \right) dx\, dy\, dz. \qquad (11.19)$$

We can now make two observations. First, the term $\exp\{jk(x^2 + y^2)/z\}$ is approximately unity near the axis and far in range (where Fraunhofer applies). Therefore, we can replace it with the value 1. Second, the expression $S^2(\cdot, \cdot)$ changes slowly with respect to z, while n_e is only a brief pulse in z. Therefore, function $S^2(\cdot, \cdot)$ can be viewed as being *indexed* by z—i.e., a different $S^2(\cdot, \cdot)$ is used for each range considered. Incorporating these two observations as approximations, we see that (11.19) can be written as the following convolution:

$$\hat{R}(x, y, z) = \left| R(x, y, z) e^{j2kz} * \left[S\left(\frac{x}{\lambda z}, \frac{y}{\lambda z} \right) \right]^2 n_e \left(\frac{z}{c/2} \right) \right|. \qquad (11.20)$$

The resulting image is a local blurring of the object, where the blurring function (the resolution cell) gets wider with increasing depth. This system is not shift invariant but is approximately so within a narrow range of depth.

From (11.20), we get a feeling for the visual characteristics of ultrasound images. First, because of the term e^{j2kz}, speckle is still present in both the Fresnel and Fraunhofer approximations. Second, we see that the range resolution remains constant as a pulse propagates through the tissue. Third, we see that the lateral (azimuth) resolution degrades with range (except when focusing is used). Finally, we see that the lateral shape of the resolution cell usually has sidelobes, caused by the form of $S(\cdot, \cdot)$, which typically looks like a sinc function (see Section 2.2.4) in a direction parallel to the transducer's face.

11.5 Ultrasound Imaging Modes

11.5.1 A-Mode Scan

The starting point for ultrasound imaging systems is the envelope detected (gain-compensated) signal, $e_c(t)$ in (11.11); this signal is called the *A-mode signal* or amplitude-mode signal. By firing the transducer on a repetitive basis, a succession of these signals can be displayed on an oscilloscope. This display is called the *A-mode scan* (or *A-mode display*); an example is shown in Figure 11.11. The time between successive firings is called the *repetition time* (or *interval*), and is denoted by the symbol T_R. This interval should be long enough so that the returning echoes have

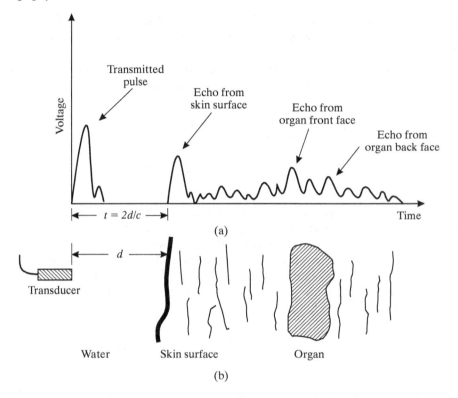

Figure 11.11
A typical A-mode display.

died out, but fast enough to capture any motion that might be present and of medical interest. For example, the A-mode display is useful when looking at heart valve motion.

11.5.2 M-Mode Scan

An *M-mode scan* is obtained by using each A-mode signal as a column in an image, the value of the A-mode signal becoming the brightness of the M-mode image. Successive A-mode signals are displayed in successive columns, wrapping around when the last column is reached. Motion of objects along the transducer axis is revealed by bright traces moving up and down across the image, as shown in Figure 11.12. This mode is most often used to image the motion of the heart valves, and it is therefore usually displayed in conjunction with the ECG.

11.5.3 B-Mode Scan

A *B-mode scan* is created by scanning the transducer beam in a plane. One way to do this is to move the transducer in the x-direction while its beam is aimed down the z-axis, as shown in Figure 11.13. Periodically firing pulses produces a succession of A-mode signals that are keyed to the x-position of the transducer. The B-mode image is created by brightness-modulating a CRT along a column using

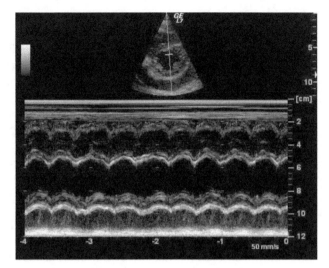

Figure 11.12
An M-mode scan. (Used with permission of GE Healthcare.)

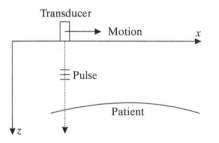

Figure 11.13
A simple B-mode scanner.

the corresponding A-mode signal. Scanning a single transducer was the dominant B-mode imaging method in early ultrasound imaging. Instead of relying upon a fixed method of translation, however, these early scanners had mechanical arms to hold the transducer and used potentiometric position sensors to determine the position and orientation of the transducer at any time. The operator could often "fix" certain directions or orientations in order to produce a tomographic scan containing data from only one cross-sectional slice. Given knowledge of the position of the transducer at the time of a pulse, it is a relatively simple matter to plot the correct line on a CRT and build up a full scan over time.

One important advantage of these manual-scan systems is that the transducer could be angled to hit the same point in the body from different directions. If a specular reflector existed at a point, a strong echo could be obtained from the direction normal to the surface. Therefore, images were typically built up by taking the maximum echo from a given point within the body. When multiple views of the same tissue are included in a single B-mode image, it is referred to as *compound B-mode scanning*. Unfortunately, images made in this way often suffer from severe artifacts due to refraction. The sound beam refracts differently depending on the incidence angle to a surface; hence, what the electronic system computes as different measurements at the same point of tissue beyond the surface is often wrong. As a result, small intensity scatters are often plotted at different points on the screen, producing object-dependent blurring. To reduce this blurring effect, these early systems often plotted only the largest amplitude returns—those corresponding to specular reflectors only.[1] These early images often looked like anatomical sketches, depicting only the outline of organs and tumors.

A further disadvantage of these early systems is that they were not real-time. Modern systems are generally real-time and are not compound scanners. Therefore, within a given image, the reflectivity value displayed for a given point within the image is generated from data obtained from a single orientation. Improvements in transducer design and pulse-echo electronics have made the display of these images in grey-scale quite acceptable, even though specular reflectors are not always imaged from their normal directions. Three types of B-mode scanners now dominate: linear scanners, mechanical sector-scanners, and phased array sector-scanners, as shown in Figure 11.14.

The *linear scanner* has a collection of transducers arranged in a line, thus mimicking the linear translation of a single transducer, but without requiring motion. Generally, several transducers are tied together on transmit and receive to synthesize larger aperture transducer which can be focused and can generate more power. The linear scanner requires a large flat area with which to maintain contact with the body; thus, abdominal imaging is its primary use, with obstetrics being a common use.

Sector scanners come in both mechanical and phased-array varieties. The *mechanical sector scanner* simply pivots a transducer about an axis orthogonal to the transducer's axis (typically through the body of the transducer, as shown in Figure 11.14). Defining the pivot point as the origin, a collection of rays emanating from the origin can be imaged. Since the rays diverge with depth, the tissue is likely to be oversampled near the origin and undersampled far away. Since all medical ultrasound imaging systems use focused transducers, however, the lateral resolution will be best near the focal region and worse both nearer and farther away from the transducer.

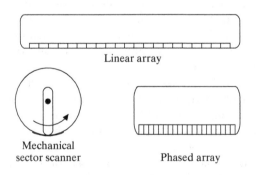

Linear array

Mechanical
sector scanner

Phased array

Figure 11.14
Three B-mode scanner types.

<hr />

[1] Also, the earliest systems used storage oscilloscopes and therefore could plot only binary images (i.e., images with only two shades of gray).

The *phased array sector-scanner* has a collection of very small transducer elements arranged in a line. The overall total linear dimension of a phased-array sector scanner is much smaller than that of a linear scanner. The ultrasound beam is electronically steered over a sector by applying differing time-delay (or phases) to the individual transducers. One very important advantage of this type of system is that the receive focus can be varied over time with the expected position of returning pulses, thus providing a dynamic focus. A disadvantage of these systems is that spurious sidelobes (called "grating lobes") of acoustic energy are generated by the phasing action and can lead to image artifacts.

Both types of sector scanners require conversion of polar coordinate data (the A-mode signals returning from along rays) to rectangular coordinates (the image memory driving the CRT). The process of converting from polar coordinates to rectangular coordinates is called *scan conversion*. Proper scan conversion reduces sampling artifacts (moiré effects, see Section 2.8) without excessive low-pass filtering (and resultant loss of resolution).

EXAMPLE 11.4

A linear transducer array is operating at 2 MHz in water. The face of each transducer element is 1 mm wide in x-direction, and 10 mm in y-direction.

Question If we want geometric approximation to be valid up to 10 cm in range, how many transducers have to be grouped together?

Answer The wavelength in water at 2 MHz is $\lambda = c/f = 1484$ m/s$/2 \times 10^6$ s$^{-1} = 0.742$ mm. The range at which the beam transitions from geometric to far field is $z = D^2/\lambda = D^2/0.742$ mm $= 10$ cm. We have

$$D = \sqrt{100 \times 0.742} = 8.61 \text{ mm.}$$

So, at least nine transducer elements must be grouped together. ∎

Depth of Penetration Ultimately, although there are many other factors involved, attenuation limits the depth of penetration of ultrasound systems. Suppose our transducer/preamplifier can handle at most an L dB loss (80 dB is typical). From Example 10.5,

$$20 \log \frac{A_z}{A_0} = -L, \tag{11.21}$$

where A_z is interpreted as the pressure amplitude after traversing a distance of z. From (10.34),

$$\alpha = -\frac{1}{z} 20 \log \frac{A_z}{A_0} \tag{11.22}$$

and

$$\alpha \approx af. \tag{11.23}$$

TABLE 11.1

Typical Depth of Penetration for Given Frequencies	
Frequency (MHz)	Depth of Penetration (cm)
1	40
2	20
3	13
5	8
10	4
20	2

The total range that a wave can travel before being attenuated below the system threshold is

$$z = -\frac{L}{af}. \tag{11.24}$$

The depth of penetration d_p can be only half this distance since a round-trip is required, giving

$$d_p = \frac{L}{2af}. \tag{11.25}$$

A short table of d_p, assuming $a = 1$ dBcm^{-1}MHz^{-1} and $L = 80$ dB, is given in Table 11.1.

Pulse Repetition Rate A new pulse can be generated only after all echoes from the previous pulse have died out. After imaging tissues at the maximum depth of penetration, all potential remaining echoes are below the level of detection. Therefore, the pulse repetition interval T_R has a lower bound given by the round-trip time to the depth of penetration:

$$T_R \geq \frac{2d_p}{c}. \tag{11.26}$$

The *pulse repetition rate* is defined as

$$f_R = \frac{1}{T_R}. \tag{11.27}$$

Assuming $a = 1$ dBcm^{-1}MHz^{-1} and $L = 80$ dB, the minimum pulse repetition interval is 0.267 ms. In other words, in this case $f_R = 3750$ pulses per second can be generated without danger of confusing the returning pulses with the wrong transmitted pulse.

B-Mode Image Frame Rate Suppose N pulses are required to generate an image (linear or sector). Then, the image frame rate F is

$$F = \frac{1}{T_R N}. \tag{11.28}$$

Continuing the example from above, if $N = 256$, then $F = 14.6$ frames/sec. Typical frame rates in commercial ultrasound systems are between 10 and 100 frames/sec.

A moving picture shown at 15 frames/sec creates a great deal of flicker, and this would be unacceptable in an ultrasound system. Scan conversion solves this problem. The A-mode data are read into a dual ported memory as fast as they come in. Simultaneously, the scan converter is reading out the data at a higher rate in order to generate around 60 frames/sec for the CRT. Aside from flicker, however, it may be desirable to capture motion occurring faster than what can be viewed at 15 frames/sec. A standard feature in most modern ultrasound systems is the capability to reduce the field of view—e.g., by reducing the angle of the sector, or the linear extent of a linear scanner—so that it may be scanned more rapidly (because N is reduced).

EXAMPLE 11.5

An ultrasound imaging system is operating in B mode, requiring 256 pulses to generate an image. Assume that the transducer is sensitive to at most 80 dB loss.

Question If the material being imaged has a speed of sound $c = 1540$ m/s, and $a = 1$ dBcm^{-1}MHz^{-1}, what should the working frequency be to achieve a frame rate of 15 frames/sec?

Answer Given the frame rate and the number of pulses needed to generate an image, we can compute the pulse repetition rate as

$$T_R = \frac{1}{FN} = 0.26 \, \text{ms}.$$

Since

$$T_R \geq \frac{2d_p}{c} = \frac{L}{afc},$$

we have

$$f \geq \frac{L}{acT_R} = 1.99 \, \text{MHz}.$$

∎

11.6 Steering and Focusing

Phased arrays use electronic steering and focusing to achieve a B-scan without transducer movement. They do this using a small linear array of transducer elements (typically between 50 and 100), with total dimension of around 1 cm (height of each element) by 3 cm. The width of each element is about 1/4 wavelength, which is typically 0.2–0.75 mm wide. The extreme rectangular shape of each element allows mechanical focusing to keep the sound largely in the image plane, but guarantees that each element will have a nearly circular field pattern within the image plane. On transmit, a directed beam is achieved by timing the firing of each element so that the sound each produces adds coherently in the desired direction and incoherently otherwise. On receive, the array is made direction-sensitive by delaying the received signals and summing them coherently. In the following two sections, we expand on these ideas and bring some elementary trigonometry to bear in the analysis of phased-array systems.

11.6.1 Transmit Steering and Focusing

As shown in Figure 11.15, adding a separate delay element to each transducer of a phased array makes it possible to steer the transducer's acoustic beam. To analyze this situation further, assume that T_0 (transducer 0) generates a pulse at $t = 0$. We want to determine what firing times are necessary for the remaining transducers in order to generate a plane wave heading off in direction θ. At time t, the leading edge of the pulse generated by T_0 has traveled distance $r_0(t) = ct$. Therefore, the line $L(t, \theta)$ representing the wavefront in the direction θ at time t is [by analogy to (2.10)] given by

$$L(t, \theta) = \{(x, z)|\ z \cos\theta + x \sin\theta = r_0(t)\}. \tag{11.29}$$

By simple geometry, the distance between T_i (the ith transducer) and this line is

$$r_i(t) = r_0(t) - id \sin\theta. \tag{11.30}$$

To generate a pulse that will also have a leading edge on $L(t, \theta)$, the pulse must be generated at time

$$t_i = \frac{r_0(t) - r_i(t)}{c} = \frac{id \sin\theta}{c}. \tag{11.31}$$

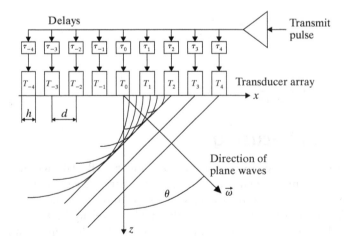

Figure 11.15
Geometry for steering a phased array.

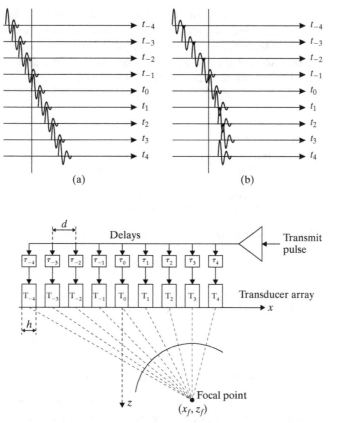

Figure 11.16
Timing diagram for (a) steering and (b) focusing.

Figure 11.17
Geometry for steering and focusing a phased array.

The sequences of pulses generated at these times is depicted in Figure 11.16(a). In practical terms, we cannot have negative times, so the resulting times must be shifted to positive time. Assume that there are $2N + 1$ transducer elements centered at the origin, and let $t_{min} = \min\{t_i, i = -N, \ldots, N\}$. Then, the time delays are given by

$$\tau_i = t_i - t_{min} \quad \text{and} \quad i = -N, \ldots, N. \tag{11.32}$$

Typically, the phased array will be focused, which is just a refinement of steering, as shown in Figure 11.17. To analyze steering and focusing together, let the focal point be at coordinate (x_f, z_f); T_i is at coordinate $(id, 0)$. Therefore, the range from T_i to the focal point is

$$r_i = \sqrt{(id - x_f)^2 + z_f^2}. \tag{11.33}$$

Assume as before that T_0 fires at $t = 0$. Then, T_i must fire at

$$t_i = \frac{r_0 - r_i}{c} = \frac{\sqrt{x_f^2 + z_f^2} - \sqrt{(id - x_f)^2 + z_f^2}}{c} \tag{11.34}$$

to reach the focal spot at the same instant, and the delays are also given by (11.32). Figure 11.16(b) depicts the pulse timings that would be used to generate a focused beam. The difference between these delays and those of steering alone is that (1) the delays are not simply a multiple of a base delay, and (2) the transducers need not be fired in the same order as their geometric order. A typical depth for the focal spot would be $z_f = d_p/2$.

EXAMPLE 11.6

A linear transducer array with 127 transducers is operating in water. Any adjacent transducers are separated by $d = 0.8$ mm (see Figure 11.18).

Question   We desire a focal point at $z = 5$ cm on the z-axis. Suppose the outmost transducers fire at $t = 0$, when does the center element fire?

Answer   The distance between the focal point and the centers of the outmost transducers is

$$r_{63} = \sqrt{(63d)^2 + (5 \text{ cm})^2} = 7.1 \text{ cm}.$$

The difference between r_{63} and r_0 is $r_{63} - r_0 = 2.1$ cm. In water, the speed of sound is 1484 m/s, so the center element fires at

$$\tau_0 = \frac{2.1 \text{ cm}}{1484 \text{ m/s}} = 14.15 \,\mu\text{s}. \qquad \blacksquare$$

11.6.2 Beamforming and Dynamic Focusing

Although the acoustic energy from a phased array pulse is timed to provide coherent summation in a direction or at a point, in fact the energy propagates quite widely due to the circular wave pattern of each element of the array. In order to achieve better sensitivity to echoes occurring along the dominant (steered) direction, two techniques are commonly employed: beamforming and dynamic focusing.

Beamforming is analogous to steering in that it increases the transducer's sensitivity to a particular direction. As shown in Figure 11.19(a), a plane wave incident upon the transducer array from a direction θ will hit one transducer at the end of the array first and then successively hit the remaining transducers. If the first

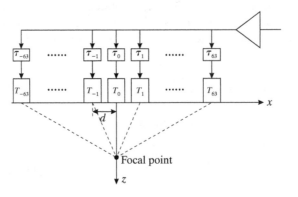

Figure 11.18
A linear transducer array with an on-axis focal point.

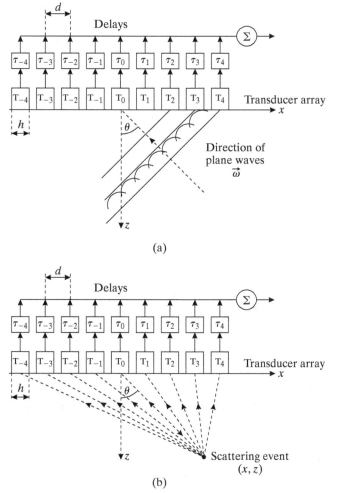

Figure 11.19
Geometry for (a) beamforming and (b) dynamic-focusing phased array.

transducer to be hit, and thereafter each successive transducer, delay their received waveforms so that they can be coherently summed, then the entire array is sensitized to the direction θ. The required receive delays are identical to the transmit delays specified in (11.31) and (11.32) for the transmission of a plane wave in direction θ. Likewise, a fixed focal point can be achieved by adopting the delays used for combined steering and focusing in (11.32) and (11.34).

Additional sophistication is achieved using *dynamic focusing*, which increases the transducer's sensitivity to a particular point in space at a particular time. Suppose that T_0 is fired at $t = 0$ in the direction θ (steered or focused), as shown in Figure 11.19(b). Assuming that a scattered spherical wave originates at position (x, z), the total distance the pulse travels from T_0 to arrive at transducer T_i is

$$r_i = \sqrt{x^2 + z^2} + \sqrt{(id - x)^2 + z^2}. \tag{11.35}$$

The time difference between the pulse's arrival at T_0 and T_i is

$$t_i = \frac{\sqrt{x^2 + z^2} - \sqrt{(id - x)^2 + z^2}}{c}. \tag{11.36}$$

In fact, as the transmit pulse propagates in the direction θ, the (x, z) position of the scattering center changes with the position of the pulse. Therefore, the time differences also change as a function of time. The position of the transmitted pulse along the θ axis is given by $(x, z) = (ct \sin \theta, ct \cos \theta)$. Therefore, the time differences are given by

$$t_i(t) = \frac{\sqrt{x^2(t) + z^2(t)} - \sqrt{(id - x(t))^2 + z^2(t)}}{c} \tag{11.37}$$

$$= t - \frac{\sqrt{(id)^2 + (ct)^2 - 2ctid \sin \theta}}{c}. \tag{11.38}$$

To avoid negative time delays, it is necessary to shift these times by subtracting $t_{min} = -Nd/c$. Hence, dynamic focusing is achieved by dynamically altering the receive time delays according to

$$\tau_i(t) = t - \frac{\sqrt{(id)^2 + (ct)^2 - 2ctid \sin \theta}}{c} + \frac{Nd}{c}. \tag{11.39}$$

EXAMPLE 11.7

Consider a dynamic focusing phased array with 127 transducers. Any adjacent transducers are separated by $d = 0.8$ mm. A point scatter on the z-axis at a range of 5 cm starts to move in the $+z$ direction with a speed of $v = 5$ cm/s at time $t = 0$.

Question   Assuming that the speed of sound is $c = 1500$ m/s, plot the difference between the delays $\tau_{63}(t)$ and $\tau_0(t)$.

Answer   Let

$$\tau_d(t) = \tau_{63}(t) - \tau_0(t).$$

The distance between the scatter and the center of the 63rd transducer at time t is

$$r_{63}(t) = \sqrt{[(63 * 0.8) \text{ mm}]^2 + [(5 + 5t) \text{ cm}]^2}.$$

The distance between the scatter and the center of the middle transducer at time t is

$$r_0(t) = (5 + 5t) \text{ cm}.$$

So, the difference in the time delay is

$$\tau_d(t) = \frac{r_{63}(t)}{c} - \frac{r_0(t)}{c} = \frac{1}{c} [r_{63}(t) - r_0(t)],$$

which is plotted in Figure 11.20.   ∎

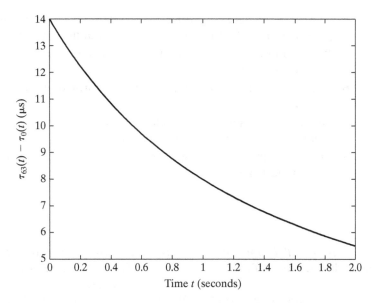

Figure 11.20
The difference in time delay between the outmost and the center transducer in a dynamic-focusing phased array with a moving scatter.

11.6.3 Three-Dimensional Ultrasound Imaging

Three-dimensional ultrasound is made possible by either scanning conventional linear or phased arrays or by using 2-D phased arrays. Real-time volumetric imaging became commercially available in 1997 using 2-D phase arrays. These systems use 256 transmit and 256 receive channels with 16:1 parallel processing in the receive mode. Such parallel processing means that the transducers can be simultaneously phased in 16 different ways on receive, beamforming 16 directions at the same time. Typically, rather than transmitting in 16 separate directions, one direction is insonified (pulsed with acoustic waves) and hyperresolved into 16 receive rays. This allows for a 16-fold increase in frame rate. New systems are under testing that have 32:1 parallel processing on receive, and another system is being developed that will have 64:1 parallel processing on receive.

11.7 Summary and Key Concepts

Ultrasound imaging is usually accomplished with a pulse-echo system, which is based on a transducer. Such systems are high performance and low cost. In this chapter, we presented the following key concepts that you should now understand:

1. The *transducer*—a primary component of an ultrasound imaging system—both generates and receives sound pulses in the so-called *pulse-echo mode* of operation.

2. Most transducers are based on a *piezoelectric crystal* and resonate at a fundamental frequency characteristic of that transducer.

3. The ultrasound beam must be *steered*, *scanned*, or *swept* across the body to produce a useful field of view.

4. The *basic imaging equation* is the *pulse-echo equation*, which arises from generalization of either the Fresnel or Fraunhofer approximation.

5. Practical ultrasound imaging typically requires *transducer motion*, which modifies the basic imaging equation.

6. The basic imaging equation can be transformed into an equation describing the *reflectivity* of the tissues.

7. *Ultrasound imaging modes* include *A-mode scanning*, *M-mode scanning*, *B-mode scanning*, and *3-D ultrasound imaging*.

8. Practical B-mode scanning requires *steering* or *beamforming* and *focusing* of the ultrasound beam.

Bibliography

Kremkau, F. W. *Diagnostic Ultrasound, Principles and Instruments*, 5th ed. Philadelphia, PA: W. B. Saunders Company, 1998.

Macovski, A. *Medical Imaging Systems*. Englewood Cliffs, NJ: Prentice Hall, 1983.

Pierce, A. D. *Acoustics: An Introduction to Its Physical Principles and Applications*. New York: McGraw–Hill, 1981.

Wells, P. N. T. *Biomedical Ultrasonics*. London, England: Academic Press, 1977.

Problems

Ultrasound Image Formation and Imaging Modes

11.1 An $L \times L$ square transducer, submersed in water, is centered at the origin and is pointed down the $+z$-axis. A large layer of homogeneous fat begins at $z = z_0$ and extends to the depth of penetration, so that the only possible returns are those from the interface between water and fat. The transducer is capable of rotating toward the x-axis by θ degrees. Assume that there is no attenuation in the water. Assume that the transmitted pulse is a gated sinusoid of duration T and height A_0.

(a) Assume that z_0 is in the far field and derive an expression for the strength of return as a function of angle. Make approximations where appropriate and do not forget the sidelobes.

(b) Assume that the transducer and preamplifier are able to detect signals 80 dB down from the transmitted pulse height. Assume also that a B-mode binary image is generated so that all signals above the detectable threshold are plotted in black and all signals below the threshold are plotted in white. Sketch the B-mode image that would result from this experiment.

11.2 Two identical square transducers with a width of 1 cm face each other. The first one is located at the origin and points in the $+x$ direction; the second one is located at (10 cm, 0, 0) and points toward the origin. The first one will be used to image the second one. Assume a homogeneous medium with $\rho_0 = 1000$ kg/m^3, $c = 1500$m/s, and $\alpha = 1$ dB/cm. Assume that the first transducer fires a perfect geometric beam with peak transmit acoustic pressure measured at its face of 12.25 N/cm^2; assume that the second transducer is a perfect reflector.

(a) Sketch the A-mode signal. Label the axes carefully, and identify the time-of-return and peak-height (as an acoustic pressure at the face) of the returning pulse.

(b) At time $t = 2$s the second transducer begins to move back and forth along the x-axis with x position $x(t) = 10 + \sin 2\pi(t - 2)$ cm, $t \geq 2$ sec. Sketch the M-mode plot for $0 \leq t \leq 5$ sec. Label axes carefully; identify key points on your plot.

(c) Now suppose the second transducer stops at its original position and we allow the first transducer to move along the y axis. Sketch the resulting B-mode image. Label axes carefully; identify key points on your plot. Make a sketch of the peak-height of the returning pulse as a function of the y position of the first transducer.

11.3 You are using a single transducer to examine a heart valve. Assume that in a given heart cycle the range of the valve is given by $z(t) = 16 + 0.5e^{-t/\tau}u(t)$cm where $\tau = 10$ ms, $u(t)$ is the unit step function, and the speed of sound is 1540 m/s.

(a) Sketch $z(t)$ over a couple of heart cycles (assume that the heart rate is 1 Hz).

(b) Assume that your transducer is fired at $t = 0$ and it has a typical transmit waveform. Carefully sketch the A-mode signal (as a function of time) that you would observe on an oscilloscope.

(c) Now assume that you repeatedly fire your transducer every 1 ms. Sketch the M-mode image that would be generated, being careful to label the axes.

(d) Suppose you wanted to image the motion of this valve by making a B-mode image of it. If it can be "covered" by 10 scan lines (given the beam size at 16 cm range), describe what steps you would take to make this image in real time. Do you think it is possible?

11.4 An ultrasound transducer is pointing down the z-axis, on which two point scatterers are located at z_1 and z_2 in a Fraunhofer field (see Figure P11.1). The transducer fires a narrowband burst with a rect shaped envelop down in the $+z$ direction. The envelop is $n_e(t) = \text{rect}[(t + T/2)/T]$, and $T = \lambda/2c$, where λ is the wavelength, and c is the speed of sound in the material. Sketch the estimated reflectivity for the following:

(a) $z_2 - z_1 = \lambda/2$.

(b) $z_2 - z_1 = \lambda/8$.

Figure P11.1
An ultrasound transducer
and two scatterers on axis.

Ultrasound Transducer Array

11.5 A linear ultrasound transducer array has 101 transducers, each of width $d = 0.1$ mm and separated by $h = 0.1$ mm. Assume the simple field pattern geometry of Figure 10.5 and $c = 1540$ m/s.

   (a) What is the maximum frequency at which the transducers can operate and still generate a plane wave at angle $\theta = 30°$?

   (b) How long will it take to sequence through the array in order to generate the plane wave in part (a)?

11.6 Consider the linear array in Figure 11.15. Suppose we want to image a cross section of a patient as shown in Figure P11.2, acquiring a scan line every $\Delta\theta = 1°$. Assume there are 101 transducers operating at a frequency of 2 MHz and $c = 1540$ m/s and $d = 0.6$ mm.

   (a) Find an expression for the time delay of the ith transducer as a function of angle of the transmitted plane wave.

   (b) How long does it take to scan the entire FOV assuming $R = 20$ cm?

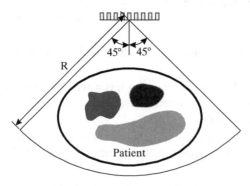

Figure P11.2
Problem 11.6

11.7 Consider a linear array of 100 flat, rectangular, 3.0 MHz transducers, each 1.5 mm wide and 2.1 cm deep. Assume, for simplicity, that they are packed tightly together so that the array is 15 cm long by 2.1 cm deep. The transducer array is tested in a medium having a speed of sound of 1500 m/s.

   (a) If 14 consecutive transducers on the transducer array are electronically grouped on each transmit and receive, what depths (or ranges) are considered to be in the far field?

   (b) At what time interval after the transmit pulse will an echo from an object at range 7 cm arrive back at the transducer array?

   (c) A B-mode image is created by successively grouping 14 transducers on the array, sliding one transducer at a time down the length of the array. Suppose the depth of penetration is 20 cm. What is the frame rate for this linear array image?

   (d) Describe two ways that the frame rate can be increased.

11.8 A phased array ultrasound transducer consists of three elements (L, C, R) each of dimensions $h = 3$ mm and $d = 8$ mm, as shown in Figure P11.3.

Each element has a square face, $s(x, y) = \text{rect}(\frac{x}{h}, \frac{y}{h})$. In the first scenario (Figure P11.3), the transducer has a silicone sheet of thickness $s = 2$ mm on its front face, which is placed on the skin. The acoustic impedances Z and the speed of sound c in silicone and in skin are given in the table below. For parts (a)–(e), assume that only one element, C, is generating and receiving ultrasound waves. The envelope of the wave is $n_e(t) = \text{rect}(\frac{t}{10})$ and its frequency is 2 MHz.

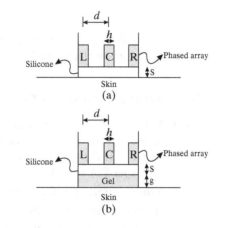

Figure P11.3
Problem 11.8

	Silicone	Skin
Acoustic impedance (Z) (kg/m^{2}sec)	1.4×10^6	1.5×10^6
Speed of sound (c) (m/sec)	1500	1550

(a) Sketch the A-mode signal from the time of the transmit pulse through the time of the echo from the silicone-skin interface. Assume a geometric field approximation and ignore attenuation and speckle. Clearly label key times and amplitudes. In order to reduce the size of the echo from the skin, an acoustic gel of width $g = 3$ mm is introduced between the silicone and the skin, as shown in the figure (Figure P11.3). The gel is selected such that $Z_{\text{gel}} = \sqrt{Z_{\text{silicone}} Z_{\text{skin}}}$.

(b) Find the speed of ultrasound in the gel, c_{gel}, so that there can be no refraction at the gel-skin interface, regardless of the angle of incidence.

(c) Sketch the A-mode signal from the time of the transmit pulse through the time of the echo from the gel-skin interface. Assume a geometric field approximation and ignore attenuation and speckle. Clearly label key times and amplitudes.

(d) Suppose you want your imaging system to be insensitive to the echo from the skin. What should the sensitivity L (in dB) be? In parts (e) and (f) ignore the silicone and gel.

(e) Is point F in Figure P11.4 in the near field or far field of the central transducer?

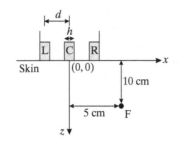

Figure P11.4
Problem 11.8.

(f) Find the time delays, τ_L, τ_C, τ_R, so that, on transmit, the phased array will focus on the point F.

Ultrasound Imaging System Design and Image Quality

11.9 A $1\text{ cm} \times 1\text{ cm}$ ultrasound transducer is placed on the $z = 0$ plane, and pointed in the $+z$-direction. Its sides are parallel to the x and y axes, and its center is at the origin. The transducer is submerged in a homogeneous media with speed of sound $c = 1540$ m/s. The working frequency is 2.5 MHz.

A line scatterer parallel to the y-axis passing point $(0, 0, 5)$ cm is being imaged.

(a) Write down an expression for the scatterer in mathematical form.

(b) What is the estimated reflectivity?

(c) If a second line scatterer parallel to the first one is placed at the same range, what is the minimal distance between them for the system to distinguish them?

(d) Repeat the above two parts for range $z = 20$ cm.

11.10 You are asked to select the best ultrasound imaging parameters for imaging a deep-lying small structure. Assume that this object is located at 20 cm depth and is embedded within a homogeneous tissue with sound speed $c = 1500$ m/s and absorption coefficient $\alpha = af$, where $a = 1.0$ dB/(cm-MHz). Also assume that the system is sensitive to at most a 100 dB signal loss.

You can choose one of three operating frequencies: $f = 1$ MHz, $f = 2$ MHz, or $f = 5$ MHz. Given any frequency, you can choose one of two transducer diameters: $D = 1$ cm or $D = 2$ cm. Assume that all transducers have flat faces, and use the split-range field pattern approximation (geometric and far-field approximations in their appropriate ranges). Which combination of f and D will give the best image of this structure? Justify your answer.

11.11 An $L \times L$ ultrasound transducer is centered at the origin and pointing in the $+z$-axis. The narrowband signal generated by the transducer has an envelope

$$n_e(t) = \text{sinc}\left(\frac{\pi t}{\Delta T}\right).$$

Two point scatterers with reflectivity of R are located at $(0, 0, z_0)$ and $(0, 0, z_0 + \Delta z)$. Ignore the attenuation and multiple reflections.

(a) Assuming that the scatterers are within the region where geometric assumption holds approximately, find the estimated reflectivity $\hat{R}(x, y, z)$;

(b) How far should the scatterers be separated in order for the imaging system to distinguish them?

(c) If the scatterers are in the far field where Fraunhofer approximation holds, what would be the estimated reflectivity $\hat{R}(x, y, z)$?

(d) Derive an expression of the depth resolution in the far field as a function of z_0.

11.12 An ultrasound imaging system is equipped with two square transducers. One operates at 5 MHz and the other at 12 MHz. The 5 MHz transducer is 2.0 cm by 2.0 cm and the 12 MHz one is 0.4 cm by 0.4 cm. The imaging system is tested in a medium having a speed of sound of 1560 m/s at both frequencies. The amplitude attenuation factor of the medium satisfies

$$\mu[\text{cm}^{-1}] = 0.04 \text{ cm}^{-1} \cdot \text{MHz}^{-1} \times f[\text{MHz}]$$

We want to image objects at ranges up to 20 cm.

(a) Find the system sensitivities $L_{5\text{MHz}}$ and $L_{12\text{MHz}}$.

(b) What are the widths of the main lobes at the range of 20 cm? (Use the simple 3 dB model of field pattern.)

(c) What is the maximal pulse repetition rate for the system?

(d) Suppose the transducers were scanned back and forth to obtain a B-mode image. What is the maximal frame rate if 128 lines are used to form an image?

(e) What will happen to the B-mode image if the speed of sound in the medium is nonuniform but we still use a uniform speed assumption?

11.13 A 1 cm × 1 cm square ultrasound transducer, submerged in oil, is centered at the origin and pointed down the +z-axis, as shown in Fig. P11.5. A large layer of fat begins at $z_0 = 20$ cm, and extends to infinity, so that the only possible returns are those from the interface between oil and fat.

Suppose the transducer operates at 1 MHz. The density, speed of sound, and absorption coefficient of oil and fat are, respectively,

$$\rho_{\text{oil}} = 950 \text{ kg/m}^3, \ c_{\text{oil}} = 1500 \text{ m/s}, \ \alpha_{\text{oil}} = 0.95 \text{ dB/cm;} \quad \text{and}$$

$$\rho_{\text{fat}} = 920 \text{ kg/m}^3, \ c_{\text{fat}} = 1450 \text{ m/s}, \ \alpha_{\text{fat}} = 0.63 \text{ dB/cm.}$$

Figure P11.5
Problem 11.13.

Assume that the transducer is sensitive to at most $L = 65$ dB pressure amplitude loss.

(a) What is the depth of penetration inside the oil?

(b) What is the approximate beamwidth at the given oil/fat interface?

(c) Suppose the peak acoustic pressure of the generated pulse is 20 N/cm^2. Compute the peak pressure of the returning pulse. Is it detectable by the transducer system? Does this result conflict with the answer in part (a)? Explain why or why not.

Now, assume that the interface is not fixed; instead, it is given by $z_0(t) = 20 - 5\cos(2\pi f_0 t)$ cm, where $f_0 = 100$ Hz.

(d) Assume that the transducer fires a pulse at $t = 0$. Sketch the A-mode signal. Label the axes carefully, and identify the time-of-return of the returning pulse.

(e) Assume that the transducer fires every 10 ms, starting from $t = 0$. Sketch the M-mode image that would be generated. Label the axes carefully.

11.14 Consider a single, flat, square, ultrasound transducer whose sides have lengths of 0.5 cm and whose resonant frequency is 2.5 MHz. Suppose it is embedded in a homogeneous medium whose density ρ is 920 kg/m^3 and whose acoustic impedance Z is 1.35×10^6 kg/m^{2}s.

(a) What range z_f characterizes the range at which the far field begins for this transducer in this medium?

(b) For a typical system (i.e., make appropriate assumption(s) about the system), what does the absorption coefficient α of this medium have to be if the depth of penetration is 20 cm?

(c) What is the lateral resolution (FWHM) of this transducer at a range $z = 10$ cm?

(d) Assume that this transducer could be scanned (slid) rapidly in the x-direction and repetitively pulsed in order to create a rectangular-shaped B-mode image. It is desired to obtain A-mode data separated by 1 mm in the x-direction and to scan a 12 cm wide region. Ignoring any potential mechanical problems with scanning the transducer, what is the maximum frame rate that can be achieved given the depth of penetration indicated in (b)?

(e) Suppose there is an interface at range 10 cm, normal to the transducer's axis, and assume that the second medium (beyond 10 cm) has density 1070 kg/m^3 and acoustic impedance 1.7×10^6 kg/m^{2}s. If the acoustic pressure of the initial pulse has amplitude A_0 and the absorption coefficient is $\alpha = 1.5$ dB/cm in the first medium, what is the amplitude of the reflected acoustic pressure pulse when it arrives at the transducer?

Magnetic Resonance Imaging

Overview

In this part of the book, we consider magnetic resonance imaging (MRI). Like computed tomography, MRI produces high resolution, high contrast cross-sectional (tomographic) anatomic images through the body. Like ultrasound imaging, MRI is noninvasive. The combination of high image quality and risk-free imaging has made MRI the fastest growing modality in recent years.

In MRI, the signals arise from the nuclear magnetic resonance properties of tissues; these properties are "stimulated" by the application of fixed magnetic and variable radio-frequency fields. Since the *pulse sequences* that govern the time-varying application of these fields can be changed by the operator, MRI—like nuclear medicine—has the potential to create a number of different images representing different underlying signals. In addition, clever combinations of pulse sequences can be used to create a dynamic series of images, which in turn can be used to estimate blood flow. *Functional magnetic resonance imaging (fMRI)* is rapidly becoming a powerful research tool, especially in understanding brain function, and it complements its clinical use for anatomic imaging. In addition, paramagnetic "contrast agents" or "tracers" akin to those used in radiography and nuclear medicine are being developed to improve image contrast and measure additional functions. Ultimately, MRI may prove to have the ideal combination of structural and functional imaging capabilities in a risk-free modality.

A typical MRI system looks like a CT system, consisting of a large gantry with a "tunnel" through which the patient, lying on a bed, is inserted. This gantry houses the apparatus for applying the magnetic and RF fields as well as the *receiver coils* from which the raw signal data arise. MRI, like CT, is a cross-sectional or tomographic imaging modality. Image reconstruction, while conceptually equivalent to that in CT, is derived directly from the nature of the raw signals in frequency space.

In clinical practice, MRI is mainly used to depict anatomy. It produces high-contrast cross-sectional images through the body [see Figure I.1(d)] and brain

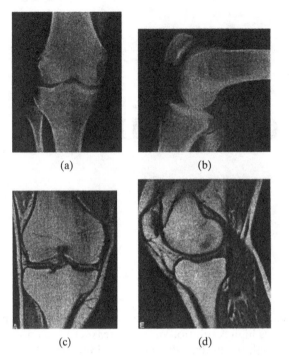

(a)   (b)

(c)   (d)

Figure V.1
A comparison of projection radiography and MRI of the knee: (a) anterior and (b) lateral projection radiographs; corresponding (c) coronal and (d) sagittal MRIs.

[see Figure I.4(b)] and has displaced CT in many clinical applications. We used a sagittal MRI of the head extensively in Chapters 2 and 3 to illustrate some basic concepts. Another clinical example is shown in Figure V.1. Here, conventional anterior and lateral projection radiographs of the knee are shown in Figure V.1(a) and (b), and the corresponding MRI coronal and sagittal slice images are shown in Figure V.1(c) and (d).

Physics of Magnetic Resonance

12.1 Introduction

Magnetic resonance imaging (MRI) is made possible by the physical phenomenon called *nuclear magnetic resonance* (NMR). In this chapter, we provide an overview of NMR in order to be able to understand MRI. It turns out that it is not necessary to understand the quantum physics aspects of NMR, although it is informative to appreciate these origins, if for no other reason than to know where to look for a deeper understanding of the physical principles of MRI. We can then rapidly progress from the quantum principles to the macroscopic principles, which can be understood from the principles of classical physics. It is macroscopic *spin systems* and their interaction with electric and magnetic fields that give rise to the signals from which MR images are made. The behavior of these spin systems is described by the so-called *Bloch equations*, which are phenomenological.[1] The Bloch equations tie three important physical properties of these spin systems—proton density, longitudinal relaxation, and transverse relaxation—to the generation of signals that can be measured by an MR scanner and turned into an image, as described in the next chapter.

12.2 Microscopic Magnetization

Nuclear magnetic resonance is concerned with the nuclei of atoms but not with radioactivity, as it is in nuclear medicine. Instead, it is the *charge* and *angular momentum* possessed by certain nuclei that are of concern in NMR. We know that all nuclei have positive charges, of course, because they are comprised of protons and neutrons. It is also true that a nucleus with either an odd atomic number or an odd mass number has an *angular momentum* Φ. Said to have *spin*, these nuclei are NMR-active. We can visualize such a nucleus as a small ball that is rotating about an axis, as shown in Figure 12.1(a).

---

[1] The term *phenomenological* refers to a process or equation that is based on an observed phenomenon rather than one that is derived from fundamental or "first" principles.

Figure 12.1
(a) Visualization of the angular momentum of a nucleus. (b) Graphic depicting the microscopic magnetization of a nucleus.

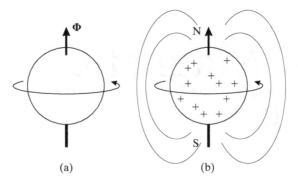

(a)   (b)

Collections of identical nuclei—regardless of the molecular environment in which they are found—are called *nuclear spin systems*. The nuclei of ^{1}H, ^{13}C, ^{19}F, and ^{31}P are important in NMR because they are prevalent in biological systems and thus provide a large enough NMR signal to be detected above the background noise. In whole-body MR imaging of anatomy, however, we are concerned only with the nuclei of ^{1}H. ^{1}H atoms are present in very high density within the body (mainly because of the water content in our bodies), and an ^{1}H nucleus gives a very strong NMR signal. Because the nucleus of the hydrogen atom is a single proton, it is sometimes said that MRI images "protons," but this is not strictly true, because protons are present in other atoms but are not imaged in conventional MRI.

Each nucleus that has a spin also possesses a microscopic magnetic field. Although the physical principles for this fact can only be rigorously derived using quantum physics, we can use a classical approach to argue for this as well. We envision each nucleus having a positive charge that is spinning around an axis, as shown in Figure 12.1(b). Since a magnetic field is present whenever there are circulating charges (as in a loop of wire), the nucleus must possess a magnetic field. The microscopic magnetic field has a magnetic moment vector $\boldsymbol{\mu}$, given by

$$\boldsymbol{\mu} = \gamma \boldsymbol{\Phi}, \tag{12.1}$$

where γ is the *gyromagnetic ratio* and has units of radians per second per tesla. It is useful to define

$$\gamma\!\!\!\!/ = \frac{\gamma}{2\pi}, \tag{12.2}$$

which has units of Hz per tesla. Some gyromagnetic ratios for common nuclei used in NMR and MRI are given in Table 12.1.

In general, there is no preferred orientation for nuclei in a given sample of material. The randomly oriented individual nuclear spins cancel each other out macroscopically, and the resulting spin system has no macroscopic magnetic field. However, spin systems do become macroscopically magnetized when they are placed in a magnetic field, because in this case the microscopic spins tend to align with the applied external magnetic field. This property of nuclear spin systems is termed *nuclear magnetism*.

TABLE 12.1

Common Gyromagnetic Ratios	
Nucleus	γ [MHz/T]
^{1}H	42.58
^{13}C	10.71
^{19}F	40.05
^{31}P	11.26

The magnitude and direction of an external magnetic field can be represented as a vector B_0, given by

$$B_0 = B_0 \hat{z}. \qquad (12.3)$$

Here, B_0 is the magnitude of the magnetic field in tesla and $\hat{z}$ is a unit vector pointing in the $+z$ direction within a laboratory (fixed) frame. Intuition suggests that the microscopic spins should all point in the same direction as the applied magnetic field. Quantum physics predicts otherwise. Each nuclear species possesses a *spin quantum number*, I, which takes on nonnegative integer multiples of 1/2. All of the nuclei in Table 12.1, in fact, have $I = 1/2$—their spins systems are called *spin 1/2 systems*. For spin 1/2 systems in equilibrium, the microscopic magnetization can point in only one of two possible orientations relative to the direction $\hat{z}$. In particular, they can be 54 degrees off $\hat{z}$ or $180 - 54 = 126$ degrees off $\hat{z}$, with a very slight preference for the 54-degree (the so-called "up") orientation, which is the low energy state. Furthermore, the phase of μ—i.e., its orientation around the z-axis—is random. Putting all these facts together, the spin system overall becomes slightly magnetized in the $\hat{z}$ direction. This gives rise to the notion of "bulk" or macroscopic magnetization, which we now study in detail in order to understand how MR imaging works.

12.3 Macroscopic Magnetization

Consider a specific spin system (e.g., the nuclei of ^{1}H) within a volume of material. When an external static magnetic field B_0 is applied, the spin system becomes magnetized. This can be modeled using a bulk *magnetization vector* M, which is really just the sum of a large number N_s of individual nuclear magnetic moments

$$M = \sum_{n=1}^{N_s} \mu_n.$$

This situation is shown in Figure 12.2. If the sample is left undisturbed in the static field B_0 for a long period of time,[2] M will reach an equilibrium value M_0 that is

---

[2]The definition of "long" and "short" times will become clear after we look at relaxation processes.

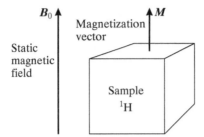

Figure 12.2
The magnetization vector M.

parallel with the direction of B_0 and has a magnitude M_0 (which is dependent only on its spatial position $r = (x, y, z)$). The magnitude of M_0 is given by

$$M_0 = \frac{B_0 \gamma^2 \hbar^2}{4kT} P_D, \qquad (12.4)$$

where k is Boltzmann's constant, T is temperature (from absolute zero), and P_D is the so-called *proton density*—i.e., the number of targeted nuclei per unit volume. In whole body imaging, the hydrogen nucleus is just a proton, which accounts for the name *proton density*. We are going to see that if M_0 is larger, the NMR signal is larger. Therefore, both larger magnets and higher proton densities produce larger NMR signals. When we learn about spatial encoding for MRI in Chapter 13, we will see that spatial differences in P_D give rise to signal differences, which in part accounts for image contrast.

It is of fundamental importance in NMR and MRI that M is a function of time. In addition to the changes in this vector as it approaches equilibrium within a magnet, it turns out that M can be manipulated in a spatially dependent fashion by using external radio-frequency excitations and magnetic fields. Thus, it should be understood that $M = M(r, t)$, where r is a three-dimensional laboratory coordinate and t is time. In analogy to the microscopic angular momentum studied at the nuclear scale, there is also a bulk *angular momentum J* corresponding to the sample, which is related to the sample's magnetization by

$$M = \gamma J. \qquad (12.5)$$

In MR imaging, the "sample" is a small volume of tissue—i.e., a voxel. The value of an MR image at a given voxel is determined by two dominant factors: the tissue properties and the scanner imaging protocol. There are several tissue properties that affect the appearance of an MR image, including two relaxation parameters T_1 and T_2, which we will define below, and the proton density. The second dominant factor is the way in which the vector M is manipulated by the scanner hardware and software using a so-called *pulse sequence*. As we shall see, the pulse sequence manipulates the magnetic field within the field of view in order to manipulate M over space and time. For now, we focus on a particular sample (voxel) and study how $M = M(t)$ changes with time. The so-called *equations of motion* for $M(t)$ are based on the *Bloch equations*, which we now begin to develop.

12.4   **Precession and Larmor Frequency**

Because $M(t)$ is a magnetic moment, it experiences a torque when an external, time-varying magnetic field $B(t)$ is applied. The equation describing this relationship is

$$\frac{dJ(t)}{dt} = M(t) \times B(t), \tag{12.6}$$

where J is the angular momentum vector associated with M. Using (12.5) to eliminate J yields

$$\frac{dM(t)}{dt} = \gamma M(t) \times B(t), \tag{12.7}$$

which is valid over a short period of time, where "short" will soon be defined.

Now suppose that $B(t)$ is a static magnetic field oriented in the z-direction—i.e., $B(t) = B_0$. If the initial magnetization vector $M(0)$ were oriented at an angle α relative to the z-axis, then the solution of (12.7) is

$$M_x(t) = M_0 \sin \alpha \cos (-\gamma B_0 t + \phi), \tag{12.8a}$$

$$M_y(t) = M_0 \sin \alpha \sin (-\gamma B_0 t + \phi), \quad \text{and} \tag{12.8b}$$

$$M_z(t) = M_0 \cos \alpha, \tag{12.8c}$$

where

$$M_0 = |M(0)|, \tag{12.9}$$

$M(t) = (M_x(t), M_y(t), M_z(t))$, and ϕ is an arbitrary angle. These equations describe a *precession* of $M(t)$ around B_0 with a frequency

$$\omega_0 = \gamma B_0, \tag{12.10}$$

which is known as the *Larmor frequency*.

As expressed in (12.10), the Larmor frequency has units of radians per second. In this book, we have expressed frequencies in cyclic rather than radial units. Therefore, we define the equivalent Larmor frequency

$$\nu_0 = \gamma B_0, \tag{12.11}$$

which has units of cycles per second, or hertz. Using (12.11) and (12.2), the equations of precession in (12.8) can be written as

$$M_x(t) = M_0 \sin \alpha \cos (-2\pi \nu_0 t + \phi) \tag{12.12a}$$

$$M_y(t) = M_0 \sin \alpha \sin (-2\pi \nu_0 t + \phi) \tag{12.12b}$$

$$M_z(t) = M_0 \cos \alpha. \tag{12.12c}$$

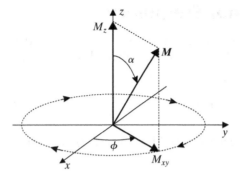

Figure 12.3
The magnetization vector M precesses about the z-axis.

Precession is an important concept in NMR and MRI. Most people have seen precession in a child's top or dreidel. A spinning top typically begins with the axis of the top, which is also its axis of rotation, pointing vertically. As the top slows, its spin axis begins to rotate about the vertical axis—this is precession. In NMR, $M(t)$ represents the axis of the top; the direction of the static magnetic field—$\hat{z}$, in this case—represents the vertical axis. This situation is shown schematically in Figure 12.3. Notice that precession occurs in the *clockwise* direction when viewed opposite the direction of the static magnetic field.

It might be assumed that the Larmor frequency is constant for a given spin system (e.g., ^{1}H) within a sample. After all, the gyromagnetic ratio is constant for a given spin system, and B_0 is supposed to be a constant. In reality, B_0 cannot be exactly constant. There are three sources of B_0 fluctuations of concern in MRI: (1) magnetic field inhomogeneities; (2) magnetic susceptibility; and (3) chemical shift. Magnet design and calibration is a very important issue in MRI. We shall see in the following chapter that *shimming* the main magnet to produce a very homogeneous field is critical to image quality. Generally, the main field inhomogeneity is kept to a few parts per million (ppm) and is slowly varying over the field of view, so it can be either ignored or compensated by postprocessing.

Magnetic susceptibility refers to the material property that decreases or increases the magnetic field within the material relative to the surrounding field. *Diamagnetic* materials slightly decrease the field, *paramagnetic* materials slightly increase the field, and *ferromagnetic* materials strongly increase the field. Both carbon and molecular oxygen (O^2) are diamagnetic, and as a result the body is mostly diamagnetic. This means that the field within the body is slightly lower than the field outside, and there are sudden changes in the magnetic field when moving from inside to outside the body. These sudden changes can have a deleterious effect on images.

Chemical shift is a measure of the change in Larmor frequency due to the chemical environment of the imaged nucleus—e.g., the molecules to which the ^{1}H atom is attached. In the chemical shift phenomenon, the local electron cloud can shield the nucleus from the full effects of the main magnetic field. Chemical shift can be modeled as a change in the magnitude of the magnetic field, by

$$\hat{B}_0 = B_0(1 - \varsigma), \tag{12.13}$$

where ς is the *shielding constant* and is a measure of the degree of chemical shift. The corresponding shifted Larmor frequency is given by

$$\hat{v}_0 = v_0(1 - \varsigma).$$ (12.14)

Chemical shift is usually expressed in ppm:

$$\varsigma(\text{ppm}) = \varsigma \times 10^6.$$ (12.15)

For example, the hydrogen nuclei in fat (CH_2) are shifted down by 3.35 ppm from those in water (H_2O). This corresponds to a shift of the Larmor frequency by -214 Hz at 1.5 T.

Groups of nuclei within a given spin system that have the same Larmor frequency (including chemical shift effects) are called *isochromats*. For example, the hydrogen nuclei in water form an isochromat, while those in fat form another isochromat. The importance of isochromats for a given spin system is that they have only very slightly different Larmor frequencies. The impact of this fact on imaging will become clear as we proceed.

12.5 Transverse and Longitudinal Magnetization

In order to obtain a deeper understanding of MRI, it is necessary to conceptualize the magnetization vector $M(t)$ as having *two* components. The first is the *longitudinal component*, which is oriented along the axis defined by the static magnetic field. The second is the *transverse component*, which is oriented in a plane orthogonal to the direction of the static magnetic field. The *longitudinal magnetization* is given by $M_z(t)$—simply the z-component of $M(t)$. The *transverse magnetization* is defined by

$$M_{xy}(t) = M_x(t) + jM_y(t),$$ (12.16)

which captures the two orthogonal components $M_x(t)$ and $M_y(t)$ as one complex quantity. The angle of the complex number M_{xy}, given by

$$\phi = \tan^{-1}\frac{M_y}{M_x},$$ (12.17)

is called the *phase* or *phase angle* of the transverse magnetization. Figure 12.3 illustrates this concept as a projection of the magnetization vector $M(t)$ against the z-axis and the xy-plane, respectively. The three Cartesian components of $M(t)$ are completely represented by just these two components, one of which is complex. Thus, understanding the evolution of $M_{xy}(t)$ and $M_z(t)$ yields, equivalently, an understanding of the temporal evolution of the three spatial components of $M(t)$.

12.5.1 **NMR Signals**

Using (12.12) and (12.16), the transverse magnetization can be written as

$$M_{xy}(t) = M_0 \sin \alpha e^{-j(2\pi \nu_0 t - \phi)} . \tag{12.18}$$

The origin of the *observed signal* in MRI can now be understood. The rapidly rotating transverse magnetization creates a radio frequency (RF) excitation within the sample. This RF excitation will in turn induce a voltage—a measurable signal—in a coil of wire located outside the sample. This signal is recorded for use in MRI. It is a common misconception that MRI uses *radio waves* to image the body. This is not true. In fact, it is *Faraday induction* that is used both to manipulate nuclear spin systems and to generate signals from active NMR samples. Thus, while it is true that the signal frequencies in whole body imaging are in the RF frequency band, radio waves themselves play very little role in MR imaging, and certainly do not contribute an irradiation dose to the patients.

Although it is common to refer to the transverse magnetization as the "signal," in order to actually receive a voltage signal from these rotating spins, there must be a coil of wire—a so-called *RF coil*—close to the sample. In Chapter 13, we will study various RF coil designs that are used in imaging; for now, we simply need to understand the abstract relationships.

Faraday's law of induction states that a time-varying magnetic field cutting across a coil of wire will induce a voltage in the wire. We can determine the amplitude of the induced voltage using the *principle of reciprocity*—that is, we first look at the magnetic field produced at a point in space as the result of a current flowing in the coil of wire. Specifically, suppose that the magnetic field at r produced by a unit direct current in the coil is given by $\boldsymbol{B}^r(r)$. Now reverse the scenario: suppose that there exists a time-varying magnetic field $\boldsymbol{M}(r, t)$ (our magnetic moment) throughout the object. Then, the voltage induced in the coil is given by

$$V(t) = -\frac{\partial}{\partial t} \int_{\text{object}} \boldsymbol{M}(r, t) \cdot \boldsymbol{B}^r(r) \, dr , \tag{12.19}$$

where · represents the dot (or inner) product.

To find a simpler expression for the induced voltage in an NMR experiment, we first assume that the object is homogeneous—i.e., $\boldsymbol{M}(r, t) = \boldsymbol{M}(t)$. We also assume that the coil produces a uniform field when excited, so $\boldsymbol{B}^r(r) = \boldsymbol{B}^r$. We also note that the z component of magnetization is a slowly changing quantity (we will see why later); hence, its temporal derivative is small and can be ignored. Using these simplifications in (12.19) and breaking each field into its component representation yields

$$V(t) = -\frac{\partial}{\partial t} \int_{\text{object}} M_x(t)B_x^r + M_y(t)B_y^r \, dr , \tag{12.20a}$$

$$= -V_s \frac{\partial}{\partial t} \left[M_x(t)B_x^r + M_y(t)B_y^r \right] , \tag{12.20b}$$

where V_s is the volume of the sample.

The transverse magnetization components are found as the real and imaginary parts of $M_{xy}(t)$ in (12.18),

$$M_x(t) = M_0 \sin \alpha \, \cos(-2\pi \nu_0 t + \phi) \tag{12.21a}$$

$$M_y(t) = M_0 \sin \alpha \, \sin(-2\pi \nu_0 t + \phi) \,. \tag{12.21b}$$

Using these expressions in (12.20) yields

$$V(t) = -2\pi \nu_0 V_s M_0 \sin \alpha \left[B_x^r \sin(-2\pi \nu_0 t + \phi) - B_y^r \cos(-2\pi \nu_0 t + \phi) \right] \,. \tag{12.22}$$

Writing the components of the reference field as

$$B_x^r = B^r \cos \theta_r \tag{12.23a}$$

$$B_y^r = B^r \sin \theta_r \,, \tag{12.23b}$$

and further simplifying (using a trigonometric identity) yields

$$V(t) = -2\pi \nu_0 V_s M_0 \sin \alpha \, B^r \sin(-2\pi \nu_0 t + \phi - \theta_r) \,. \tag{12.24}$$

The basic NMR signal is a sinusoid at the Larmor frequency.

The magnitude of the NMR signal in (12.24),

$$|V| = 2\pi \nu_0 V_s M_0 \sin \alpha \, B^r, \tag{12.25}$$

is of central interest, since we would ordinarily want to maximize it. Recall from (12.4) that M_0 is proportional to B_0, and ν_0 is also proportional to B_0. Therefore (12.25) reveals that signal strength is proportional to B_0^2, which means that higher field magnets will yield larger signals. Current whole-body scanners use field strengths of 1.5 tesla, but scanners with fields up to 4 tesla are available and approved for human use by the FDA, and these scanners yield much higher signal strengths. We see also that the maximum signal is produced when $\alpha = \pi/2$. The angle α is called the *tip angle* (or sometimes the flip angle), and is something we can control when acquiring MR data. We will find that it is sometimes beneficial to use smaller tip angles in order to image faster, as will be discussed in the next chapter, but this practice comes at the expense of signal strength. The last factor of interest is the sample volume V_s, a quantity that is, to some degree, under our control. In NMR, we can derive a larger signal by putting more of the substance under study into the scanner. We will discover in MRI that V_s will represent *voxel size*, and that larger signal strength is achieved by having larger voxels.

EXAMPLE 12.1

From (12.25), we can see that in order to improve the resolution, while keeping the signal strength at the same level, we need to increase B_0. Now consider a scanner with $B_0 = 1.5$ tesla.

Question   If we want to double the resolution in all three dimensions, what B_0 should we use to keep the signal strength unchanged?

Answer   By doubling the resolution in all three dimensions, we halve the voxel size in all directions. This change reduces the volume of a voxel by a factor of $1/2^3 = 1/8$. From (12.25), we know that in order to keep the signal strength at the same level, we must have B_0 increased by a factor of $\sqrt{8}$. So, in order to double the resolution, we need to use

$$B_0 = 1.5 \times \sqrt{8} = 4.24 \text{ tesla.}$$ ∎

12.5.2 Rotating Frame

It is sometimes convenient to express and visualize the evolution of the magnetization vector in a frame of reference, called the *rotating frame*, that is rotating at the Larmor frequency ν_0. The coordinates in the rotating frame are related to those in the stationary frame by

$$x' = x \cos(2\pi \nu_0 t) - y \sin(2\pi \nu_0 t) , \tag{12.26a}$$

$$y' = x \sin(2\pi \nu_0 t) + y \cos(2\pi \nu_0 t) , \tag{12.26b}$$

$$z' = z . \tag{12.26c}$$

In this frame of reference, (12.18) becomes

$$M_{x'y'}(t) = M_0 \sin \alpha e^{j\phi} . \tag{12.27}$$

In other words, in the rotating frame, $M_{x'y'}$ is a *stationary vector* in the rotating complex plane with magnitude $M_0 \sin \alpha$ and phase angle ϕ.

We now look at the method used to tip $M(t)$ away from B_0 in order to elicit an NMR signal.

12.6   RF Excitation

We have seen that the magnetization vector M will precess if it is initially oriented away from B_0, and the transverse component of this precession will induce a current in an antenna surrounding the sample. By putting RF current through an antenna surrounding the sample, the spin system can be deliberately excited and thereby control the behavior of M. This is how we can produce magnetization vectors that are not parallel to B_0, for example. In other words, we can excite the spin system using RF signals so that the stimulated system will in turn induce RF signals as output.

Equation (12.7) can be used to understand how RF excitation will allow us to control the magnetization vector. Consider a system in equilibrium [so that $M(t)$ lines up with B_0]. If a small magnetic field $B_1 = B_1 \hat{x}$ oriented in the x-direction is turned on (adding to the main field), then (12.7) predicts a small motion of $M(t)$ in the $+y$ direction. One understands this as an incremental precession around the x-axis due to the presence of the additional field B_1. This is the beginning of a process that can lead, ultimately, to the magnetization vector $M(t)$ resting on the transverse plane, and to process about the z-axis once B_1 is turned off.

There is a difficulty with the above approach to spin system excitation, however. As soon as the incremental precession around the x-axis begins, the spin system is no longer in equilibrium, and the magnetization vector $M(t)$ begins to precess (also) around the z axis, which has a much larger field strength, and therefore a much larger rate of precession. Once that precession begins, an analysis of the location of $M(t)$ (see Problem 12.4) reveals that the desired effect of B_1 is lost. In fact, if we are to continually push $M(t)$ toward the transverse plane (the desired goal in most instances), we will need to track the position of $M(t)$ as it precesses around the z-axis, and apply a B_1 field whose orientation will produce the correct motion, in accordance with (12.7).

Since $M(t)$ is precessing at the Larmor frequency, the first step in tracking the position of $M(t)$ is to apply the B_1 field at the Larmor frequency rather than keeping it constant. In this way, the excitation becomes an *RF excitation*. This has a desirable effect, since the precessing vector $M(t)$ is pushed down toward the transverse plane whenever it coincides with the $\pm y$ axis (see Problem 12.4). Such an RF excitation is said to be *linearly polarized* because its B_1 field is oriented along only one linear axis. It is possible to improve this approach by adding another RF field oriented in the y direction (still orthogonal to the main field). By applying a quadrature excitation (sine instead of cosine) to this y-oriented B_1 field, the magnetization vector is continuously pushed toward the transverse plane. Such an RF excitation is said to be *circularly polarized* since the direction of the B_1 field traces out a circle in the transverse plane.

Circularly polarized RF excitations are produced using quadrature RF coils (see Chapter 13), which are in common use today. The circularly polarized RF field can be modeled as a complex magnetic field in the transverse plane

$$B_1(t) = B_1^e(t)e^{-j(2\pi \nu_0 t - \varphi)}, \qquad (12.28)$$

where $B_1^e(t)$ is the envelope of $B_1(t)$ and φ is its initial phase. The simplest envelope is a rectangular pulse with amplitude B_1 and duration τ_p, yielding a simple RF burst. In the rotating frame, the RF field is given by

$$B_1(t) = B_1^e(t)e^{j\varphi}. \qquad (12.29)$$

For simplicity, we will assume in the following that $\varphi = 0$, so that B_1 is oriented in the x' direction of the rotating frame. This excitation causes M to precess in the clockwise direction in the $y'z'$-plane (in analogy to the precession we observed earlier for the static B_0 field), as shown in Figure 12.4(a). This type of motion of M is called *forced precession*, because it occurs in response to a deposition of RF energy that comes from outside the sample. The frequency of this precession is given by $\nu_1 = \gamma B_1$, where $B_1 = |B_1^e(t)|$. Of course, the actual precession must include the rotation in the xy-plane. The actual evolution of $M(t)$ is a spiral from the z-axis toward the xy-plane in a clockwise orientation when viewed from the $+z$-axis, as shown in Figure 12.4(b).

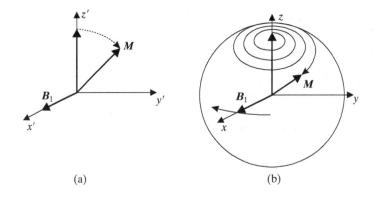

Figure 12.4
(a) In the rotating frame, the magnetization vector M precesses around the x' axis in response to an RF excitation. (b) In the laboratory frame, the actual motion of M is a spiral.

The final tip angle and phase of M depends on both the amplitude and duration of $B_1^e(t)$. If we turn off the RF after M has precessed down into the transverse plane, then the pulse is called a $\pi/2$ (pronounced "pi over two") pulse. A $\pi/2$ pulse is commonly used as an *excitation pulse*, because it elicits the maximum signal from a sample that is in equilibrium. An RF pulse twice as long as this is called a π pulse; it will place M along the $-z$ axis. This is often called an *inversion pulse*, because it inverts the orientation of M. In general, the final tip angle after an RF excitation pulse of duration τ_p is given by

$$\alpha = \gamma \int_0^{\tau_p} B_1^e(t)\,dt\,. \tag{12.30}$$

For the special case of a rectangular pulse, M is tipped through an angle

$$\alpha = \gamma B_1 \tau_p\,. \tag{12.31}$$

Because the RF excitation pulse changes α, such a pulse is referred to as an α-*pulse*.

EXAMPLE 12.2
We apply an RF pulse to a sample of protons. The sample is in equilibrium with the B_0 field in the $+z$ direction.

Question   We need to tip the magnetization vector M into x-y plane in 3 ms. What should the strength of RF excitation be?

Answer   We need a $\pi/2$ pulse to tip the magnetization vector M aligned in z direction into x-y plane. The gyromagnetic ratio for protons is $\gamma = 42.58$ MHz/T. The tip angle is

$$\alpha = \pi/2 = 2\pi\gamma B_1 \tau_p = 2\pi \times 42.58\,\text{MHz/T} \times B_1 \times 3\,\text{ms}$$

So the strength of the RF excitation should be

$$B_1 = 1.96 \times 10^{-6}\,\text{tesla} = 0.0196\,\text{gauss}. \qquad \blacksquare$$

12.7 Relaxation

After application of an α-pulse, and assuming that $\alpha \neq \pi$, M will precess in response to the presence of the main magnetic field B_0, as described earlier. According to (12.7), this precession will never end. If this were true, then RF waves would emanate from the sample and could be detected forever as a sinusoidal voltage in an external antenna. Obviously, this situation cannot be true; there must be some mechanism to dampen this (otherwise perpetual) motion. In fact, there are two independent relaxation processes that together cause the received signal to vanish: longitudinal relaxation and transverse relaxation. We now describe each of these in detail.

Transverse relaxation acts first to cause the received signal to decay. Also known as *spin-spin relaxation*, transverse relaxation is caused by perturbations in the magnetic field due to other spins that are nearby. This interaction, heavily influenced by random microscopic motion, causes spins to momentarily speed up or slow down, changing their phases relative to other nearby spins, as illustrated in Figure 12.5. This *dephasing* causes a loss of coherence of the RF wave produced by the spin system, and a concomitant loss of signal in the receiver antenna. The resultant signal is known as a *free induction decay* (FID), as illustrated in Figure 12.6.

The signal decay in a free induction decay is well-modeled as an exponential decay. The time constant of this decay—called the *transverse relaxation time*—is given the symbol T_2 and, like a conventional time constant, has units of time. Accordingly, rather than modeling the transverse magnetization as a never-ending

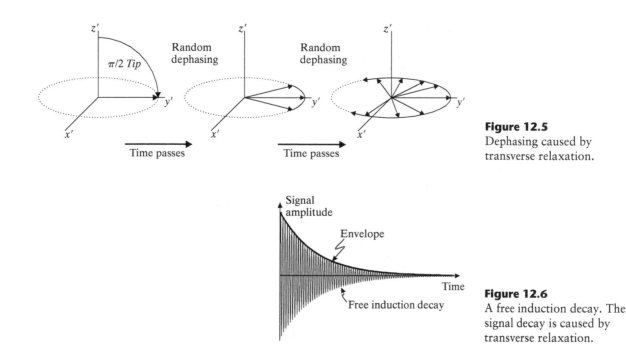

Figure 12.5
Dephasing caused by transverse relaxation.

Figure 12.6
A free induction decay. The signal decay is caused by transverse relaxation.

complex exponential as in (12.18), a more accurate representation is given by

$$M_{xy}(t) = M_0 \sin \alpha e^{-j(2\pi \nu_0 t - \phi)} e^{-t/T_2} . \tag{12.32}$$

T_2 is generally different for various types of tissues and is one of the important physical properties of the body that gives rise to contrast in MR images. We will see how this works in the next chapter. The decay of signal strength $|M_{xy}(t)|$ due to T_2 relaxation is shown graphically in Figure 12.7(a).

After this careful presentation of T_2, we are now going to "pull a fast one on you," and reveal that the received signal actually decays more rapidly than T_2. In fact, local perturbations in the static field B_0 cause the received signal to decay exponentially with a time constant T_2^* (pronounced "tee two star"), which satisfies $T_2^* < T_2$. This situation is illustrated in Figure 12.7(a). It is useful to model the decay associated with these "external" field effects using a time constant T_2'. The relationship between the three transverse relaxation constants is

$$\frac{1}{T_2^*} = \frac{1}{T_2} + \frac{1}{T_2'} . \tag{12.33}$$

It is natural to ask why we need the concept of T_2 if it does not fit reality. We will find that although the initial signal decays with T_2^*, there remains an underlying magnetization coherence that lasts longer—it decays with time constant T_2, in fact. This is consistent with the fact that the relaxation due to T_2' effects is reversible. In

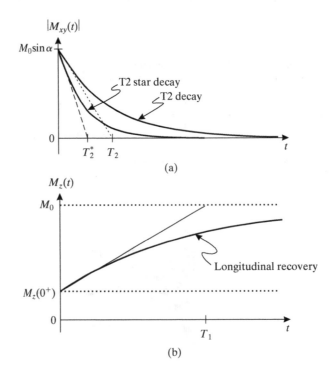

Figure 12.7
(a) Transverse and (b) longitudinal relaxation.

making images, we will crucially depend on the *refocusing* of this latent coherence through the concept of *echoes*. So, although the initial signal disappears after a few T_2^*'s, there is an underlying signal potential lasting a few T_2's.

The second relaxation mechanism that causes a loss of signal is called *longitudinal relaxation* or *spin-lattice relaxation*. This process concerns the longitudinal magnetization $M_z(t)$, which recovers back to its equilibrium value M_0 as a rising exponential, as depicted in Figure 12.7(b). Although transverse relaxation is a decreasing exponential and longitudinal relaxation is an increasing exponential, both processes lead to a loss of NMR signal.

Suppose that an α-pulse $(\alpha \neq \pi)$ is applied at $t = 0$. Then, the longitudinal magnetization obeys

$$M_z(t) = M_0(1 - e^{-t/T_1}) + M_z(0^+)e^{-t/T_1} , \qquad (12.34)$$

where T_1 is a material property called the *longitudinal relaxation time*. The notation $M_z(0^+)$ refers to the longitudinal magnetization immediately after the α-pulse and is given by

$$M_z(0^+) = M_0 \cos \alpha . \qquad (12.35)$$

It is implied by this notation that the duration of the α-pulse is negligible with respect to the longitudinal relaxation time. This is a good assumption since, typically, α-pulses are a few milliseconds long while longitudinal recovery lasts several hundred milliseconds.

This recovery of longitudinal magnetization is illustrated in Figure 12.7(b). Like T_2, T_1 is generally different for various types of tissue and is also responsible for generating contrast in MR images, as we shall see. For tissues in the body, the relaxation times are in the ranges 250 ms $< T_1 < 2500$ ms and 25 ms $< T_2 < 250$ ms. Usually, $5T_2 \leq T_1 \leq 10T_2$ (although there are some exceptions), and for all materials $T_2 \leq T_1$.

Although we have been using the term "equilibrium" throughout the chapter, our understanding has been limited to some notion of an "unchanging" or "quiescent" system. We can now understand what equilibrium means mathematically. A sample is said to be in *equilibrium* if its longitudinal magnetization $M_z(t)$ is essentially equal to its final value M_0 throughout the sample. In practice, a sample is considered to be in equilibrium if there have been no external excitations for at least $3T_1^{max}$, where T_1^{max} is the largest T_1 in the sample. In the next chapter, we will encounter the term *steady-state*; it is worth noting now that this is a different notion than equilibrium. Equilibrium implies that there has been no excitation for some time into the past; steady-state describes a spin system that has been *periodically excited* so that the spin system is undergoing a periodic longitudinal relaxation process.

EXAMPLE 12.3
Suppose a sample is in equilibrium, and a $\pi/2$ pulse is applied.

Question What happens to the longitudinal magnetization of the sample?

Answer The longitudinal magnetization immediately after the α-pulse is $M_z(0^+) = M_0 \cos(\pi/2) = 0$. Therefore, (12.34) yields

$$M_z(t) = M_0(1 - e^{-t/T_1}),$$

where M_0 is the equilibrium value of longitudinal magnetization. ∎

EXAMPLE 12.4

Suppose a sample is in equilibrium, and an (unspecified) α-pulse is applied.

Question What are the transverse and longitudinal magnetizations of the sample, expressed in the rotating frame?

Answer Since the sample is initially in equilibrium, the longitudinal magnetization immediately after the α-pulse is given by (12.35). Substituting (12.35) into (12.34) yields the longitudinal magnetization

$$M_z(t) = M_0(1 - e^{-t/T_1}) + M_0 \cos\alpha \, e^{-t/T_1}.$$

This expression is the same whether expressed in the laboratory coordinate frame or the rotating frame. The transverse magnetization, expressed in the rotating frame, is just $M_{xy}(t)$ in (12.32) multiplied by $e^{+j2\pi \nu_0 t}$, which yields

$$M_{xy}(t) = M_0 \sin\alpha \, e^{j\phi} e^{-t/T_2}.$$

∎

12.8 The Bloch Equations

Putting together both the *forced* and *relaxation* behavior of a magnetic spin system yields the *Bloch equations*

$$\frac{dM(t)}{dt} = \gamma M(t) \times B(t) - \mathrm{R}\{M(t) - M_0\}, \tag{12.36}$$

which describe the behavior of M in the laboratory frame.[3] Here, $B(t)$ is composed of the static field and the RF field,

$$B(t) = B_0 + B_1(t),$$

and R is the *relaxation matrix*, given by

$$\mathrm{R} = \begin{pmatrix} 1/T_2 & 0 & 0 \\ 0 & 1/T_2 & 0 \\ 0 & 0 & 1/T_1 \end{pmatrix}. \tag{12.37}$$

The Bloch equations are used to construct models of the behavior of magnetization vectors during excitation. From this behavior, the NMR signal can be inferred by

---

[3]This equation is referred to in the plural because it is a vector equation comprising three scalar equations.

computing the transverse magnetization. For most applications, these equations are transformed into the rotating frame of reference (see the problems at the end of this chapter).

EXAMPLE 12.5

The Bloch equations describe the behavior of M in the laboratory frame.

Question   Find the equations for the components of M in x-y plane, and verify that the transversal relaxation after a $\pi/2$ pulse (in x direction) satisfies the equations.

Answer   The Bloch equations are given above in (12.36). By expanding the cross product, we have

$$\frac{d}{dt}\begin{pmatrix} M_x(t) \\ M_y(t) \\ M_z(t) \end{pmatrix} = \gamma \begin{pmatrix} M_y(t)B_z(t) - M_z(t)B_y(t) \\ -M_x(t)B_z(t) + M_z(t)B_x(t) \\ M_x(t)B_y(t) - M_y(t)B_x(t) \end{pmatrix} - \begin{pmatrix} \frac{1}{T_2}M_x(t) \\ \frac{1}{T_2}M_y(t) \\ \frac{1}{T_1}(M_z(t) - M_{0z}) \end{pmatrix}.$$

After a $\pi/2$ pulse, the RF field B_1 is shut down, and only B_0 is nonzero. Therefore, $B_x(t) = B_y(t) = 0$, and the equations for $M_x(t)$ and $M_y(t)$ are simpler:

$$\frac{d}{dt}\begin{pmatrix} M_x(t) \\ M_y(t) \end{pmatrix} = \gamma \begin{pmatrix} M_y(t)B_z(t) \\ -M_x(t)B_z(t) \end{pmatrix} - \begin{pmatrix} \frac{1}{T_2}M_x(t) \\ \frac{1}{T_2}M_y(t) \end{pmatrix}.$$

The transversal relaxation after a $\pi/2$ pulse is given by

$$M_x(t) = M_0 \cos[-(2\pi \nu_0 t - \pi/2)]e^{-t/T_2} = -M_0 \sin(2\pi \nu_0 t)e^{-t/T_2}, \qquad \text{and}$$

$$M_y(t) = M_0 \sin[-(2\pi \nu_0 t - \pi/2)]e^{-t/T_2} = -M_0 \cos(2\pi \nu_0 t)e^{-t/T_2}.$$

The initial phase is $\pi/2$ because the RF pulse is applied in x direction, the magnetization vector M is tipped into y direction. By substituting the relaxation equations into the Bloch equations for $M_x(t)$, we get

$$\frac{dM_x(t)}{dt} = \frac{d}{dt}\left[-M_0 \sin(2\pi \nu_0 t)e^{-t/T_2}\right]$$

$$= -2\pi \nu_0 M_0 \cos(2\pi \nu_0 t)e^{-t/T_2} + \frac{1}{T_2}M_0 \sin(2\pi \nu_0 t)e^{-t/T_2}$$

$$= \gamma B_0 M_y(t) - \frac{1}{T_2}M_x(t).$$

Similarly, we have

$$\frac{dM_y(t)}{dt} = -\gamma B_0 M_x(t) - \frac{1}{T_2}M_y(t). \qquad \blacksquare$$

12.9   Spin Echoes

Pure transverse relaxation, characterized by the time constant T_2, is a random phenomenon. The fact that the FID decays faster, with time constant T_2^*, is due to fixed perturbations in the magnetic field. These fixed perturbations cause the precession of some spins to speed up and others to slow down (relative to the nominal rate of rotation predicted by the Larmor frequency). As described in Section 12.7, in a very brief period, nearby spins are largely dephased—i.e., they begin to point in different directions in the transverse plane—as illustrated by the top row of Figure 12.8.

The existence of *spin echoes* is due to the fact that the faster (slower) spins, which now lead (lag) the spin system, can be made to lag (lead) the spin system using a short-duration 180-degree pulse, as shown in Figure 12.8(d). From this new phase position, the fast spins "catch up" and the slow spins "fall back," forming a *spin echo*, as shown in the bottom row of Figure 12.8. Thus, a spin echo is the signal that is generated by the transverse spins recovering their coherence after loss of coherence followed by a deliberate 180-degree RF pulse. The time interval from the initial $\pi/2$-pulse to the formation of the spin echo is known as the *echo time* and is given the symbol T_E. This time is under our control because we specify the application time of the 180-degree pulse, which is at $T_E/2$.

Until this section, we have not had to consider the timing of NMR signals; we have simply generated an α pulse and watched what happens. In the generation of spin echoes, however, we see that it is the succession of two pulses that matters, and the timing of the second pulse relative to the first determines when the signal (echo) occurs. As we will discover in Section 12.10 and in even greater detail in Chapter 13, it is the type and ordering of excitations and their relative timings that gives rise to different tissue contrasts, as well as a host of other image characteristics such as resolution, noise, and artifacts. In the case of spin echoes, Figure 12.9 shows a simple timing diagram, known as a *pulse sequence*, that can be followed to generate the echo signal.

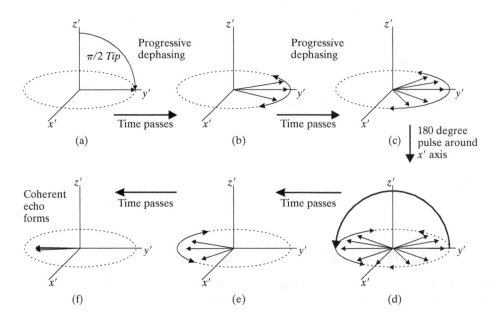

Figure 12.8
Formation of a spin echo.

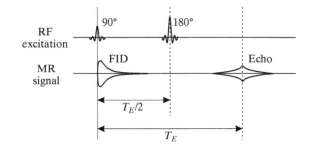

Figure 12.9
Pulse sequence diagram for generating spin echoes.

There are two mechanisms that combine to make spin echoes decrease in amplitude over time. First, because time elapses during the spin echo process, longitudinal relaxation causes the magnitude of the transverse magnetization to decrease. This idea is evident in Figure 12.8, where the transverse component of the magnetization vector is shown to be getting smaller with time. Second, because of the random effects of transverse relaxation, the phase of the coherent echo is never perfectly aligned, as illustrated in Figure 12.8(f). Since T_2 is often much smaller than T_1, the first effect can often be ignored; therefore, the amplitude of an echo is approximately given by the ideal T_2 exponential decay of the tissue at T_E. This means that we can elicit *multiple echoes* using multiple 180-degree pulses, and the signal strength of these echoes will decay exponentially with time constant T_2, until they disappear at approximately $3T_2$.

EXAMPLE 12.6

Suppose two ^{1}H isochromats are in different locations in a 1.5 T (tesla) magnet, and the fractional difference in field strength is 20 ppm.

Question   How long will it take before these isochromats are 180 degrees out of phase?

Answer   Suppose one sample has field strength B_0 and the other B_0'. The magnitude of the field strength difference can be found as follows:

$$B_0' = B_0(1 - 20 \times 10^{-6})$$

$$|B_0' - B_0| = (20 \times 10^{-6})B_0$$

The difference in Larmor frequency is then

$$\Delta \nu = \gamma |B_0' - B_0|$$

$$= \gamma (20 \times 10^{-6})B_0$$

$$= 42.58 \text{ MHz/T} \times (20 \times 10^{-6}) \times 1.5 \text{ T}$$

$$= 1277.4 \text{ Hz}.$$

Since 180° is half a cycle, the time it will take before these isochromats are 180° out of phase is

$$\Delta t = \frac{1/2 \text{ cycle}}{1277.4 \text{ cycle/s}} = 391 \ \mu\text{s}.$$

Question   What will be their phase difference at $T_E/2$ if the echo time is 4 ms?

Answer T_E is defined to be the echo time; therefore, $T_E/2 = 2$ ms. The number of cycles occurring in 2 ms is

$$\text{Number of cycles} = 1277.4 \text{ cycles/s} \times 2 \text{ ms}$$

$$= 2.555 \text{ cycles}$$

The phase difference is therefore

$$\text{Phase difference} = 2.55 \text{ cycles} \times 2\pi \text{ radians/cycle}$$

$$= 16 \text{ radians}.$$

Thus, after this relatively short amount of time, the two signals have completely lost coherence.   ∎

12.10   Contrast Mechanisms

We now understand that it is the transverse magnetization $M_{xy}(t)$ that produces the measurable MR signal; the larger its magnitude, the larger the measured signal. If we are to see a contrast between tissues—e.g., different image intensities in the gray matter and white matter of the brain—the measured signal must be different in those tissues. In MRI, our ability to generate tissue contrast depends on both the intrinsic NMR properties of the tissues—i.e., P_D, T_2, and T_1—and the characteristics of the externally applied excitations. So far, we have seen that it is possible to control the *tip angle* α and the *echo time* T_E of the RF excitation. It is also possible to control the interval between successive α-pulses; this is the so-called *pulse repetition interval*, and is given the symbol T_R. In this section, we consider the response of small volumes containing different tissues to the same external application of RF excitations. We will see how tissue contrast can be manipulated by externally controllable parameters.

Figure 12.10 shows three images of the same slice through a human skull. These clearly show quite different tissue contrasts in the three images. The contrast in the images are classified as (a) P_D-weighted, (b) T_2-weighted, and (c) T_1-weighted. This does not mean that the brightness of these images are proportional to P_D, T_2, and T_1, but merely that the differences in intensity seen between different tissues are largely determined by the differences in P_D, T_2, and T_1, respectively, of the tissues. In the brain, there are two dominant tissue types, gray matter (GM) and white matter (WM), which are surrounded by cerebrospinal fluid (CSF). The NMR properties of these three "tissues" are given in Table 12.2.

Proton Density Weighted Contrast   In proton density weighted images, the image intensity should be proportional to the number of hydrogen nuclei in the sample. This weighting can be obtained with the kind of NMR experiment we have already considered. We merely need to start with the sample in equilibrium, apply an excitation RF pulse, and image quickly, before the signal has a chance to decay from T_2 effects. Thus, a P_D-weighted contrast can be obtained by using a long T_R (which

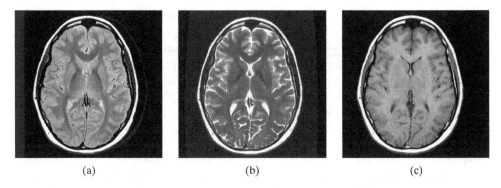

(a)   (b)   (c)

Figure 12.10 Three images of the same slice through the skull. Contrast between the tissue types are classified as (a) P_D-weighted, (b) T_2-weighted, and (c) T_1-weighted. (Used with permission of GE Healthcare.)

TABLE 12.2

Typical Brain Tissue Parameters Measured at 1.5 T			
Tissue Type	Relative P_D	T_2 (ms)	T_1 (ms)
White matter	0.61	67	510
Gray matter	0.69	77	760
Cerebrospinal fluid	1.00	280	2650

Source: Adapted from Liang and Lauterbur, 2000.

allows the tissues to be in equilibrium) and either no echo or a short T_E (in order to minimize T_2 decay). The preferred tip angle is $\pi/2$, in order to obtain the maximum signal. The image shown in Figure 12.10(a) was obtained using $T_R = 6000$ ms, $T_E = 17$ ms, and $\alpha = \pi/2$. It directly reflects the contrast expected from Table 12.2. In practice, 6000 ms is an unusually long repetition time. It is generally impractical in MRI to use more than about $T_R = 3500$ ms because the images take too long to acquire.

T_2-weighted Contrast   To reveal T_2 contrast, differences in the transverse relaxation times of different tissues must be apparent. We know that due to T_2^* effects, the FID decays more rapidly than pure transverse relaxation would predict, and this is not sufficient time to observe T_2 differences between tissues. To obtain T_2-weighted images, therefore, echoes must be used. But what should T_E be? We have already seen that if T_E is small, we get P_D-weighting. On the other hand, if T_E is large, the signal strength would be too small to detect above the noise level. In practice, for T_2 contrast, T_E should be selected to be approximately equal to the T_2 values of the tissues being imaged. Figure 12.10(b) is the result of using $T_R = 6000$ ms, $T_E = 102$ ms, and $\alpha = \pi/2$. The use of a large T_R here is consistent with obtaining maximum signal strength, but its use also reduces intermingling of T_1 contrast (which will become clear in the next section). The appearance of Figure 12.10(b) is consistent with the data in Table 12.2, especially when one realizes that the image intensities are related to T_2 through a decaying exponential. In particular, we realize that GM and WM have completely decayed (greater than three time constants) by the time the CSF has decayed by only about a third of its starting intensity. This means that the GM and WM should have small contrast with respect to each other, and large contrast with respect to CSF. White matter is slightly darker than gray matter because its NMR signal has decayed slightly faster.

T_1-weighted Contrast   To obtain T_1-weighted contrast, differences in the longitudinal component of magnetization must be emphasized. This is done by exciting the tissue repeatedly before it has had a chance to fully recover its longitudinal magnetization. We know that if the sample is in equilibrium and excited using an α pulse, then the transverse magnetization is given by

$$M_{xy}(t) = M_0 \sin \alpha e^{-j(2\pi v_0 t - \phi)} e^{-t/T_2} . \tag{12.38}$$

This (potential) signal dies away with a time constant T_2, and since $T_1 \gg T_2$, it is already negligible when $t \approx T_1$. On the other hand, the longitudinal component, given by

$$M_z(t) = M_0(1 - e^{-t/T_1}) + M_z(0^+)e^{-t/T_1} , \tag{12.39}$$

is not negligible when $t \approx T_1$. So, if we set $T_R \approx T_1$ (for some tissue in the field of view), we have created a situation where the transverse component has vanished (and therefore no signal is available, even from an echo), and the longitudinal component has not returned to equilibrium.

Now suppose an α-pulse is applied at $t = T_R$, where $T_R \approx T_1$ for some tissue. In this case, (12.38) no longer applies, because the sample was not in equilibrium to begin with. Instead, the longitudinal magnetization follows (12.39) and has longitudinal magnetization $M_z(T_R)$ just prior to excitation. For generality, we denote the longitudinal magnetization just prior to an RF excitation by $M_z(0^-)$. Then, the transverse magnetization after excitation follows

$$M_{xy}(t) = M_z(0^-) \sin \alpha \, e^{-j(2\pi v_0 t - \phi)} e^{-t/T_2} \tag{12.40a}$$

$$= M_{xy}(0^+)e^{-j(2\pi v_0 t - \phi)} e^{-t/T_2} , \tag{12.40b}$$

where $M_{xy}(0^+)$ denotes the magnitude of the transverse magnetization immediately after excitation.

In this scenario, $M_{xy}(0^+) < M_0 \sin \alpha$ for most tissues, and the exact value of $M_{xy}(0^+)$ depends directly on the longitudinal relaxation time of the particular tissue. This process is illustrated in Figure 12.11 for two tissues, one with a short T_1 (top row) and one with a long T_1 (bottom row). Starting in the left column at an early time when $t > 3T_2$, there is no transverse magnetization, and the longitudinal magnetization is starting to recover. In the middle column, the tissue with the short T_1 has undergone much more recovery than that with the long T_1. In the right

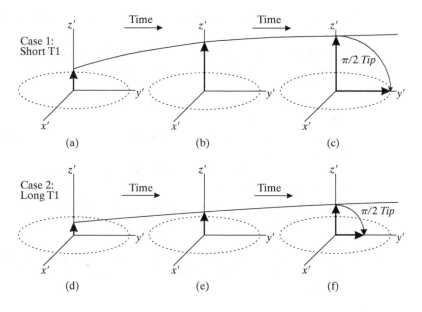

Figure 12.11
The principle behind T_1-weighted contrast.

column, a $\pi/2$-pulse tips the longitudinal magnetization into the transverse plane. The signal strength is proportional to the length of the magnetization vector in the transverse plane, and this length is partially determined by the T_1 value of the tissue.

The image shown in Figure 12.10(c) was obtained using $T_R = 600$ ms, $T_E = 17$ ms, and $\alpha = \pi/2$. From the data in Table 12.2, we see that this choice of T_R falls in-between the T_1 values for GM and WM but is much smaller than that of CSF. Therefore, both GM and WM will have recovered approximately two-thirds of their longitudinal magnetization, whereas CSF will have recovered relatively little. As a result, the CSF signal should be very small, while that of GM and WM should be larger. This situation is reflected well in Figure 12.10(c), where the GM and WM are relatively bright, while the CSF is dark.

12.11 Summary and Key Concepts

Nuclear magnetic resonance is the phenomenon behind magnetic resonance imaging. This phenomenon is well-known in chemistry and is now exploited for medical imaging. In this chapter, we presented the following key concepts that you should now understand:

1. *Magnetic resonance imaging* is based on the phenomenon of *nuclear magnetic resonance*.
2. A nucleus with an odd Z or A has *angular momentum* or *spin*; collections of such nuclei are called *nuclear spin systems*.
3. Such systems become magnetized when an external magnetic field is applied, producing a *bulk magnetization vector* that is a function of time.

4. The so-called *equations of motion* for this magnetization vector are based on the *Bloch equations*.

5. These equations describe a *precession* of the magnetization vector around the external magnetic field direction, with a frequency known as the *Larmor frequency*.

6. An *RF pulse* will cause the magnetization vector to precess around the external magnetic field.

7. The magnetization vector has two components: *transverse magnetization* and *longitudinal magnetization*.

8. The *observed signal* in MRI is an RF pulse produced by the rapidly rotating transverse magnetization.

9. *Relaxation* describes the gradual dampening of the precession (and associated signal); it has two components: *transverse* or *spin-spin relaxation* and *longitudinal* or *spin-lattice relaxation*.

10. *Contrast* in MR images is produced by manipulating the RF excitation pulse-sequence to produce MR signals that are influenced by different weighted contributions of proton density, T_1, and T_2.

Bibliography

Ernst, R. R., Bodenhausen, G., and Wokaun, A. *Principles of Magnetic Resonance in One and Two Dimensions*. Oxford, England: Oxford University Press, 1987.

Haacke, E. M., Brown, R. W., Thompson, M. R., and Venkatesan, R. *Magnetic Resonance Imaging: Physical Principles and Sequence Design*, NY: John Wiley and Sons, 1999.

Henkelman, R. M., "Methods of relaxation measurements." In *NMR in Medicine: The Instrumentation and Clinical Applications*, Woodbury, NY: American Institute of Physics, 1985.

Liang, Z. P. and Lauterbur, P. C. *Principles of Magnetic Resonance Imaging: A Signal Processing Perspective*, Piscataway, NJ: IEEE Press, 2000.

NessAiver, M. *All You Really Need to Know About MRI Physics*, Baltimore, MD: Simply Physics, 1997.

Sprawls, P. and Bronskill, M. J., Eds. *The Physics of MRI: 1992 AAPM Summer School Proceedings*. Woodbury, New York: American Institute of Physics, 1992.

Stark, D. and Bradley, W., Eds. *Magnetic Resonance Imaging*. St. Louis, MO: Mosby, 1992.

Problems

Magnetization

12.1 A nonuniform magnetic field B pointing in the z-direction is applied to a sample of protons. The field B (in tesla) varies as a function of z (in cm):

$$B(z) = 1 + 0.5z.$$

The magnetization vector M precesses around z-axis. Suppose at time $t = 0$, all magnetization vectors have same phase. At what time will the magnetization vector M at $z = 1$ cm and that at $z = 0$ have same phase again?

12.2 Prove that equations (12.12) are solutions to (12.7).

12.3 Protons that are in different chemical species, such as the CH_2 and CH_3 groups in fat, will resonate at slightly different frequencies due to local "shielding" of the static field by the electronic environment. This change in the resonance frequency is called *chemical shift*; it has made NMR an indispensable tool for chemists and physicists. If N species are present in the sample, and we neglect all interactions between nuclei, the FID is composed of a set of N decaying oscillators. Write down an equation that gives an expression for the transverse magnetization under these circumstances.

RF Excitation and Relaxation

12.4 A sample of ^1H in equilibrium condition in a static magnetic field B_0 is excited by a short circularly polarized RF pulse. The RF pulse is a magnetic field in the transverse plane

$$B_1(t) = B_1^e(t)e^{-j2\pi v_0 t} \text{ gauss,}$$

where v_0 is the Larmor frequency of the sample. The envelope of the RF pulse is a triangle function with parameter T:

$$B_1^e(t) = \begin{cases} \dfrac{1}{10}\left(1 - \dfrac{|t - T|}{T}\right), & 0 \le t \le 2T \\ 0, & \text{otherwise} \end{cases}.$$

(a) Find the tip angle of the magnetization vector as a function of t for $0 \le t \le 2T$.

(b) What is the value of T to make $B_1(t)$ a $\pi/2$ pulse?

12.5 The longitudinal and transverse relaxation of a magnetization vector after an RF excitation is governed by the following differential equations (in rotating framework):

$$\frac{dM_z}{dt} = -\frac{M_z - M_0}{T_1}$$

$$\frac{dM_{xy}}{dt} = -\frac{M_{xy}}{T_2}$$

Solve the above equations for $M_z(t)$ and $M_{xy}(t)$, assuming $M_z(0)$ and $M_{xy}(0)$ are known.

12.6 The following equations from Example 12.4 give the components of **M** after an α pulse, when the system is in equilibrium just before the α pulse:

$$M_z(t) = M_0(1 - e^{-t/T_1}) + M_0 \cos \alpha \, e^{-t/T_1}$$

$$M_{xy}(t) = M_0 \sin \alpha \, e^{j\phi} e^{-t/T_2} \, .$$

Suppose that we now excite the sample with a train of α pulses, separated by a time T_R. The equilibrium condition is true when T_R is long compared to T_1 and we can assume that M_z just before the pulse is equal to M_0. Derive a more general formula for $M_z(t)$. You can assume that the transverse magnetization has completely dephased before each RF pulse—i.e., $M_{xy}(T_R) = 0$. [*Hint*: In this more general formula, M_0 will be replaced with the *steady-state* value of the longitudinal magnetization. Define M_z after the $(n+1)^{th}$ pulse to be M_z^{n+1}, and M_z after the nth pulse to be M_z^n. Relate these two quantities with an equation. Derive another (very simple) equation from the steady-state condition. You now have enough information to solve the problem.]

12.7 In this problem, we answer the question: What is the flip angle that maximizes MR signal intensity for a pulse sequence with given T_R and T_E values? Solve this problem for given T_1, T_2, and T_2^* values. Assume that the magnetization is in the steady-state condition—i.e., that the magnetization is a periodic function of time with a period of T_R. Also assume that the transverse magnetization completely dephases before each RF pulse, i.e., $M_{xy}(0^-) = 0$.

(a) Calculate $M_z(0^-)$, the steady-state magnetization value just before the RF pulses.

(b) Calculate $M_{xy}(T_E)$, the signal level in the steady-state condition.

(c) Calculate the optimum flip angle.

12.8 (a) For the pulse sequence given in Figure P12.1, write down the expression for the FID signal.

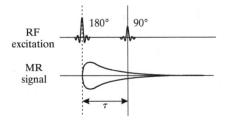

Figure P12.1
Pulse sequence for
Problem 12.8.

(b) Explain how we can use this pulse sequence to measure T_1.

Bloch Equations and Spin Echoes

12.9 Transform the Bloch equations into the rotating frame.

12.10 (a) Explain why a π pulse is applied at $T_E/2$ in order to get an echo at $t = T_E$.

(b) Suppose that a $\pi/2$ pulse is applied to a sample at $t = 0$ and that a sequence of π pulses is applied to the sample at $t = \dfrac{2k+1}{2}T_E$, $k = 0, 1, \ldots$. If T_E is small compared with T_2, find an expression for the magnitude of $M_{xy}(kT_E)$ for $k = 1, 2, \ldots$.

12.11 A rectangular RF pulse given by

$$\vec{B}_1(t) = A\,\text{rect}(t/10^{-5})[\cos(2\pi v_0 t)\hat{x} + \sin(2\pi v_0 t)\hat{y}]$$

is used to tip the magnetization vector away from its equilibrium position along the z-axis by 15 degrees. What is the value of A required to achieve this tip angle?

12.12 Assume that a hard $\pi/2$ pulse sends the longitudinal magnetization of a sample into the transverse plane in the direction of the x-axis. At that instant a spatially varying (but temporally constant) magnetic field $\Delta B(r)$ oriented in the z-direction is added to the static field B_0. The spatially varying Larmor frequency is then given by

$$v_0(r) = \gamma(B_0 + \Delta B(r)),$$

and the phase of the precessing sample is given by

$$\phi(r, t) = -\gamma(B_0 + \Delta B(r))t.$$

(The minus sign is required because of the usual convention that positive angles in the x-y plane are made in the counterclockwise direction, and the physics requires that precession occurs in the clockwise direction given a magnetic field oriented in the $+z$-direction.) A hard π pulse is applied at time τ.

(a) What is the phase $\phi(r, t)$ immediately before and immediately after the π pulse?

(b) What is the phase $\phi(r, t)$ at time $T_E = 2\tau$?

(c) Make a conclusive statement about the use of spin echoes when there are spatially varying gradients in use.

Contrast Mechanism

12.13 Explain proton density weighted contrast. Describe how to select imaging parameters to obtain P_D-weighted images. Explain why a large T_E cannot be used.

12.14 Suppose we are imaging a human brain (see Table 12.2 for the NMR properties of different brain tissues). We use the strategy described above in Problem 12.6, and acquire signals after the system is in the steady state. For simplicity, we assume that the intensity of the reconstructed MR image is directly proportional to the magnitude of the transverse magnetic field right after the RF excitations. In order to get maximal signal strength, we use $\pi/2$ pulses.

(a) In order to get the best contrast between gray matter and CSF, what is the best value of T_R?

(b) Using the T_R value computed in part (a), what is the contrast between gray matter and CSF? What is the contrast between gray matter and white matter?

12.15 We are using the pulse sequence shown in Figure P12.2 to image the human brain. A $\pi/2$ pulse is applied to the brain while in equilibrium. Suppose the duration of the RF pulse is negligible. The FID signal is sampled at time τ. The pulse repetition interval is T_R.

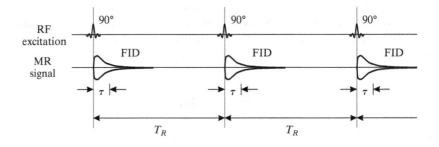

Figure P12.2
Pulse sequence for
Problem 12.15.

We use $T_R = 6000$ ms, $\tau = 20$ ms. Is this a T_2-weighted contrast? If not, what modification is needed to obtain T_2-weighted images? Sketch the pulse sequence if it is different from the one shown in Figure P12.2. Provide some reasonable values of τ, T_R, flip angle, and other parameters that may be added in the new pulse sequence.

Magnetic Resonance Imaging

CHAPTER

13

In the previous chapter, we studied nuclear magnetic resonance (NMR), the physics behind magnetic resonance imaging (MRI). We saw that it is possible to manipulate nuclear spin systems in such a way that a radio frequency signal is generated by the object in the form of a free induction decay (FID). We also saw that spin echoes can be used to refocus an otherwise very brief or transient FID. We even saw that a variety of different types of tissue contrasts can be created by manipulation of the timing of the various excitations. What we did not see, however, is how the spatial dependency of the underlying object can be encoded so that images can be created.

In fact, looking back at the previous chapter, you will notice that we always assumed there was a single object, the sample, and a single coherent signal, the FID or echo, generated by that object. A single object and corresponding single signal was the basic viewpoint of the first 40 years or so of NMR. Then, in the early 1970s, Paul Lauterbur had the idea to spatially encode the NMR signal in order to create images. The first MR scanners were built in the late 1970s, and there has been a continuous growth in technical development and clinical use since that time.

In this chapter, we first explore the instrumentation necessary to create MR images. We then present the image formation process, starting from the imaging equations and ending with the computer algorithms that use these equations to generate MR images. Finally, we discuss the factors affecting image quality in order to get a sense of the limitations of MR imaging and of the possibilities for the future.

13.1 Instrumentation

13.1.1 System Components

As depicted in Figure 13.1, there are five principal components comprising an MRI scanner: (1) the main magnet; (2) a set of coils to provide a switchable spatial gradient in the main magnetic field; (3) resonators or "coils" for the transmission and reception of radio-frequency pulses; (4) electronics for programming the timing of transmission and reception of signals; and (5) a console for viewing, manipulating,

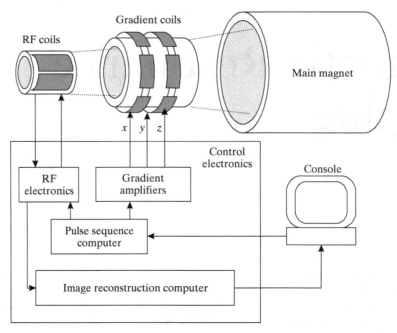

Figure 13.1
Block diagram of MR scanner components.

and storing images. Figure 13.2(a) shows a photograph of an MRI magnet with patient table and Figure 13.2(b) shows the operator console.

The magnet, gradient coils, and RF coils must be isolated from the electronic noise of the outside world—e.g., radio stations—in order to prevent interfering signals. Accordingly, these elements are placed in a copper-lined room, which acts as a *Faraday cage*. All electronic signals that enter the scan room, such as gradient currents and power outlets, must first go through band-pass filters to ensure that no electrical noise in the range of the receiver electronics is present inside the room. A rack of electronics, including the gradient amplifiers and transmit/receive electronics, are typically housed in a room adjacent to the scanner room.

Like CT and PET scanners, the majority of systems on the market today have the patient lying on a sliding table inside a cylinder. This geometry requires large (> 1 meter diameter) solenoidal coils of superconducting wire to be used to provide a strong (> 1 tesla) main magnetic field. Other geometries have also been produced, such as "open" magnets with two pole pieces arranged either vertically or horizontally. These systems give easier access to the patient, but have lower field strengths (< 1 tesla) and hence lower amplitude signal. We focus our attention on the common 1.5 T MR scanners, which dominate the market.

13.1.2 Magnet

The most common type of magnet used in MRI systems is the cylindrical superconducting magnet (typically with a 1 meter bore size). In a superconducting magnet, as depicted in Figure 13.3, coils of niobium-titanium wire are immersed in liquid

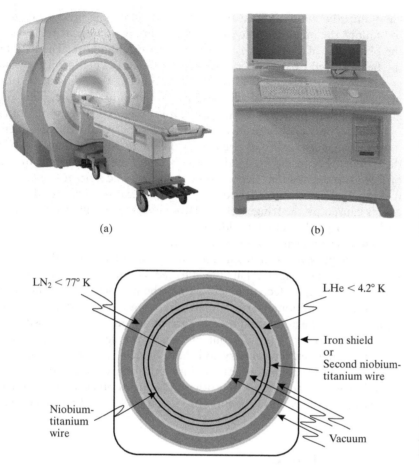

Figure 13.2
The main magnet with the patient table (a), and the console for operating the scanner (b). (Used with permission of GE Healthcare.)

(a)   (b)

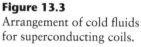

Figure 13.3
Arrangement of cold fluids for superconducting coils.

helium held at about 4° K. Niobium-titanium is a superconductor at temperatures less than 9.5° K, and helium boils at 4.2° K. A dewar, also known as a *cryostat*, is used to keep the helium from boiling. The cryostat is essentially a "sandwich" that contains liquid helium between layers of liquid nitrogen and a vacuum, which help to insulate the helium.

Field strengths in MR scanners vary from 0.5 T (21 MHz) to 3.0 T (128 MHz), with a few systems operating at 9 T (383 MHz). By far the most common magnet used for whole body systems operates at 1.5 T. This field strength has no special or optimal characteristics; it was simply the highest field achievable at the time the MRI market was being established (in the late 1970s). Because these magnets use superconducting wire, the current is on all of the time. For a 1.5 T magnet, the stored energy in this constant current is approximately 2.8 million joules; put another way, if the solenoid carrying the current was a single wire, that wire would carry 740 amperes of current.

There are two major challenges in the design and maintenance of superconducting magnets. First, the homogeneity of the magnetic field within the bore (field of view) must be maintained at better than ±5 ppm. The process of *shimming* is used to "tailor" the magnetic field in order to improve homogeneity. *Passive shimming* is the placement of small pieces of metal just inside the bore. The metal pieces disturb the field created by the superconducting wire and, when properly placed, the result is a more homogeneous field. *Active shimming* is the adjustment of currents in as many as 30 small coils of wire placed just within the bore. This process is more automated than passive shimming, but it is more expensive to manufacture.

The second challenge in the design of superconducting magnets is the minimization of the so-called *fringe field*—the magnetic field that is outside the bore of the magnet. Because these fields are very large, the fringe field can cause significant problems in devices that depend on either magnetic storage—e.g., credit cards—or moving currents—e.g., cathode ray tubes. Because of these kinds of detrimental effects caused by the fringe fields, MR scanners generally cannot be sited arbitrarily within a hospital or clinic building.

Two mechanisms can be used to reduce fringe fields, again one passive and one active. The passive mechanism is to simply put a large iron shield around the entire superconducting magnet. This tends to reduce the fringe field beyond the iron, but also detrimentally affects the field within the magnet's bore. The active mechanism is to add an additional set of superconducting coils *outside* the primary coils (see Figure 13.3). The additional set has current moving in the opposite direction, which significantly lowers the field outside. Since the second set also reduces the field within the magnet, the primary coils have to operate at a higher current than would otherwise be the case, in order to maintain the desired magnetic field strength. Active shielding has the advantage that other large metal objects outside the scanner—e.g., beams within the walls of the building—will have less of an effect on the field within the bore.

13.1.3 Gradient Coils

As depicted in Figure 13.1, the *gradient coils* fit just inside the bore of the magnet (after any active shimming coils, if present). The function of the gradient coils is to provide a temporary change in the magnitude B_0 of the main magnetic field as a function of position in the magnet bore. There is no process described in Chapter 12 that requires such a function; in fact, we have gone to great lengths to ensure that the main field is constant. So, why would we want to deliberately perturb this "perfection" by creating an additional field that deliberately causes the main field to vary spatially? It turns out that this is the key to spatially encoding the NMR signal—the key to creating images.

Gradient coils provide the means to choose slices of the body for selective imaging. In this way, MRI can be *tomographic*—i.e., it can image slices. Gradient coils also provide the means to spatially encode the pixels within a given image slice, so that the individual FIDs and echoes coming from each one of thousands of pixels can be unraveled and turned into an image. We will see how all this works in Section 13.2. For now, let's study the gradient coils themselves.

There are usually three orthogonal gradient coils, one for each of the physical x, y, and z-directions, as shown in Figure 13.4 . For a cylindrical magnet, these gradient

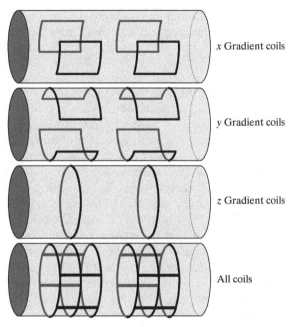

x Gradient coils

y Gradient coils

z Gradient coils

All coils

Figure 13.4
Arrangement of the
gradient coils.

coils are wound on a cylindrical former and stabilized by "potting" the windings in epoxy. Circulating water or air is used to cool the current-carrying elements. Gradient coils are subject to large forces—sometimes called *Lorentz forces*—since they carry currents in a strong magnetic field. Furthermore, in order to acquire images, the gradient currents are repeatedly turned on and off, leading to significant vibration. In addition to the potential to break down the epoxy and thereby weaken electrical connections, this vibration is the source of the rather loud "knocking" sounds associated with MR scans.

The purpose of each of the gradient coils is to add or subtract a spatially dependent magnetic field to the main field. In particular, if all three coils are turned on at the same time with strengths G_x, G_y, and G_z, respectively, then the main field is (ideally) given by

$$\boldsymbol{B} = (B_0 + G_x x + G_y y + G_z z)\, \hat{\boldsymbol{z}}. \tag{13.1}$$

Notice that the gradient coils do not change the direction of the magnetic field; instead, they add to (or subtract from) the magnitude of the main field. This is a common point of confusion when first learning MRI, and should be clearly understood before moving on. Ideally, the change in field strength resulting from currents in the gradient coils is a linear function of spatial position. This helps considerably in the development of algorithms to reconstruct images, as we shall see. The *gradient amplitude* is defined by the constants G_x, G_y, and G_z, which have units of Gauss per centimeter. It is often written in vector form as

$$\boldsymbol{G} = (G_x, G_y, G_z). \tag{13.2}$$

Using the notation $r = (x, y, z)$ to denote a vector spatial position, we find that (13.1) can be written using a dot product notation as

$$B = (B_0 + G \cdot r)\hat{z}. \tag{13.3}$$

The gradient coils are designed to produce a linear spatial perturbation of the magnetic field, as described above. As shown in Figure 13.4, the x and y gradients can be produced by pairs of *saddle coils* and the z gradient can be produced using two *opposing coils*, each wound around the circumference of the cylindrical bore. The maximum gradient amplitude is determined by the maximum current that the coil can carry. This, in turn, is limited by the gradient amplifier, the heating of the coil, and the forces on the coil. Typically, 100–200 amperes of current are available for the x, y, and z coil each. The maximum gradient amplitude in clinical scanners is usually on the order of 1–6 gauss/cm (or 10–60 mT/m).

We will see that the faster a gradient can "turn on," the faster we will be able to acquire an image. The switching time from zero to the maximum gradient amplitude is also dependent on the coil and amplifier design. Typical switching times are on the order of 0.1–1.0 ms. The so-called *slew rate* is more commonly used to characterize the overall performance of a gradient coil and amplifier pair. The *slew rate* is the maximum achievable rate of change of the gradient value and is given in units of mT/m/msec. Typical values range from 5–250 mT/m/msec.

The gradient coils must also have an auxiliary *shielding coil* on the outside of the gradient coil cylinder in order to minimize the change in the magnetic field outside the gradient housing. Without this shield, the changing flux through surrounding metal components such as the magnet housing would cause significant eddy currents, which exist on time scales from milliseconds to seconds. These induced eddy currents completely change the temporal profile of the magnetic field gradients. Some of the most innovative engineering that has emerged from the development of MRI has been the design of gradient coils that can be switched rapidly without inducing large eddy currents in the metallic components of the magnet housing.

In MR imaging methods, the gradients are switched on and off very rapidly. Ideally, this change in field strength would occur instantaneously; however, there are physical limitations that prevent this. The first limitation is the self-inductance of the gradient coil, which can be overcome with high voltage amplifiers, or by redesigning the coil to be smaller (usually this means shorter in the z-direction). The tradeoff when making the gradients smaller is that the region of linearity in the change of the magnetic field becomes smaller, and hence the maximum achievable field of view becomes smaller. The second limitation is the induction of eddy currents in the patient, causing peripheral nerve stimulation (muscle twitching). The Food and Drug Administration (FDA) has set a limit of 40 tesla/second on the exposure to magnetic field switching. Above this limit, the probability of peripheral nerve stimulation is not negligible.

EXAMPLE 13.1

Although the bore of a magnet is one meter, the human body normally takes up considerably less room. For the sake of concreteness, let's say the diameter of a human body is d, and

$d \approx 0.5$ m—this is the field of view (FOV). In the center of the magnet, the additional field caused by a gradient is zero, rising to its maximum values on the edges of the FOV, which is at a radius of $d/2$.

Question How rapidly can an MR scanner legally establish a gradient of $G = 40$ mT/m in the human body, and what is the slew rate required to achieve this?

Answer The maximum contribution of the magnetic field in the human body due to the gradient coils is

$$G_{\max} = 40 \text{ mT/m} \times \frac{d}{2} = 10 \text{ mT}.$$

We assume that the gradient waveform is linear until it reaches the desired final value. Therefore, the minimum time to reach the maximum value is

$$t_{\min} = \frac{G_{\max}}{40 \text{ Tesla/s}} = \frac{10 \text{ mT}}{40 \text{ Tesla/s}} = 0.25 \text{ ms}.$$

The slew rate required to achieve this is

$$\text{SR} = \frac{40 \text{ mT/m}}{0.25 \text{ ms}} = 160 \text{ mT/m/ms}.$$

∎

13.1.4 Radio-Frequency Coils

In the previous chapter, we saw the importance of radio-frequency (RF) induction in NMR. Currents applied around the sample that are oscillating at the Larmor frequency cause the nuclear spins to precess, tipping them toward the transverse plane. Once a spin system is excited, coherently rotating spins can induce RF currents (at the Larmor frequency) in nearby antennas, yielding measurable signals associated with the FID and echoes. Thus, RF *coils* (sometimes called *resonators*) in MRI systems, as shown in Figure 13.1, serve to both induce spin precession and to have currents induced in them by the spin system.

There are two basic types of RF coils: volume coils and surface coils. *Volume coils* are designed to (mostly) surround the object being imaged, while *surface coils* are designed to be placed on the surface in very close proximity to the object to be imaged. Different types of RF coils are depicted in Figure 13.5. Volume coils are preferable to surface coils in most instances because their sensitivity (field) patterns are very uniform within the body. This means that transmitted energy is uniformly distributed throughout the sample; so, for example, tip angles intended to be $\pi/2$ are fairly close to being $\pi/2$ radians everywhere. In addition, FIDs generated throughout the sample in response to an excitation are received in a relatively uniform manner using volume coils. Surface coils, in contrast, are very sensitive to sources close to the coil, but their sensitivity degrades rapidly away from the coil. Arrays of surface coils can be used to improve the coverage—e.g., "belts" containing four coils can be strapped around the torso—but the uniformity is still quite inferior to the carefully designed geometrically perfect volume coils.

All scanners have a *body coil*, which fits just inside the gradient coils, as shown in Figure 13.1 (labeled as RF coils in the figure). Usually the body coil is a *birdcage*

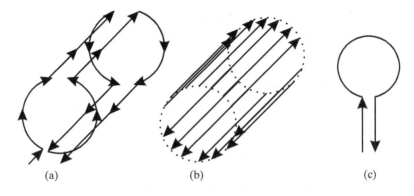

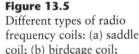

Figure 13.5
Different types of radio frequency coils: (a) saddle coil; (b) birdcage coil; (c) surface coil.

(a)   (b)   (c)

resonator as depicted in Figure 13.5(b). Most scanners come with a *head coil*, which is a much smaller coil—birdcage, saddle coil, or Aldermann-Grant coil (not shown)—designed to fit closely around the human head. Other specialty volume coils, such as knee coils, neck coils, and small extremity coils can be purchased for most MRI scanners. Surface coils are individual loops [see Figure 13.5(c)] placed in close proximity to the organ being imaged or they can be used in combination—so-called *phased arrays*—in order to improve their homogeneity and sensitivity over that of single coils.

During RF transmission, relatively large currents are produced in the coil elements from an RF amplifier, with a power requirement of approximately 2 kW for human imaging. Ideally, a transmission coil produces a relatively uniform B_1 field throughout the entire imaging volume. On reception, the coil must pick up very low amplitude magnetic fields, which produce very small voltages in the coil. Because transmission and reception require currents of much different amplitude in the coil, these functions are often split into two separate subsystems. In practice, a body coil located just inside the gradient coils is used to transmit RF into the patient, and another coil—e.g., head coil, surface coil, or phased array coil—positioned close to the volume of interest is used for signal reception. This is a very good way to increase the signal-to-noise ratio of the imaging signal, as will be discussed later. For both the transmit and receive coils, it is important to realize that only the transverse components of the magnetic field are used for imaging.

13.1.5 Scanning Console and Computer

The console [see Figures 13.1 and 13.2(b)] in a typical MRI system is used by the operator to select the scanning protocol, set the gating to the patient's electrocardiogram (ECG) or breathing (to synchronize acquisition to the appropriate periodic physiologic process), graphically select the orientation of the scan planes to image, review images obtained, and change variables in the pulse sequence to modify the contrast between tissues. The operator's console is connected to a compute engine, such as an array processor, which performs the image reconstruction. Current scanners reconstruct about 10 to 50 images per second, which is adequate for real-time scanning of slices. The speed of reconstruction is often limited by the data transfer rate from the receiver electronics to the array processor. Some scanners now have real-time image feedback much like ultrasound, so that the operator can use the real-time images from a single slice to maneuver the scan plane.

13.2 MRI Data Acquisition

Radiography encodes the spatial position of objects using the position of x-rays that hit a detector. Factors affecting the encoding of spatial position in radiography are the magnification caused by the diverging beam and the superposition of overlaying structures. In radiography, the superposition of structures is never "decoded;" radiologists learn to read radiographs with the knowledge that images are corrupted in this fashion. CT encodes the spatial position of objects by observing their effect in many projections, each from a different orientation. The position of an object in CT is actually spread across the observed data set in a rather complicated way. CT reconstruction algorithms decode the spatial position by filtering and integrating the data using a mathematical algorithm—e.g., convolution backprojection.

So far, we understand only that MRI will use the gradient coils to encode spatial position. But how will this be done? How will the data be decoded? The "trick" is to use both the Larmor frequency and the phase of the transverse magnetization to encode spatial position. In this section, we present both frequency and phase encoding of spatial position and develop a fundamental MR image formation technique.

13.2.1 Encoding Spatial Position

Before discussing the encoding of spatial position, we should ask, "What is the spatial coordinate system?" It is customary for $+z$ to be oriented along the direction of the B_0 field and for the $+y$-direction to be oriented up. The $+x$-direction must therefore be horizontal and oriented in such a way to create a right-handed coordinate system. When the patient goes in head-first and supine, this means that the $+z$-direction is from the head to the feet; $+y$ is oriented posterior (back) to anterior (front); and $+x$ is oriented right to left, as shown in Figure 13.6. In this scenario, we see that if we were able to image a slice whose z-coordinate is constant, we would get a so-called *axial* image; we would get a *coronal* image by holding y constant; and we would get a *sagittal* image by holding x constant.

Although an MR scanner can create images at arbitrary location and orientations, for simplicity, we will describe only the formation of an axial image. In this case, the z-direction corresponds to the *through-plane* direction, and the x- and y-directions correspond to the *in-plane* directions. It is straightforward to acquire both coronal and sagittal images and so-called *oblique* images as well. Oblique images are those that are not orthogonal to any laboratory coordinate. The methods we develop for the formation of axial images will carry over with relatively little modification to the acquisition of images in these other orientations.

In order to encode spatial position using frequency, the Larmor frequency should vary as a function of spatial position. Recall that the frequency at which a magnetization vector $M(t)$ precesses about a magnetic field B is given by $\nu = \gamma B$,

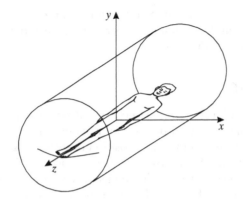

Figure 13.6
Laboratory coordinates in an MR scanner.

where $B = |\mathbf{B}|$. Now suppose that we "turn on" a constant gradient field $\mathbf{G} = (G_x, G_y, G_z)$ by applying a constant current to all three gradient coils. Then, the total magnetic field is given by (13.3), and the Larmor frequency is given by

$$\nu(\mathbf{r}) = \gamma(B_0 + \mathbf{G} \cdot \mathbf{r}), \qquad (13.4)$$

where the dependence of Larmor frequency $\nu(\mathbf{r})$ on spatial position $\mathbf{r} = (x, y, z)$ is made explicit.

This concept, called *frequency encoding*, is the first of two concepts used to encode spatial position in MRI, the other being called *phase encoding*. Frequency encoding is used both for slice selection, as we describe in Section 13.2.2, and for reading out samples in Fourier space, as we describe in Section 13.2.3. Phase encoding is used to select a position in Fourier space, as described in Section 13.2.6.

EXAMPLE 13.2
It is very common to use only one gradient component when applying the concept of frequency encoding.

Question If a sample is put in a magnetic field with $B_0 = 1.5$ tesla, and a z gradient with strength $G_z = 3$ gauss/cm is applied, what is the Larmor frequency for the protons on the $z = 0$ plane? If we are to image a slab with a thickness of 0.5 m centered at $z = 0$, what is the range of Larmor frequency of the protons in the slab?

Answer In this case, $\mathbf{G}$ has only one nonzero component

$$\mathbf{G} = (0, 0, G_z).$$

Using (13.4) gives

$$\nu(\mathbf{r}) = \gamma(B_0 + \mathbf{G} \cdot \mathbf{r}), \qquad (13.5a)$$

$$= \gamma(B_0 + G_z z). \qquad (13.5b)$$

For protons, the gyromagnetic ratio is $\gamma = 42.58$ MHz/tesla. So on the $z = 0$ plane, the Larmor frequency is

$$\nu_0 = \gamma B_0 = 63.87 \, \text{MHz}.$$

With the z gradient, the strength of the magnetic field is

$$B(z) = B_0 + G_z z.$$

Its range within the slab of sample is between

$$B_{\min} = B_0 - G_z \times 0.25\,\text{m} = 1.5\,\text{Tesla} - 7.5\,\text{mTesla} \qquad \text{and} \qquad (13.6\text{a})$$

$$B_{\max} = B_0 + G_z \times 0.25\,\text{m} = 1.5\,\text{Tesla} + 7.5\,\text{mTesla}. \qquad (13.6\text{b})$$

So the range of the Larmor frequency within the slab of sample is

$$63.55\,\text{MHz} \le \nu \le 64.19\,\text{MHz}. \qquad \blacksquare$$

13.2.2 Slice Selection

We have already studied several medical imaging techniques that image 2-D *slices* of the human body: CT, SPECT, PET, and ultrasound. In CT and ultrasound, the energy used to image the selected slice is restricted to the slice itself. In that way, the physiological property giving rise to image contrast arises from within only the selected slice—there is no other part of the body from which signal can arise. In SPECT and PET, the entire body is a potential source, but the observed signal is selected by collimation so that it belongs to a specific slice.

In MRI, the basic principles of both of these approaches can be used. Specifically, it is possible to excite only a selected slice so that the received signal can arise only from within the selected slice; it is also possible to excite the whole volume and then to extract images of selected slices. The first technique is called 2-D MR imaging and the second is called 3-D MR imaging. We confine our detailed presentation in this chapter to 2-D imaging, and only comment briefly on 3-D imaging. The first step in 2-D imaging is to perform *slice selection*, the selective excitation of a nuclear spin system in a slice. As noted above, for convenience we describe the selection—and then imaging—of axial slices. The principles carry over quite directly to the imaging of arbitrary slices.

Principle of Slice Selection In Example 13.2, we found that application of the gradient $G = (0, 0, G_z)$ yields a Larmor frequency that is a function of z. This spatially varying Larmor frequency can be written as

$$\nu(z) = \gamma(B_0 + G_z z), \qquad (13.7)$$

which is illustrated in Figure 13.7. If a pure sinusoidal RF excitation at a specific frequency were applied, then an infinitesimally thin slice of the body would undergo forced precession. That slice corresponds to all points whose z position has the specific Larmor frequency being excited. It is not possible to create such a sinusoidal waveform in practice, however, nor is it practically desirable. Instead, we create a waveform that excites a range of frequencies, which in turn excites a range of tissues, corresponding to a "thick slice" or a *slab*.

Figure 13.8 shows two slice selection scenarios. In the first, Figure 13.8(a), a z-gradient $G_z = G_A$ is used and a signal containing the range of frequencies

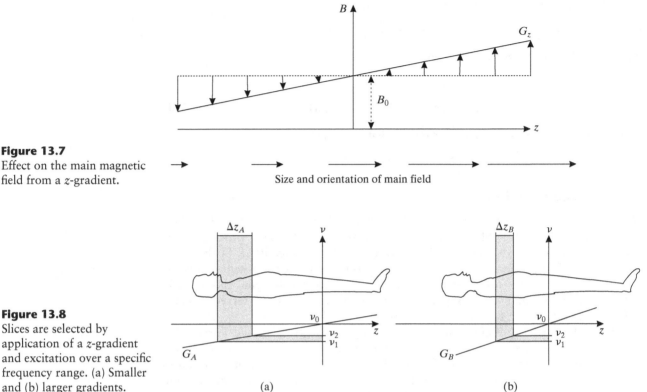

Figure 13.7
Effect on the main magnetic field from a z-gradient.

Figure 13.8
Slices are selected by application of a z-gradient and excitation over a specific frequency range. (a) Smaller and (b) larger gradients.

$\nu \in [\nu_1, \nu_2]$ is generated. (We talk about what this signal actually looks like below.) This combination causes a forced precession of all the spins around the neck area of the subject in this example. The slice is thicker than would ordinarily be desirable, however. In the second scenario, Figure 13.8(b), a larger z-gradient $G_z = G_B$ is used and the same RF signal is applied. Because the z-gradient is larger in magnitude than that in Figure 13.8(a)—i.e., $|G_B| > |G_A|$—the same RF waveform having frequencies in the range $[\nu_1, \nu_2]$ causes the forced precession of a thinner slice. Notice also that the slice is in a different position—at the lower chest—in this case.

By studying Figure 13.8 we see that there are actually three parameters that are used together to select slices: *z-gradient strength* G_z, *RF center frequency*,

$$\overline{\nu} = \frac{\nu_1 + \nu_2}{2}, \qquad (13.8)$$

and *RF frequency range*,

$$\Delta\nu = |\nu_2 - \nu_1|. \qquad (13.9)$$

For example, let's assume that we want to image a thinner slice at the neck region using one of the two z-gradients in Figure 13.8. There are two ways to achieve this.

Suppose we use the smaller z-gradient, G_A, as in Figure 13.8(a). We see that the RF center frequency is fine, but the RF frequency range would have to be made smaller in order to make the excited slice narrower. This is the first approach. The second approach uses the larger z-gradient, G_B, as in Figure 13.8(b). In this case, the RF frequency range is fine—it already produces a thin slice—but the RF center frequency must be made smaller in order to shift the slice from the chest region to the neck region.

With these three parameters, it is possible to control both *slice position*, $\bar{z}$, and *slice thickness* Δz. Solving (13.7) for z, we find that the lowest and highest excited frequencies, ν_1 and ν_2, yield the slice boundaries,

$$z_1 = \frac{\nu_1 - \gamma B_0}{\gamma G_z} \quad \text{and} \tag{13.10a}$$

$$z_2 = \frac{\nu_2 - \gamma B_0}{\gamma G_z}, \tag{13.10b}$$

where $\nu_1 = \nu(z_1)$ and $\nu_2 = \nu(z_2)$. Slice position $\bar{z}$ is therefore given by

$$\bar{z} = \frac{z_1 + z_2}{2} \tag{13.11a}$$

$$= \frac{\bar{\nu} - \nu_0}{\gamma G_z}, \tag{13.11b}$$

where (13.11) follows after some algebra. Slice thickness Δz is given by

$$\Delta z = |z_2 - z_1| \tag{13.12a}$$

$$= \frac{\Delta \nu}{\gamma G_z}. \tag{13.12b}$$

EXAMPLE 13.3

Suppose that we desire a slice thickness of 2.5 mm and that the z gradient strength is $G_z = 1$ G/cm.

Question What RF frequency range should be excited?

Answer Solve (13.12) for frequency range

$$\Delta \nu = \gamma G_z \Delta z$$

$$= \gamma \times 1 \text{ G/cm} \times 2.5 \text{ mm}$$

$$= 4.258 \frac{\text{kHz}}{\text{G}} \times 1 \frac{\text{G}}{\text{cm}} \times 0.25 \text{ cm}$$

$$= 1.06 \text{ kHz}.$$

This gives the frequency range. The selection of the specific frequencies themselves—i.e., ν_1 and ν_2 —depends on the specific slice to be selected relative to the $z = 0$ origin. ∎

We note that thinner slices have fewer nuclei, which makes the NMR signal smaller. At some point with progressively thinner slices, the received NMR signal—FID or echo—is too small to detect above the ambient antenna noise, and thinner slices are not practically achievable. In whole-body imaging today, slices thinner than about 1.5 mm require special pulse sequences and longer imaging times.

Practical RF Waveforms We learned in the previous section that slice selection uses a constant gradient together with an RF excitation over a range of frequencies. We now consider what RF waveform to use in exciting the range of frequencies $[v_1, v_2]$. We desire a signal whose frequency content is

$$S(v) = A \text{ rect}\left(\frac{v - \bar{v}}{\Delta v}\right). \tag{13.13}$$

According to Fourier transform theory (see Section 2.2.4 and Example 2.7), we see that the signal itself should be

$$s(t) = A\Delta v \text{ sinc}(\Delta v t)e^{j2\pi \bar{v} t}. \tag{13.14}$$

This analysis is valid provided both that the gradient is constant during RF excitation and that the RF excitation is short. An illustration of $s(t)$ is provided in Figure 13.9(a); its envelope is illustrated in Figure 13.9(b).

Let us consider the exact effect that the RF signal $B_1(t) = s(t)$ has on the spin system. Equation (12.30) gives the final tip angle α after an RF excitation pulse of duration τ_p and is repeated here:

$$\alpha = \gamma \int_0^{\tau_p} B_1^e(t) \, dt, \tag{13.15}$$

where $B_1^e(t)$ is the envelope of the RF excitation evaluated in the rotating coordinate system. For isochromats whose Larmor frequency is v, the excitation signal in the rotating coordinate system is

$$B_1^e(t) = s(t)e^{-j2\pi v t}. \tag{13.16}$$

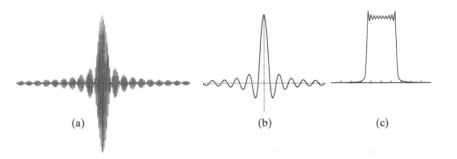

Figure 13.9
A slice selection waveform (a) and its envelope (b) and its Fourier transform (c).

(a)   (b)   (c)

After some algebraic manipulation (see Problem 13.5), it can be shown that

$$\alpha(z) = \gamma A \text{rect}\left(\frac{z - \bar{z}}{\Delta z}\right), \qquad (13.17)$$

which shows that the tip angles excited by the waveform in (13.14) define a perfect slab excitation, exactly as expected by the intuitive argument above. In principle, the value of constant A could be selected to produce an arbitrary tip angle within the slab, such as a $90°$ maximum signal excitation.

Although it is useful to know what the slice profile is in terms of tip angle, it is somewhat more important to know what the signal strength will be as a function of z. Here, we recall that the amplitude of the transverse magnetization is proportional to $\sin\alpha$. In the case of small tip angle excitation, $\sin\alpha \approx \alpha$, which means that $|M_{x'y'}(t)|$ will (initially) have a rect profile as well. Often, the small tip angle approximation is used for initial design purposes all the way up to $\alpha = 90°$ excitation. Fine-tuning of the actual excitation pulses can be made by using the Bloch equations to simulate all excitations.

In reality, it is not possible to synthesize a perfect sinc excitation. The preceding analysis can be carried out again, however, in the case of a truncated sinc excitation. In this case, the resulting tip angle distribution will be

$$\alpha(z) = \gamma A \tau_p \text{rect}\left(\frac{z - \bar{z}}{\Delta z}\right) * \text{sinc}\left(\tau_p \gamma G_z (z - \bar{z})\right), \qquad (13.18)$$

where τ_p is the duration of the pulse (truncated beyond $[-\tau_p/2, \tau_p/2]$). An illustration of a slice profile resulting from a truncated sinc excitation is shown in Figure 13.9(c); it is essentially the Fourier transform of the slice selection waveform. Because of truncation, the slice profile is not perfectly rectangular in this example. This means that the edges of the slice will be somewhat blurry and there will be "ripples" caused by the convolution of the (ideal) rect profile with the sinc function associated with truncation. In general, there is a tradeoff between the duration and shape of the RF waveform and the profile edge "crispness" and ripples within the slice profile. A shorter excitation RF pulse multiplied by a Hamming window is generally desirable even though it will not produce a "crisp" rectangular slice profile.

EXAMPLE 13.4

From Example 13.3, a 2.5 mm thick slice can be excited using a gradient strength of $G_z = 1$ G/cm and an RF pulse bandwidth of 1.06 kHz.

Question What is the duration of the main lobe and first two side lobes of the RF pulse required to achieve this slice thickness?

Answer The envelope of the waveform is $\text{sinc}(\Delta\nu t)$. The time of the first zero t_1 satisfies

$$t_1 = 1/\Delta\nu = 943 \; \mu s.$$

The duration of main lobe and two side lobes T is four times this value. Therefore,

$$T = 3.77 \text{ ms}.$$

A pulse that comprises only the main lobe and its two sidelobes is called a *two-period sinc approximation*.   ∎

Refocusing Gradients   During RF excitation, the spin system within the excited slab is undergoing forced precession. The slice profile reveals differences in the final tip angles and hence implies different transverse magnetizations experienced at different z positions. Assuming reasonable design of the RF excitation waveform—e.g., some windowing to reduce spurious ripples—these differences are negligible in the final images. The strength of the gradient and RF waveform can be tuned so that the center of the slice experiences a desired tip angle, often 90° for imaging.

There is another effect that takes place during RF excitation, however, that is not negligible—that is the effect of *slice dephasing*. During forced precession, the spins at the "lower" edge of the slice are precessing slower than those at the "higher" edge; that is simply the result of having different Larmor frequencies. As a result of this, the spins become out of phase with each other across the slice. To a good approximation, the phase that is introduced is linear and is equal to $\gamma G_z (z - \bar{z}) \tau_p / 2$, where τ_p is the duration of the RF pulse. To see this mathematically, it is necessary to solve the Bloch equations. We can get a sense that this is correct, however, by noticing that the greatest impact of the excitation is halfway through the pulse, since that is where the $B_t(t)$ field is strongest. If the entire RF pulse were concentrated at the center of the gradient pulse, then the phase would accumulate over duration $\tau_p / 2$, rather than the whole pulse duration. Since the resulting FID or echo relies on in-phase precession, it is necessary to *rephase* the spins within the slice.

Spin rephasing is accomplished using a so-called *refocusing lobe* z-gradient waveform. One possible refocusing lobe waveform is simply a constant negative gradient of strength $-G_z$ for a duration of $\tau_p / 2$, applied immediately after the initial RF excitation. Having no accompanying RF excitation, this negative gradient simply causes a change in the Larmor frequency as a function of z. The accumulated phase relative to that at $\bar{z}$ over the duration of the refocusing lobe is $\gamma (-G_z)(z - \bar{z}) \tau_p / 2$. Adding this to the phase accumulated over the initial RF excitation pulse yields zero, which means that the spins are now in phase over the slice and they will give an FID and can be used to form an echo in the usual way. The refocusing lobe can be shaped differently than a rectangle, however, since all that matters is that its integral is equal to half the integral of the slice selection gradient, but with opposite sign. Most pulse sequences choose waveforms that minimize the duration so that the pulse sequence can be as fast as possible.

A Simple Pulse Sequence   We now understand the basic elements of slice selection. A constant z-gradient is applied during which an RF waveform of duration τ_p is applied. After the RF waveform is completed, another gradient is applied to refocus the spins within the slice. After this, we should expect to find an FID arising from the excited spins in the slice that was selected.

Figure 13.10 shows a simplified pulse sequence, illustrating the concepts of slice selection. In this diagram, the gradient amplitudes are shown to change instantaneously as step functions, which is physically impossible given the minimum switching times and finite slew rates of the gradient coils in an MR scanner. In reality,

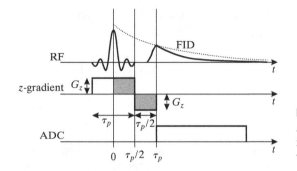

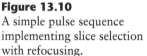

Figure 13.10
A simple pulse sequence
implementing slice selection
with refocusing.

the gradient pulses typically look like trapezoids with rising and falling edges having slopes equal in magnitude to the maximum slew rate of the gradient amplifiers. The mathematics becomes a bit more complicated when trapezoidal gradient pulses are used, but the principles of pulse sequence design are the same. Here, we stick to the unrealizable rectangular pulses for didactic purposes.

At the completion of the refocusing gradient pulse, the phase angle of all magnetization vectors in the selected slice will be the same, and the signal from these magnetization vectors will add constructively. This permits the reception of an FID and the creation of echoes in the usual way. If no dephasing were present across the selected slice, then we would expect the FID to begin at the center of the RF pulse. As shown in Figure 13.10, this is where we place the time origin $t = 0$ in the slice selection pulse sequence. Because of dephasing, the appearance of the FID is delayed until near the conclusion of the refocusing lobe.

Although there may be brief FIDs or echoes throughout a pulse sequence, the location of the analog-to-digital converter (ADC) window in a pulse sequence indicates where data are sampled for the creation of an image. In Figure 13.10, the ADC is turned on immediately after slice refocusing in order to capture the FID.

13.2.3 Frequency Encoding

Basic Signal Model We learned in Chapter 12 [see (12.40)] that the transverse magnetization for a uniform sample (and uniform magnetic field) after an α excitation is given by

$$M_{xy}(t) = M_{xy}(0^+)e^{-j(2\pi v_0 t - \phi)}e^{-t/T_2}, \tag{13.19}$$

where

$$M_{xy}(0^+) = M_z(0^-)\sin\alpha. \tag{13.20}$$

Suppose we have a heterogeneous isochromat within an excited slice. We model the spatial distribution of proton density, longitudinal relaxation, and transverse relaxation using functions of x and y only—i.e., $P_D(x, y)$, $T_1(x, y)$, and $T_2(x, y)$—assuming that the slice is fairly thin so there is no z variation. This implies that there will be a spatial variation of transverse magnetization immediately after

RF excitation, which can be written as $M_{xy}(x, y; 0^+)$. The received signal is then an integral over the slice, given by

$$s(t) = A \int_{-\infty}^{\infty} \int_{-\infty}^{\infty} M_{xy}(x, y; 0^+) e^{-j2\pi v_0 t} e^{-t/T_2(x,y)} \, dx \, dy \qquad (13.21\text{a})$$

$$= e^{-j2\pi v_0 t} \int_{-\infty}^{\infty} \int_{-\infty}^{\infty} A M_{xy}(x, y; 0^+) e^{-t/T_2(x,y)} \, dx \, dy, \qquad (13.21\text{b})$$

where A is a constant representing many different gain terms arising from both physics and instrumentation and ϕ is assumed to be zero without lack of generality.

There are a couple of details about (13.21) that we must make clear. First, we know from Chapter 12 that the FID decays more rapidly than T_2; in fact, it decays with time constant $T_2^* < T_2$. Therefore, we must view (13.21) either as an idealized signal model, or one that applies only for very short time intervals, where the difference in decay rates is negligible. Why not replace T_2 with T_2^* in (13.21)? While it is true that we would then get a more accurate representation of the true signal, we would then find that it is not as easy to understand the generation of signals using echoes, a vitally important concept in imaging, which we address in subsequent sections. We will see that the use of T_2 is most appropriate for spin-echo acquisitions, which is what we have been discussing so far. When we introduce gradient-echo acquisition in Section 13.2.5, we will use T_2^*.

Second, it should be noted that $t = 0$ represents the center of the slice selection RF waveform, as shown in Figure 13.10. This is consistent with the time at which the spins would begin transverse and longitudinal relaxation as the pulses were made infinitesimally narrow. Third, it should be noted that this equation ignores the short time τ_p it takes for the FID to actually appear after the refocusing lobe of the slice select gradient. Nearly all practical imaging approaches record echoes rather than FIDs (as we shall see), so this small "dead period" is not usually important.

For clarity, let us define the *effective spin density* as

$$f(x, y) = A M(x, y; 0^+) e^{-t/T_2(x,y)}, \qquad (13.22)$$

which represents the MR quantity that is being imaged here. The variable t is not included in $f(x, y)$ because it is assumed that the signal acquisition period is small relative to T_2. Using the definition given by (13.22) in (13.21), we find that the received signal is given by

$$s(t) = e^{-j2\pi v_0 t} \int_{-\infty}^{\infty} \int_{-\infty}^{\infty} f(x, y) \, dx \, dy. \qquad (13.23)$$

The received signal is always demodulated in MRI hardware, yielding the baseband signal

$$s_0(t) = e^{+j2\pi v_0 t} s(t) \qquad (13.24\text{a})$$

$$= \int_{-\infty}^{\infty} \int_{-\infty}^{\infty} f(x, y) \, dx \, dy, \qquad (13.24\text{b})$$

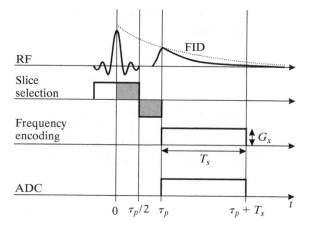

Figure 13.11
Frequency encoding with a constant gradient.

which is a constant, independent of x and y and even t (assuming a short data acquisition period). Equation (13.24) verifies that this procedure integrates out all the spatial dependency in the signal $s_0(t)$. Therefore, position is not encoded in this signal in any way other than the selective excitation of slice selection.

Readout Gradient The first concept required for spatially encoding MR signals is called *frequency encoding*. In frequency encoding, a gradient is turned on during the FID, as shown in Figure 13.11, causing the Larmor frequencies to be spatially dependent [see (13.4)]. (This same concept will apply to echoes, as we will see in a later section.) The direction of the frequency encoding gradient is called the *readout direction* because the signal that is "read out" (i.e., digitized) is spatially encoded in that direction. The readout direction is arbitrary (except that it should be orthogonal to the slice selection gradient); for didactic purposes, we will associate it with the x-direction. Accordingly, the Larmor frequencies during a frequency encode gradient are given by

$$\nu(x) = \gamma(B_0 + G_x x), \qquad (13.25)$$

which should be compared to the Larmor frequencies during a slice selection gradient, given in (13.7).

The response during an FID under slice selection is still an integration of all the spins within the excited slice, but the Larmor frequency must be written as in (13.25). Therefore,

$$s(t) = A \int_{-\infty}^{\infty} \int_{-\infty}^{\infty} M_{xy}(x, y; 0^+) e^{-j2\pi(\nu_0 + \gamma G_x x)t} e^{-t/T_2(x,y)} \, dx \, dy \qquad (13.26a)$$

$$= e^{-j2\pi\nu_0 t} \int_{-\infty}^{\infty} \int_{-\infty}^{\infty} A M_{xy}(x, y; 0^+) e^{-t/T_2(x,y)} e^{-j2\pi\gamma G_x x t} \, dx \, dy. \qquad (13.26b)$$

Using the definition of effective spin density in (13.22), the demodulated (*baseband*) signal is given by

$$s_0(t) = \int_{-\infty}^{\infty}\int_{-\infty}^{\infty} f(x,y)e^{-j2\pi\gamma G_x xt}\, dx\, dy. \tag{13.27}$$

Equation (13.27) reveals an important concept in MR imaging, and deserves a bit more explanation. In particular, the double integral on the right-hand side can be interpreted as a 2-D Fourier transform of $f(x,y)$, provided that the frequency variables are properly identified. First, we identify the spatial frequency variable in the x-direction as

$$u = \gamma G_x t, \tag{13.28}$$

which has units of inverse length (typically cm^{-1}). We next realize that the spatial frequency variable in the y-direction must be identically zero,

$$v = 0. \tag{13.29}$$

Denoting $F(u,v)$ as the 2-D Fourier transform of $f(x,y)$, we can now make the identity

$$F(u,0) = s_0\left(\frac{u}{\gamma G_x}\right), \tag{13.30}$$

which shows that the demodulated FID represents a certain "scan" of the 2-D Fourier space of the effective spin density.

In magnetic resonance imaging, Fourier space is usually referred to as *k-space*. This practice arises from the convention in physics where the wave number k represents a spatial frequency. Usually, the wave number has units of radians per unit length—i.e., it is a radial frequency—and this was the convention in early MRI as well [see also discussion around (10.10)]. More recently, MRI researchers associate the units of inverse length with k. In this case, the k-space variables can be identified with our Fourier frequencies,

$$k_x = u, \tag{13.31a}$$

$$k_y = v. \tag{13.31b}$$

Although somewhat unconventional, in order to avoid confusion between radial and cyclic frequencies and to maintain consistency throughout the book, we will use our customary Fourier frequencies u and v to describe MRI. When referencing a modern book on MRI, the identification in (13.31) will permit direct comparison.

Scanning Fourier Space   The recognition that MR imaging can be interpreted as a "scanning" of 2-D Fourier space is a tremendously important simplifying

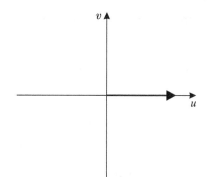

Figure 13.12
Fourier trajectory for a simple frequency encoded FID readout.

and unifying concept. Like CT, which has the projection slice theorem to tie 1-D projections to central slices in 2-D Fourier space, viewing pulse sequences as a type of scanning in Fourier space allows us to understand why a certain set of simple pulse sequences, like that in Figure 13.11, needs to be augmented or repeated with a changing parameter in order to acquire enough information to make a 2-D tomographic image.

For example, let's take a look at the Fourier information we obtain using the pulse sequence in Figure 13.11. From (13.30), this single readout provides Fourier information only on the horizontal u-axis in Fourier space, because this is where $v = 0$. Furthermore, since observations are taking place only where $0 < t < T_s$, we are only able to observe the positive u-axis over a particular interval. This Fourier scan, usually called a *Fourier trajectory*, is shown in Figure 13.12. It is quite evident from this diagram that only a small amount of Fourier space is scanned using this one small pulse sequence; somehow, we must find a way to scan more. How do we return to the origin again so that we might scan in some other direction? How do we scan in another direction?

In the next several sections, we study different techniques that will allow us to traverse Fourier space in a remarkably flexible manner, acquiring Fourier data as we "drive." As in CT, the particular reconstruction algorithm that we use must match the particular geometry of the acquired data. MR is much more flexible in this regard than CT, and both rectilinear (as described above) and polar scanning are commonly used in MR imaging, requiring the fast Fourier transform algorithm and convolution backprojection, respectively, for reconstruction. The following sections are ordered for didactic purposes and do not necessarily present an order based on historical development or increasing practical importance. Furthermore, some of the pulse sequences we present are never or rarely used in practice; therefore, we will specifically identify those that are widely used and worthy of more detailed study.

EXAMPLE 13.5

For a sample of protons, we want to scan a line segment in the k space from $u = 0$, $v = 0$ to $u = 0.5 \, \text{cm}^{-1}$, $v = 0$.

Question Assuming that the readout gradient has magnitude of $G_x = 1$ gauss/cm, how long should the duration of the readout gradient be?

Answer   From (13.28), we have

$$u = \gamma G_x t.$$

In order to scan the line segment from $u = 0$, $v = 0$ to $u = 0.5 \, \text{cm}^{-1}$, $v = 0$, we must have

$$\gamma G_x T_s = 0.5 \, \text{cm}^{-1}.$$

The gyromagnetic ratio for proton is

$$\gamma = 42.58 \, \text{MHz/Tesla} = 4.258 \, \text{kHz/gauss}.$$

The duration of the readout gradient should thus be

$$T_s = \frac{0.5}{4.258 \times 10^3 \times 1} = 0.117 \, \text{ms}.$$ ∎

13.2.4 Polar Scanning

Changing the direction of the Fourier space scan simply requires a different frequency encoding gradient. In the previous section, the x-direction was identified as the readout direction and an x-gradient was used to achieve this particular encoded FID signal. However, a more general gradient involving both an x- and a y-component can be used to encode the Larmor frequency

$$\nu(x, y) = \gamma(B_0 + G_x x + G_y y), \tag{13.32}$$

as shown in Figure 13.13(a). Following an analogous development as in the previous section [see (13.27)] leads to a baseband signal given by

$$s_0(t) = \int_{-\infty}^{\infty} \int_{-\infty}^{\infty} f(x, y) e^{-j2\pi\gamma(G_x x + G_y y)t} \, dx \, dy. \tag{13.33}$$

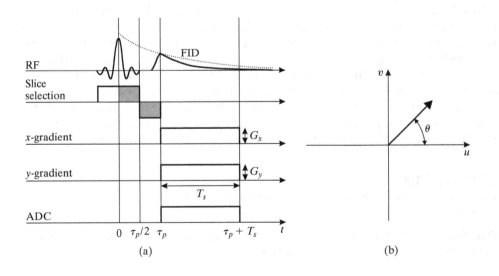

Figure 13.13
(a) A pulse sequence for arbitrary polar scan; (b) a Fourier trajectory for this polar scan.

From this, we identify the Fourier frequencies

$$u = \gamma G_x t \quad \text{and}$$ (13.34a)

$$v = \gamma G_y t,$$ (13.34b)

and find that the implied Fourier trajectory is a ray emanating from the origin in the direction

$$\theta = \tan^{-1} \frac{G_y}{G_x},$$ (13.35)

as depicted in Figure 13.13(b).

   This more general scan strategy has the potential to be used in a complete imaging scenario. What seems to be required is the repetition of the basic pulse sequence enough times to cover Fourier space in polar rays. This is a good idea, except for a still unanswered question: How do we move back to the origin in order to start another polar scan? Actually, the simplest answer is to simply wait for the transverse magnetization to decay. There are two other useful concepts, however, and we begin to develop one of these in the next section.

13.2.5 Gradient Echoes

We studied the concept of spin echoes in Chapter 12, and we will return to those in a bit. First, we introduce another mechanism to create an echo, called a *gradient echo*. This idea can be readily connected to both the Fourier trajectories we have just introduced and the intuitive idea of spins realigning themselves, as in spin echoes and rephasing in slice selection.

   Consider the pulse sequence shown in Figure 13.14(a). The first part of this pulse sequence is very recognizable. In fact, up through time $t = T_E - T_s/2$, the sequence looks like the simple frequency encoded pulse sequence introduced above. In this case, however, the x-gradient is negative, which simply means that the trajectory in

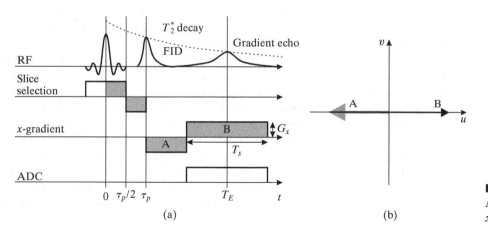

(a)   (b)

Figure 13.14
A simple gradient echo in the x-direction.

Fourier space from the origin moves in the negative u-direction, rather than in the positive u-direction that we studied above. This is illustrated by the "A" gradient lobe in Figure 13.14(a) and the "A" Fourier space trajectory in Figure 13.14(b).

The FID signal is ignored in this pulse sequence (as there is no ADC window during the FID). Instead, the purpose of the negative x-gradient is to move the Fourier space position onto the negative u-axis, so that when the positive readout gradient is applied immediately thereafter, as shown using the "B" lobe in Figure 13.14(a), the Fourier trajectory traverses back in the positive u-direction, crossing the origin and continuing onto the $+u$-axis, as shown by the "B" trajectory in Figure 13.14(b).

It may not be immediately apparent why we should expect to see an echo peaking at time T_E. So, let us consider the effect of the negative x-gradient lobe with strength $-G_x$, applied immediately after slice selection. Of course, we realize that this lobe encodes spatial frequency, precisely as if it were a readout gradient. Therefore, spins are rotating faster on the $-x$ axis than they are on the $+x$ axis. These faster spins accumulate phase as

$$\phi(x, t) = \gamma \int_{\tau_p}^{t} -G_x x \, dt = -\gamma G_x x(t - \tau_p), \quad \tau_p < t < \tau_p + T_s/2. \quad (13.36)$$

This loss of phase coherence, by the way, occurs more rapidly than even T_2^* because of the applied gradient. At time $t = \tau_p + T_s/2$, a positive x-gradient with strength G_x is applied. During this applied gradient, the phase is accumulated according to

$$\phi(x, t) = -\gamma G_x x T_s/2 + \gamma \int_{\tau_p + T_s/2}^{t} G_x x \, dt, \quad (13.37a)$$

$$= -\gamma G_x x T_s/2 + \gamma G_x x(t - \tau_p - T_s/2). \quad (13.37b)$$

By direct substitution, we see that at time $t = \tau_p + T_s$, the accumulated phase is identically zero regardless of the actual value of G_x or the position x. This realignment of spins occurs at the center of the readout gradient, identified as the *echo time* T_E in Figure 13.14(a).

A gradient echo can therefore be viewed in two ways. It can be seen as the natural consequence of an increased signal at the Fourier origin when "driving around Fourier space" or as a realignment of spins at $t = T_E$ due to a realignment of phase accumulations. There is one key feature of the gradient echo approach that encourages the "driving around Fourier space" viewpoint rather than that of "realignment of spins." That feature has to do with the dotted line in Figure 13.14(a), denoting the decay of the signal amplitude in this gradient echo pulse sequence. In particular, the signal strength decays with time constant T_2^* in the gradient echo approach, not with time constant T_2 as we saw in the spin echo approach in Chapter 12. The gradient echo refocuses the phase deviations that we have deliberately introduced through gradient application, not those deviations that are physically present due to static field inhomogeneities. This is a fundamental difference between gradient echoes and spin echoes, and it must always be kept in mind when designing or analyzing pulse sequences.

13.2.6 Phase Encoding

With frequency encoding and gradient echoes, it is clear that a wide variety of schemes (pulse sequences) can be designed to cover 2-D Fourier space and yield enough information to reconstruct a picture of the slice. There is a second important mechanism, however, that is used to encode spatial information in MRI: phase encoding. If we view frequency encoding as a mechanism to read out Fourier data in the u-direction, then phase encoding is viewed as the mechanism to position our readout line in the v-direction in Fourier space. After our discussion of frequency encoding and gradient echoes, phase encoding should seem very intuitive.

Basic Concept   Consider the pulse sequence shown in Figure 13.15(a). The pulse sequence includes the usual slice-selective RF pulse followed by a refocusing gradient. The very next action is a y-gradient pulse ("A" in the figure) with strength G_y and duration T_p. This pulse achieves what is referred to as *phase encoding*, which can be interpreted as a polar scan in Fourier space in the vertical direction, as shown in Figure 13.15(b) ("A" in the figure). Although there is no readout during this pulse (because there is no ADC window at this time), from (13.33) we see that the phase accumulated during this pulse is given by

$$\phi_y(y) = -\gamma G_y T_p y. \qquad (13.38)$$

The very next action in this pulse sequence is a standard x-oriented readout gradient ["B" in Figure 13.15(a)], which acquires data in Fourier space along the B trajectory in Figure 13.15(b). The addition of the phase encode step to (13.27), the imaging equation we previously derived for the FID readout pulse sequence, yields the baseband signal

$$s_0(t) = \int_{-\infty}^{\infty}\int_{-\infty}^{\infty} f(x,y)e^{-j2\pi\gamma G_x xt}e^{-j2\pi\gamma G_y T_p y}\,dx\,dy. \qquad (13.39)$$

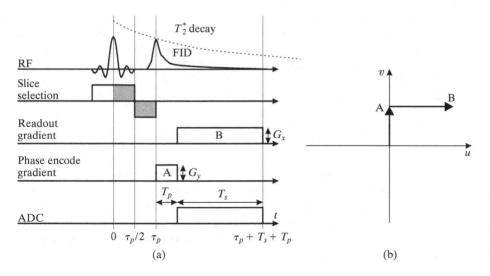

(a)   (b)

Figure 13.15
(a) A simple pulse sequence showing the phase encoding of an FID; (b) its Fourier trajectory.

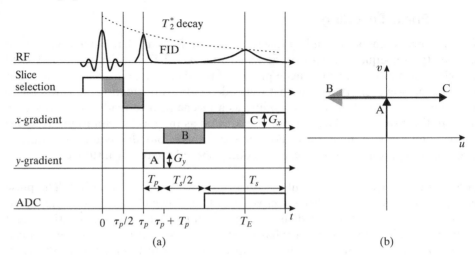

Figure 13.16
(a) A gradient-echo pulse
sequence and (b) its Fourier
trajectory.

Comparing this to the 2-D Fourier transform yields the following identifications:

$$u = \gamma G_x t, \tag{13.40a}$$

$$v = \gamma G_y T_p, \tag{13.40b}$$

and

$$F(u, \gamma G_y T_p) = s_0 \left(\frac{u}{\gamma G_x} \right), \quad 0 \le u \le \gamma G_x T_s. \tag{13.41}$$

A 2-D Gradient-Echo Pulse Sequence   It is common to combine phase encoding, gradient echoes, and frequency encoding in a single pulse sequence. Figure 13.16(a) shows a pulse sequence identical to that of Figure 13.15(a), except that the FID readout has been replaced with a gradient echo readout. In Fourier space [Figure 13.16(b)], there are three distinct movements corresponding to the A, B, and C segments in the pulse sequence. Only the C segment corresponds to actual data collection. This pulse sequence is capable of acquiring an entire "line" of Fourier space from negative to positive frequencies. Analogous analysis to the previous pulse sequence reveals that the Fourier frequencies scanned are given by

$$F(u, \gamma G_y T_p) = s_0 \left(\frac{u}{\gamma G_x} \right), \quad -\gamma G_x T_s/2 \le u \le \gamma G_x T_s/2. \tag{13.42}$$

The pulse sequence depicted in Figure 13.16(a) would work, but it can be easily improved. A better pulse sequence would overlap the refocusing lobe, the phase encoding gradient, and the gradient echo formation lobe. These, it turns out, can all be done at the same time, as their phase preparation processes are independent. The advantage is that the amplitude of the echo is significantly larger since T_E is shorter. This possibility is explored further in Problem 13.15.

EXAMPLE 13.6

A rectangular-shaped phase encoding gradient is not practical because it has discontinuities at both ends (and large slew rates are required even to just approximate it). Since phase encoding depends on the area of the pulse rather than its specific shape, we can use waveforms with more practical shapes.

Question Suppose we want to use a sine-shaped gradient to achieve the same phase accumulation as a rectangular one with $G_y = 1$ gauss/cm, and $T_p = 0.5$ ms. Assuming the same duration is required, what is the expression of the phase encoding gradient and what is its maximum slew rate?

Answer The duration of the gradient is $T_p = 0.5$ ms, exactly half the period of a sine function. So the gradient can be expressed as

$$G_y(t) = G_{y\max} \sin(2\pi t).$$

In order to achieve the same phase accumulation, we need

$$G_y T_p = \int_0^{0.5\,ms} G_{y\max} \sin(2\pi t)\, dt$$

$$0.5\,\text{gauss} \cdot \text{ms/cm} = G_{y\max} \int_0^{0.5\,ms} \sin(2\pi t)\, dt$$

$$= \frac{1}{\pi} G_{y\max}.$$

So, the maximum value for the phase encoding gradient is $G_{y\max} = \pi/2$ gauss/cm. Therefore,

$$G_y(t) = \frac{\pi}{2} \sin(2\pi t) \text{ gauss/cm},$$

where t is in ms. The slew rate is the maximum value of $\left|\dfrac{dG_y(t)}{dt}\right|$, which can easily be computed to be

$$SR = \pi^2 \text{ gauss/cm/ms}. \qquad \blacksquare$$

13.2.7 Spin Echoes

We saw in Chapter 12 that spin echoes could be used to refocus spin systems that had been dephased due to T_2^* effects (static magnetic field inhomogeneities). By using spin echoes, an MR signal could be generated until lost due to T_2 effects (random perturbations of the magnetic field), which are not reversible. Because $T_2 > T_2^*$, the spin echo can produce a measurable signal long after the initial FID decays away. In this section, we will see how the spin echo can be used in an imaging sequence, and its interpretation in Fourier space scanning.

Recall that a spin echo is generated by applying a 180° RF pulse shortly after the initial α-pulse. Spins that were lagging behind are now ahead and vice versa. Consider what this means in terms of phase accumulation. Suppose we had deliberately applied a phase encode with gradient strength G_y just prior to the 180° pulse, which means that phases that had advanced due to their y-position would now be lagging and vice versa. It would be as if we had applied a phase encode with the gradient strength $-G_y$. Now, suppose that we had prepared for a gradient echo by applying an x-gradient with strength $-G_x$ and then applied a 180° RF pulse. Again, the spins that were leading would now be lagging, and it would be as if we had applied an x-gradient with strength G_x. In both cases, the effect of the 180° pulse is to interchange the meaning of lead and lag, and this happens in both the x- and y-directions at the same time.

The effect of a 180° pulse in Fourier space can be understood by carefully interpreting some of our words from the previous paragraph. We stated that the position in Fourier space after the 180° pulse would be as if both the phase encode and the gradient echo preparation pulse had the opposite sign. Therefore, the 180° pulse causes a sudden change of sign of the frequencies corresponding to both the x- and y-directions, which is a reflection through the origin in Fourier space. If the position had been (u, v) prior to the pulse, it would now be $(-u, -v)$. This is a very rapid repositioning in Fourier space, costing only the time it takes to execute a 180° pulse.

Given this description, it is straightforward to construct a pulse sequence incorporating a spin echo. One possibility is shown in Figure 13.17. Section 1 in panel (a) performs the usual slice selection. Section 2 creates a phase encode, moving the Fourier frequency straight up from the origin, as shown in panel (b). Section 3 performs an x prephasing gradient, moving the Fourier location into the first quadrant, as shown in panel (b). Section 4 is a 180° RF pulse executed during a slice selection gradient. Refocusing of this slice selection gradient is not necessary because of a peculiarity of the 180° pulse. In particular, phase accumulated during

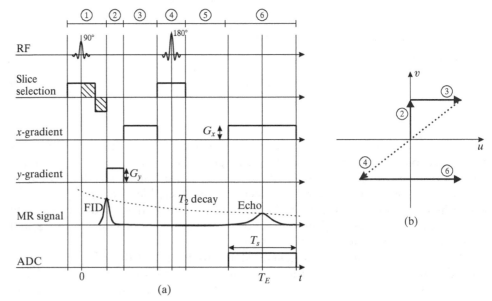

Figure 13.17
A pulse sequence diagram for a spin echo image acquisition.

the first half of the pulse is exactly balanced in the second half of the pulse, because the spins are reversed. Section 5 is a required waiting period, a dead-time during which an echo is formed. Section 6 is the conventional frequency-encoded readout, scanning across Fourier space while collecting data.

EXAMPLE 13.7

Spin echo pulse sequences (Figure 13.17) can be used to measure T_2 values of a sample of a homogeneous tissue. A simple modification to Figure 13.17 is that $G_x = 0$.

Question What is the baseband signal at time $t = T_E$ for a sample of a homogeneous tissue?

Answer From equations (13.22) and (13.27), we can determine that the baseband MR signal is

$$s_0(t) = \int\limits_{-\infty}^{\infty} \int\limits_{-\infty}^{\infty} A M_{xy}(x, y; 0^+) e^{-t/T_2(x,y)} e^{-j2\pi\bar{\gamma} G_x x t} \, dx \, dy.$$

By the assumption that the sample is homogeneous, we have $M_{xy}(x, y; 0^+) = M_{xy}$, and $T_2(x, y) = T_2$. Since $G_x = 0$, the baseband signal at time $t = T_E$ is

$$s_0(T_E) = A M_{xy} e^{-T_E/T_2} \Delta A,$$

where ΔA is the area of the cross section of the sample in x-y plane. Now, suppose we use the same pulse sequences with different echo time T_E' to image the same sample. The baseband signal at time $t = T_E'$ measured is

$$s_0'(T_E') = A M_{xy} e^{-T_E'/T_2} \Delta A.$$

The value of T_2 can be solved as

$$T_2 = (T_E' - T_E) \ln \left[\frac{s_0(T_E)}{s_0'(T_E')} \right]. \qquad \blacksquare$$

13.2.8 Pulse Repetition Interval

The gradient echo and spin echo pulse sequences have the potential to acquire all of 2-D Fourier space (at least in a sampled fashion). In order to do so, however, the basic pulse sequence must be repeated with different scan parameters. For example, the polar scan method must be repeated using different readout orientations θ. Both the gradient echo and spin echo techniques must be repeated using different phase encoding gradient values G_y. The duration of the interval between such repetitions is called the *pulse repetition time* and is given the symbol T_R.

Our development of MR imaging equations (Section 13.3) will depend on whether the sequence is presumed to be either a *slow imaging sequence* or a *fast imaging sequence*. The slow imaging regime assumes $T_R \gg T_2$. In this case, the transverse magnetization has completely disappeared before application of the next α-pulse, and there is no possibility of producing an echo from a previous excitation. The Fourier analysis of successive pulse sequence can correctly treat each one as if it starts from the Fourier origin. This is a little like ultrasound, where T_R must be large enough so echoes cannot be received from previous pulses and thereby confused with echoes coming from present pulses.

The fast imaging regime, however, breaks down this "barrier" using several different techniques, including the acquisition of multiple spin echoes or gradient echoes from a single initial RF excitation, and *spoiling* the transverse magnetization prior to subsequent excitation. Based on the previous discussion, we can concoct several schemes to acquire multiple spin echoes or gradient echoes. Basically, the "driving around in Fourier space" idea works fine, provided that we think correctly about compensating for spins that need to be rephased and continuing loss of signal due to T_2 and T_2^* effects. We do not have space to cover these techniques within the main body of the text, but offer several examples that can be worked out in the problems.

The idea of spoiling is something new, however. In this approach, a special additional gradient pulse called a *spoiler pulse* is used just prior to each successive excitation in order to guarantee that no echoes will form from previous excitations. The basic idea is apply a z gradient in order to dephase the spins in the z-direction, so that spins integrated over the thickness of the slice will add destructively so that no signal can be produced. A different value of the spoiler gradient is used with each successive T_R interval so that there will be no possibility that the spins would actually be forced to line up again at some point later in the sequence. Gradient echo sequences that use this technique are called *spoiled gradient echo* (SPGR) image sequences.

13.2.9 Realistic Pulse Sequences

We now present three pulse sequences that might be implemented in practice. These pulse sequences are based on the principles developed in the previous section, but show adherence to slew rate limitations, take advantage of overlap, and use alternate wave shapes where area is the only consideration.

Two-Dimensional Gradient-Echo Pulse Sequence The prototype 2-D gradient-echo pulse sequence is shown in Figure 13.16. There are several aspects of this pulse sequence that are unrealizable, impractical, or undesirable. Figure 13.18 gives a more realistic pulse sequence for the following reasons. First, in this pulse sequence, all gradient waveforms are represented as trapezoidal pulses, which reflects the fact that there is a limited slew rate on the gradient amplifiers. Second, the slice selection refocusing lobe, the phase encode pulse, and the readout prefocusing pulse are all done at the same time. This does not cause any problem since these processes are independent. This practice saves time and puts the initial scan position in the correct location in Fourier space (although the exact Fourier trajectory taken from the origin to the final position might be quite complicated).

The third reason that Figure 13.18 is a more realistic pulse sequence is that it acknowledges the fact that the basic pulse sequence must be repeated with different phase encode values. The start of the second pulse sequence is shown after the breaks appearing on the time axes, and the repetition interval is explicitly shown on the pulse diagram as T_R. The breaks in the time axes indicate that time may elapse

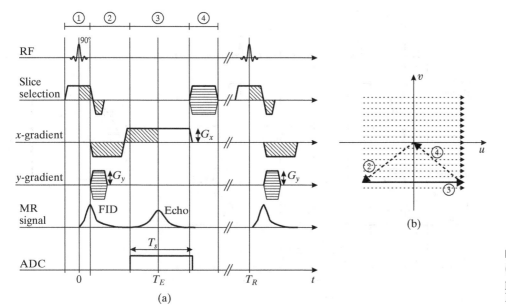

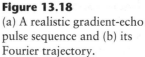

(b)

Figure 13.18
(a) A realistic gradient-echo pulse sequence and (b) its Fourier trajectory.

before starting the next excitation. The different phase encode values are indicated using a glyph that shows the basic waveform as a bold trapezoidal envelope and the other waveforms as lighter trapezoids. The Fourier diagram in panel (b) indicates the Fourier trajectory in a single excitation as well as additional trajectories for multiple excitations. It is clear that this pulse sequence is capable of covering Fourier space after multiple excitations.

A fourth aspect of Figure 13.18 that is different than Figure 13.16 is the additional z-gradient during time period 4. This is the spoiler gradient discussed above, and it is often used to produce a faster imaging sequence. Like the phase encode gradient, the spoiler uses a different value with each repetition, as depicted by the bold and lighter lines.

Two-Dimensional Spin-Echo Pulse Sequence   Having seen the elements that make up a more realistic 2-D gradient echo pulse sequence, the more realistic spin-echo sequence shown in Figure 13.19 should come as no surprise. As in the gradient-echo sequence, gradient waveforms are shown as trapezoids, reflecting the actual limitations of MR gradient amplifiers. We depict a repetition of the basic pulse sequence using breaks in the time axes and indicating the pulse repetition interval T_R. There are also special glyphs to depict the use of multiple phase encodes over successive excitations.

The spin-echo sequence cannot use a spoiler since successive excitations need to create true echoes. Fast imaging techniques using spin echoes exist, and some discussion is presented in the problems.

Two-Dimensional Polar Imaging   A realistic 2-D polar imaging pulse sequence is shown in Figure 13.20. We show a spin-echo sequence here, which is typical

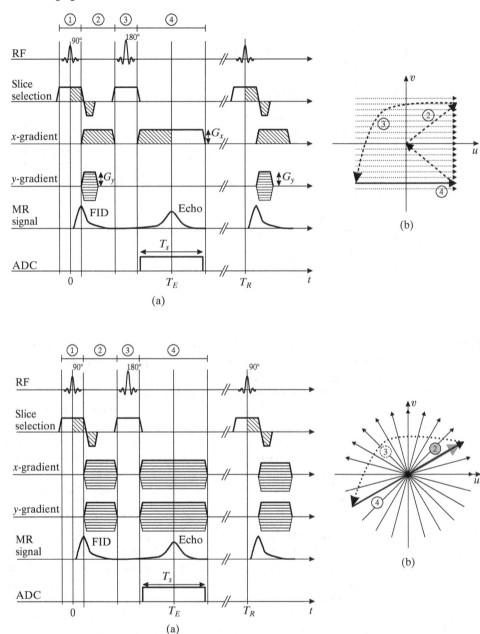

Figure 13.19
(a) A realistic spin-echo pulse sequence and (b) its Fourier trajectory.

Figure 13.20
(a) A realistic spin-echo polar pulse sequence and (b) its Fourier trajectory.

for polar imaging, but we could design a gradient-echo or spoiled gradient-echo sequence as well. As in both of the realistic sequences shown above, we have indicated the required repetition of the basic pulse sequence using both breaks in the time axes and glyphs that indicate the changing waveforms with each excitation.

The x- and y-gradients, which are used together for both phase encoding and frequency encoding (readout), in a polar pulse sequence must be chosen so that the correct orientation is scanned. In order to scan angle θ_n during the nth excitation, following (13.35), we see that the gradients must be selected as follows:

$$\theta_n = \tan^{-1} \frac{G_{y,n}}{G_{x,n}} \,. \tag{13.43}$$

13.3 Image Reconstruction

Since the data acquired from 2-D MR imaging pulse sequences can be interpreted as scans of Fourier space, the image reconstruction algorithm in MRI is the inverse Fourier transform. Since we have presented both rectilinear and polar data acquisition, we present the corresponding reconstruction algorithms.

13.3.1 Rectilinear Data

For conventional pulse sequences that acquire data in a rectilinear fashion, such as the gradient-echo or spin-echo sequences outlined above, the baseband signal is a temporal waveform that is dependent on the area A_y of the phase encode gradient (which can be negative). With reference to (13.39), the baseband signal can be written as

$$s_0(t, A_y) = \int_{-\infty}^{\infty} \int_{-\infty}^{\infty} f(x, y) e^{-j2\pi\gamma G_x x t} e^{-j2\pi\gamma A_y y} \, dx \, dy \,, \tag{13.44}$$

where the dependence on both t and A_y has been made explicit. We identify the Fourier frequencies as [see (13.40)]

$$u = \gamma G_x t \,, \tag{13.45a}$$

$$v = \gamma A_y \,, \tag{13.45b}$$

which allows us to identify the Fourier transform of $f(x, y)$ as

$$F(u, v) = s_0 \left(\frac{u}{\gamma G_x}, \frac{v}{\gamma} \right), \quad 0 \le u \le \gamma G_x T_s \,. \tag{13.46}$$

An MR image is reconstructed as the inverse 2-D Fourier transform of $F(u, v)$, which from (13.46) can be written as

$$f(x, y) = \int_{-\infty}^{\infty} \int_{-\infty}^{\infty} s_0 \left(\frac{u}{\gamma G_x}, \frac{v}{\gamma} \right) e^{+j2\pi(ux+vy)} \, du \, dv \,. \tag{13.47}$$

This equation is truly fundamental in MRI. Its simplicity, however, hides so many practical aspects of MRI that we can be easily lulled into a false sense of understanding. For example, we do not really know what $f(x, y)$ represents at this stage in our presentation. The meaning of $f(x, y)$ is not as conceptually straightforward as the linear attenuation coefficient in CT, the radioactivity concentration in SPECT, or even the reflectivity in ultrasound. In fact, $f(x, y)$ is determined by several tissue NMR properties as well as certain pulse sequence parameters. We will have more to say about this in Section 13.4.

Another aspect of (13.47) that must be addressed is the fact that both frequency variables, u and v, must be sampled. This leads to the practical requirement that (13.47) be computed using a 2-D discrete inverse Fourier transform algorithm [typically based on the fast Fourier transform or (FFT)]. In this book, we do not require a background in digital signal processing, so we will not discuss the specifics of algorithm implementation. However, the fact that Fourier space is *sampled* in MRI is critical to image quality and will be discussed in Section 13.4.

EXAMPLE 13.8

A 2-D MR image is reconstructed from data collected by sampling Fourier space. Assume that a data acquisition using a spin-echo pulse sequence (Figure 13.19), and a square region of Fourier space centered at $(u, v) = (0, 0)$ is sampled in a rectilinear fashion.

Question   The parameters are set to be $T_R = 50$ ms, $T_s = 3$ ms, $G_x = 1$ gauss/cm, $\Delta G_y = 0.1$ gauss/cm, $\gamma = 4.258$ kHz/gauss, and $T_p = 0.3$ ms is fixed. How many lines are acquired in Fourier space? How long does it take to acquire the entire image?

Answer   The Fourier frequency is given by

$$u = \gamma G_x t.$$

Since $G_x = 1$ gauss/cm and $T_s = 2$ ms, the range of u acquired is

$$u_{max} - u_{min} = \gamma G_x T_s = 12.774 \, \text{cm}^{-1}.$$

The square coverage in Fourier space requires $v_{max} - v_{min} = 12.774 \, \text{cm}^{-1}$. Since we have a fixed T_p and $\Delta G_y = 0.1$ gauss/cm, the adjacent lines in Fourier space are separated by

$$\Delta v = \gamma \Delta G_y T_p = 0.128 \, \text{cm}^{-1}.$$

In order to cover the square in Fourier space, we need to acquire

$$N = \frac{12.774 \, \text{cm}^{-1}}{0.128 \, \text{cm}^{-1}} = 100 \, \text{lines}.$$

The imaging time is

$$T = NT_R = 5 \, \text{s}. \qquad \blacksquare$$

13.3.2 Polar Data

Pulse sequences that acquire polar data in Fourier space have baseband signals that depend on both time and orientation, $s_0(t, \theta)$, where θ is given by (13.35). In this case, the frequency variable is identified in polar coordinates as

$$\varrho = \gamma \sqrt{G_x^2 + G_y^2} \; t. \tag{13.48}$$

From the development in CT, we know that the projection slice theorem equates a "slice" of $F(u, v)$ to the Fourier transform of $g(\ell, \theta)$, which is a projection of $f(x, y)$. That is,

$$G(\varrho, \theta) = F(\varrho \cos \theta, \varrho \sin \theta), \tag{13.49}$$

where

$$G(\varrho, \theta) = \mathcal{F}_{1D}\{g(\ell, \theta)\}, \tag{13.50}$$

and

$$g(\ell, \theta) = \int_{-\infty}^{\infty} \int_{-\infty}^{\infty} f(x, y)\delta(\ell - x \cos \theta - y \sin \theta) \, dx \, dy. \tag{13.51}$$

The baseband signal is a polar scan in Fourier space, which yields

$$G(\varrho, \theta) = s_0 \left(\frac{\varrho}{\gamma \sqrt{G_x^2 + G_y^2}}, \theta \right). \tag{13.52}$$

From the 2-D Radon transform theory developed in Chapter 6, we know that $f(x, y)$ can be reconstructed using filtered backprojection [see (6.23)], which is given by

$$f(x, y) = \int_0^{\pi} \left[\int_{-\infty}^{\infty} |\varrho| G(\varrho, \theta) e^{j2\pi\varrho\ell} \, d\varrho \right]_{\ell = x \cos \theta + y \sin \theta} d\theta. \tag{13.53}$$

We know that convolution backprojection is an equivalent approach [see (6.26)], and polar-to-rectangular interpolation methods could also be used to form a rectilinear array of data. The reconstruction method would thus be a standard 2-D inverse Fourier transform.

13.3.3 Imaging Equations

We introduced the concept of MRI *contrast mechanisms* in Section 12.10. We explained that the basic tissue parameters P_D, T_2, and T_1 are responsible for contrast and that the pulse sequence parameters—α, T_E, and T_R—can be manipulated to change the image contrast and, hence, the appearance of the reconstructed MR images. Table 13.1 summarizes our findings from the previous chapter, describing the basic way to create contrast that is primarily dependent on P_D, T_2, and T_1. Please look back at Figure 12.10 to remind yourself how dramatically different the same cross section can look depending on the contrast mechanism used to create the image.

Although the list in Table 13.1 is valid, the given scanner parameters are not the only way to produce the target contrast. In particular, the changes in contrast due to tip angle α are not reflected in this table. To get a much more complete understanding of the possibilities, it is necessary to derive an *imaging equation* for a given pulse sequence, which we now develop. It should be noted that these three

TABLE 13.1

Contrast Generation in Basic MRI	
Contrast	Scanner Parameters
P_D	Long T_R, read FID or use short T_E
T_2	Long T_R, $T_E \approx T_2$,
T_1	Read FID or use short T_E, $T_R \approx T_1$

contrasts are not the only possible contrast mechanisms in MRI. Other key contrast mechanisms in MRI include T_2^*, flow, diffusion, and chemical shift. These topics are left for independent study; a good starting point is the bibliography given at the end of the chapter.

We have established that MRI reconstructs an image of the effective spin density $f(x, y)$, which is given by

$$f(x, y) = AM(x, y; 0^+)e^{-t/T_2(x,y)} \,. \tag{13.54}$$

The first fact we need to realize is that t should be replaced by the time at which the signal is sampled. In the case of FID sampling $t = 0$, while for echo sampling $t = T_E$. Now consider a pulse sequence that issues a steady succession of α RF excitation pulses and assume that $T_R \gg T_2$ (which implies that the transverse magnetization is gone by the time each successive RF excitation is applied). If $T_R \approx T_1$ (at least for some tissues), then the sample will not be in equilibrium; that is, the longitudinal magnetization will not reach its equilibrium value before the next excitation pulse. But, each pixel will establish a steady-state z magnetization in a certain sense. We now explore the concept of steady-state pulse sequences.

From (12.34), we know that the first α-pulse will yield the longitudinal magnetization

$$M_z(t) = M_z(0^+)e^{-t/T_1} + M_0(1 - e^{-t/T_1}), \tag{13.55}$$

where it is understood in the context of imaging that M_0, M_z, and T_1 are functions of x and y within an excited image slab. If an α-pulse is used, we have

$$M_z(0^+) = M_z(0^-) \cos\alpha, \tag{13.56}$$

and $M_z(0^-) = M_0$ for the first pulse. So,

$$M_z(t) = M_0 \cos\alpha e^{-t/T_1} + M_0(1 - e^{-t/T_1}), \tag{13.57}$$

which at the moment of the second excitation has the value

$$M_z(T_R) = M_0 \cos\alpha e^{-T_R/T_1} + M_0(1 - e^{-T_R/T_1}). \tag{13.58}$$

This longitudinal magnetization will be flipped by α in the next excitation, and will then undergo further recovery.

It is straightforward to show that the magnetization just prior to the nth α-pulse must obey

$$M_z(nT_R) = M_z([n-1]T_R)\cos\alpha e^{-T_R/T_1} + M_0(1 - e^{-T_R/T_1}). \qquad (13.59)$$

Equation (13.59) is a difference equation that can be solved for its steady-state value $M_z^\infty(0^-)$. In particular, the steady-state value must obey

$$M_z^\infty(0^-) = M_z^\infty(0^-)\cos\alpha e^{-T_R/T_1} + M_0(1 - e^{-T_R/T_1}), \qquad (13.60)$$

which yields, after some algebra,

$$M_z^\infty(0^-) = M_0 \frac{1 - e^{-T_R/T_1}}{1 - \cos\alpha e^{-T_R/T_1}}. \qquad (13.61)$$

Looking back at the effective spin density [(13.54)], we realize that the longitudinal magnetization in (13.61) is tipped by α during RF excitation in order to generate a transverse component—the source of the NMR signal. Therefore, the reconstructed spin density for a steady-state pulse sequence can be written as

$$f(x, y) = AM_z^\infty(x, y; 0^-)\sin\alpha e^{-T_E/T_2(x,y)}, \qquad (13.62)$$

where $T_E = 0$ in the case of FID imaging. Incorporating (13.61) yields

$$f(x, y) = AM_0 \sin\alpha e^{-T_E/T_2(x,y)} \frac{1 - e^{-T_R/T_1}}{1 - \cos\alpha e^{-T_R/T_1}}, \qquad (13.63)$$

which is a very common imaging equation for MR imaging. It should always be remembered that it applies for steady-state imaging when $T_R \gg T_2$.

EXAMPLE 13.9

From (13.63), we can see that the parameters, T_E, T_R, and α, can be optimized to obtain the best contrast between different tissues.

Question   Suppose that two tissues have the same proton density and T_2 value but have different T_1 values, T_1^b and T_1^f. What is the optimal T_R value that provides the best local contrast of effective spin density, given that $\alpha = \pi/2$?

Answer   The flip angle is $\alpha = \pi/2$. So the effective spin density for two tissues are

$$f^b(x, y) = AM_0 e^{-T_E/T_2}(1 - e^{-T_R/T_1^b}),$$

$$f^f(x, y) = AM_0 e^{-T_E/T_2}(1 - e^{-T_R/T_1^f}).$$

The local contrast of the effective spin density is

$$C = \frac{f^f(x, y) - f^b(x, y)}{f^b(x, y)} = \frac{e^{-T_R/T_1^b} - e^{-T_R/T_1^f}}{1 - e^{-T_R/T_1^b}}.$$

By assuming $T_R \gg T_1^{\text{b}}$, we have

$$C \approx e^{-T_R/T_1^{\text{b}}} - e^{-T_R/T_1^{\text{f}}}.$$

By taking the derivative of C with respect to T_R and setting it to zero, we have

$$\frac{dC}{dT_R} = -\frac{1}{T_1^{\text{b}}} e^{-T_R/T_1^{\text{b}}} + \frac{1}{T_1^{\text{f}}} e^{-T_R/T_1^{\text{f}}} = 0.$$

The optimal T_R for best local contrast is

$$\hat{T}_R = \frac{T_1^{\text{b}} T_1^{\text{f}}}{T_1^{\text{b}} - T_1^{\text{f}}} \ln\left(\frac{T_1^{\text{f}}}{T_1^{\text{b}}}\right). \qquad \blacksquare$$

13.4 Image Quality

In MRI, there are many issues that must be understood in order to get a good sense of image quality. In this section, we will address the concepts of sampling, resolution, noise, SNR, and artifacts. We will find that the issues are a bit more involved than most medical imaging modalities, and we can pursue only the most basic concepts in this book. Further details can be found in several of the references listed at the end of this chapter.

13.4.1 Sampling

We now understand that MR imaging can be interpreted as a procedure that samples the 2-D spatial Fourier space of the effective spin density in a plane. The most basic part of this acquisition process is the sampling of the baseband signal $s_0(t)$—an FID or echo—during a readout gradient using an analog-to-digital converter (ADC). Suppose N_a samples are acquired T seconds apart during a readout that has gradient strength G_x. Then, the ADC will acquire data for duration

$$T_s = N_a T, \qquad (13.64)$$

a variable that is identified in all of our pulse sequences. The sampling rate for this process is

$$f_s = \frac{1}{T}, \qquad (13.65)$$

and it is given the name *receiver bandwidth* (RBW) in MRI. This is because the ADC uses an antialiasing filter that cuts off frequencies outside the interval $[-f_s/2, f_s/2]$ in order to avoid temporal aliasing. The size of this interval is the receiver bandwidth—i.e., RBW $= f_s/2 - -f_s/2 = f_s$.

Recall that the readout gradient encodes the Larmor frequencies of the spin system in the x-direction. Accordingly, on one side of the field of view (FOV), the spins have a higher frequency during the readout interval, and on the other side they have a lower frequency. We know that the received signal $s(t)$ is demodulated using the Larmor frequency $\nu_0 = \gamma B_0$, which corresponds to the center of the FOV, where the Larmor frequency is ν_0. Therefore, relative to the center of the FOV, the baseband signal frequencies on one side of the FOV will have positive frequencies and the other side will have negative frequencies during a readout interval.

Since the sampling process uses an antialiasing filter, the sampled baseband signal cannot represent frequencies outside the interval $[-f_s/2, f_s/2]$. This means that spins with higher (negative or positive) frequencies will be invisible—the FOV is cut off in the x-direction by the readout ADC. Therefore, the size of the FOV in the x-direction FOV_x is specified by the spatial range of the spin system with Larmor frequencies within the RBW. Accordingly,

$$\gamma G_x \text{FOV}_x = f_s, \tag{13.66}$$

which leads to the relationship

$$\text{FOV}_x = \frac{f_s}{\gamma G_x} \tag{13.67a}$$

$$= \frac{1}{\gamma G_x T}, \tag{13.67b}$$

where the last equality follows from (13.65). By comparing the denominator of (13.67a) with (13.40a), it is evident that the quantity $\gamma G_x T$ represents a "step" Δu in Fourier space in the readout direction. Accordingly,

$$\Delta u = \gamma G_x T \tag{13.68}$$

is the separation of samples in k space in the u-direction corresponding to a sampled MRI readout.

Sampling in the v-direction is determined by the sequence of phase encode gradients that are used. Typically, a step size in phase encode gradient area ΔA_y is selected, and any particular phase encode waveform has an area that is an integer multiple of ΔA_y. Accordingly, we find that

$$\Delta v = \gamma \Delta A_y \tag{13.69}$$

and

$$\text{FOV}_y = \frac{1}{\gamma \Delta A_y} \tag{13.70a}$$

$$= \frac{1}{\Delta v}. \tag{13.70b}$$

Despite the analogous expressions for FOV in the readout and phase-encode directions, there is a key difference between them that requires careful interpretation and implementation. In the readout direction, an antialiasing filter is used to remove frequencies higher than those in the readout FOV. This has the effect of obliterating the signal coming from tissues outside the readout FOV. In the phase encode direction, however, there is no antialiasing filter; the Fourier space lines that are sampled are essentially "point samples" with respect to the phase encode direction. So, if tissues actually exist outside the calculated phase-encode FOV, they will be aliased and will cause wraparound in the phase encode direction in the reconstructed images.

The aliasing problem caused by a too-large step size in phase encode gradients is analogous to the concept of aliasing presented in Chapter 2. Here, we are sampling Fourier space rather than image space, and the aliasing causes wraparound in image space rather than in Fourier space. For example, in a conventional axial brain image in which the phase encode direction corresponds to the back-to-front direction, too-large phase-encode sampling will cause the back of the brain to appear at the front and vice versa.

EXAMPLE 13.10

The field of view of an MR image is related to the sampling steps in Fourier space. Suppose the reconstructed MR image has same resolution and size in both x and y directions.

Question If we set $T_s = 3$ ms, $G_x = 1$ gauss/cm, and $f_s = 85.33$ kHz, how many samples are acquired for each readout gradient? What is ΔA_y? What is the size of the FOV?

Answer The duration of a readout gradient is $T_s = 3$ ms, and RBW is $f_s = 85.33$ kHz. So there are $N_a = T_s f_s = 256$ samples taken for each readout gradient. The sampling period is $T = 1/f_s = 11.72\,\mu$s. Since the reconstructed image has same resolution and size in both directions, we have

$$\Delta v = \Delta u = \gamma G_x T = \gamma \Delta A_y.$$

This means that, $\Delta A_y = G_x T = 11.72$ gauss·µs/cm. The FOV has the same size in both directions, which is

$$\text{FOV}_x = \frac{f_s}{\gamma G_x} = \frac{85.33\,\text{kHz}}{4.258\,\text{kHz/gauss} \times 1\,\text{gauss/cm}} = 20.04\,\text{cm}. \qquad \blacksquare$$

13.4.2 Resolution

Viewing MRI as a Fourier imaging method makes a basic discussion of image resolution fairly straightforward. We know that pulse sequences acquire data from a region in Fourier space using a sequence of readouts [see Figures 13.18(b), 13.19(b), and 13.20]. If there were any Fourier information outside the region acquired by the pulse sequence—i.e., higher frequency information—then it is not imaged, and it is reconstructed as if it were identically zero. Therefore, a pulse sequence inherently represents a low-pass filtering process applied to the underlying image. This limits the achievable resolution of MRI.

Let us consider the conventional acquisition approaches that use rectilinear scanning, as in Figures 13.18 and 13.19. The low-pass filter implied by these methods is a rectangle in Fourier space with dimensions

$$U = N_x \gamma G_x T \tag{13.71a}$$

$$V = N_y \gamma \Delta A_y, \tag{13.71b}$$

where N_x and N_y are the number of readout and phase encode samples acquired, respectively. The implied low-pass filter is therefore given by

$$H(u, v) = \text{rect}\left(\frac{u}{U}\right) \text{rect}\left(\frac{v}{V}\right), \tag{13.72}$$

and the equivalent spatial PSF is

$$h(x, y) = UV \, \text{sinc}(Ux) \, \text{sinc}(Vy). \tag{13.73}$$

It is sufficient to consider the main lobe of the 2-D sinc function in (13.73) in order to define resolution. As we have often done throughout this book, we will approximate the FWHM of a sinc function as half of the interval between the first two zeros. Accordingly,

$$\text{FWHM}_x = \frac{1}{U} = \frac{1}{N_x \gamma G_x T} \tag{13.74a}$$

$$\text{FWHM}_y = \frac{1}{V} = \frac{1}{N_y \gamma \Delta A_y}. \tag{13.74b}$$

Incorporating the definitions in (13.68) and (13.69) yields

$$\text{FWHM}_x = \frac{1}{N_x \Delta u} \tag{13.75a}$$

$$\text{FWHM}_y = \frac{1}{N_y \Delta v}, \tag{13.75b}$$

which are called the *Fourier resolutions* of MRI.

When polar sampling is used, the concept of resolution is conceptually the same as that in CT (see Section 6.4.1). In CT, two effects degraded spatial resolution: detector size and the ramp filter window function $W(\varrho)$. In MR, the limited trajectories being scanned in Fourier space—not detector size—are relevant in degrading resolution. Suppose $S(\varrho)$ represents a rectangular window describing the frequency scan (assumed to be symmetric about the origin) of each polar scan. Then the reconstructed effective spin density is given by

$$\hat{f}(x, y) = f(x, y) * h(r), \tag{13.76}$$

where $r = \sqrt{x^2 + y^2}$. The point spread function is circularly symmetric and given by the inverse Hankel transform [see (2.109)]

$$h(r) = \mathcal{H}^{-1}\{S(\varrho)W(\varrho)\}. \qquad (13.77)$$

It is important to distinguish what is sometimes called the *pixel size* from the Fourier resolution. Our formulas for reconstructing $f(x, y)$ are stated in terms of continuous variables—both spatial and Fourier variables—although we know that they will be implemented digitally. Without assuming prerequisite knowledge of digital signal processing, it is difficult to get too deeply into the details of implementation; but the concept of pixel size can be clarified easily enough.

Observations at discrete u and v locations will be obtained by MR scanning, and the reconstruction formula, (13.47) or (13.53), will be discretized to accept only those observed values. Similarly, the x and y locations at which $f(x, y)$ is computed will also be discretized. The separation of the x and y samples, Δx and Δy, define the dimensions of a pixel, and together they define the pixel size. These dimensions are completely arbitrary. Although we may have 256 by 256 Fourier space samples, we can choose to reconstruct J by J pixels in the image space, and J can be absolutely anything. Of key importance, however, is that the underlying Fourier resolution does not change as a function of J, Δx, and Δy, but only as a function of the image acquisition Fourier window.

EXAMPLE 13.11

The resolution of an MR image is related to the k space coverage by Fourier transform. Suppose we want to acquire a 256×256 image, with resolution of $1\,\text{mm} \times 1\,\text{mm}$.

Question   Suppose the readout gradient is $G_x = 1$ gauss/cm, what is the receiver bandwidth f_s? What is ΔA_y? What is the k space coverage? If T_R is 50 ms, what is the imaging time?

Answer   The Fourier resolution in the x direction is

$$\text{FWHM}_x = \frac{1}{N_x \gamma G_x T} = 1\,\text{mm}.$$

So the receiver bandwidth is

$$f_s = \frac{1}{T} = N_x \text{FWHM}_x \gamma G_x = 109\,\text{kHz}.$$

The Fourier resolution in the y direction is

$$\text{FWHM}_y = \frac{1}{N_y \gamma \Delta A_y} = 1\,\text{mm}.$$

So we have

$$\Delta A_y = \frac{1}{N_y \gamma \text{FWHM}_y} = 9.17\,\text{gauss·}\mu\text{s/cm}.$$

The k space coverage is $V = U = 1/\text{FWHM}_x = 10\,\text{cm}^{-1}$. The imaging time is $T = 50\,\text{ms} \times 256 = 12.8\,\text{s}$. ∎

13.4.3 Noise

Noise in MRI arises from statistical fluctuations of the signal sensed by the receiver coils—e.g., FID or echo. The dominant source of this noise is *Johnson noise*, which is generically caused by the thermal agitation of electrons or ions in a conductor. In MRI, Johnson noise arises predominantly from the electrolytes in the patient, but it can also arise within the receive coil and the electronics attached to it. The variance of the received noise can be characterized as

$$\sigma^2 = \frac{2k\mathcal{T}R}{T_A},$$ (13.78)

where k is Boltzmann's constant, $\mathcal{T}$ is temperature, R is the effective electrical resistance that the receiver coil "sees," and T_A is the total acquisition time. It is important to note that T_A is the sum of all the readout times during which data is collected; this time is less than the total time it takes to complete a given pulse sequence since there are many intervals in which the ADC is not active.

Normally, it is desirable to reduce noise as much as possible. We do not have control over temperature $\mathcal{T}$ (in ordinary human imaging scenarios, anyway), but we can control both R (to some extent) and T_A. The effective resistance in both the RF coil and the electronics is ordinarily much smaller than that of the body, and it can be ignored. Therefore, to reduce R, we need to reduce the amount of patient "seen" by the RF coil. We learned earlier that there are different RF coil designs—e.g., body coils, head coils, and surface coils. Therefore, choosing a coil whose sensitivity pattern encompasses a smaller volume will inevitably reduce the noise in the receiver. For example, it is highly desirable to use a head coil when imaging the head, since the body coil (attached within the bore of the scanner) would "see" a large part of the thorax, thereby contributing noise to the signal. Coils for the extremities are even smaller, and they are therefore better for imaging small body parts. Surface coils, while typically seeing a very small volume, have an uneven sensitivity pattern. In such coils the MR intensity will vary across the image, producing a "shading artifact" that is usually undesirable.

For a body or head coil, which can be approximated as a solenoid (a multiturn loop of wire), an equation for R can be worked out (see Problem 13.22). Suppose the solenoid has N turns per unit length, a total length of L, and each turn has radius r_0. We can then determine that

$$R = \frac{\pi^3 \mu_0^2 \nu_0^2 N^2 L r_0^4}{2\rho}.$$ (13.79)

Here, ν_0 is the Larmor frequency, $\mu_0^2 = 4\pi \times 10^{-7}$ weber/amp-meter is the permeability constant, and ρ is the resistivity of the body. From (13.78) and (13.79), it seems that noise can be reduced by decreasing (1) the solenoid's length, (2) the number of turns per unit length, (3) the solenoid's radius, or (4) ν_0. If we reduce the length of the solenoid, then its imaging volume is made smaller, which may be undesirable in volumetric imaging. If we reduce its radius, then only a smaller object will fit within the coil. This is essentially the difference between a head coil and a body coil; although either coil can be used to create an image of the head, a head coil has lower noise than a body coil and is therefore preferable.

According to (13.79), reducing the number of coils per unit length seems like a good idea, but it turns out that this will also have the effect of reducing the signal and will not actually improve the image quality. Finally, reducing v_0 is easily accomplished by using less current in the superconducting magnet, which in turn reduces B_0 and hence $v_0 = \gamma B_0$. It turns out, however, that the NMR signal will be reduced faster, causing an actual degradation of image quality. To understand these two facts, we need to understand the parameters affecting signal strength in MRI and then look at signal-to-noise ratio.

EXAMPLE 13.12

Noise is sometimes thought to be controlled by the sampling rate (receiver bandwidth), and that higher sampling rates cause an increase in the image noise.

Question   Find an expression for MRI noise variance in terms of the sampling rate (receiver bandwidth) f_s.

Answer   The total acquisition time is

$$T_A = MT_s,$$

where M is the number of readouts and T_s is the total time of a readout (where we assume that all readouts have the same duration in a given pulse sequence). Assuming there are N_a samples in each readout with a time separation of T per sample,

$$T_s = N_a T.$$

Since

$$f_s = \frac{1}{T},$$

we can write

$$T_A = \frac{MN_a}{f_s},$$

and using this in (13.78) yields

$$\sigma^2 = \frac{2k\mathcal{T}Rf_s}{MN_a}. \tag{13.80}$$

It therefore appears that increasing f_s leads directly to an increase in image noise. But what has really happened is that the larger f_s leads to a faster scanning of Fourier space and a reduction in overall acquisition time; this leads to a higher image noise.

In this example, had we simultaneously changed the readout gradient G_x while increasing the sampling rate in such a way that T_s remained constant, the noise would have remained the constant as well. Alternatively, if we doubled f_s but scanned Fourier space twice and averaged the results, then the total acquisition time would be the same and the image noise would remain unchanged. Thus (13.80) should not be used, even though it is technically correct, as it hides the true mechanism governing image noise in MRI.   ∎

13.4.4 **Signal-to-Noise Ratio**

We are interested to know what factors can be adjusted to reduce the noise, while simultaneously keeping the signal strong. This concept is embodied in signal-to-noise ratio (SNR), of course, and since we already have an expression for the noise, we need only find an expression for the signal. Under the assumption of a homogeneous sample and uniform RF field, (12.25) gives the result we need. For convenience, we repeat this equation for signal strength here:

$$|V| = 2\pi \nu_0 V_s M_0 \sin \alpha \, B^r \, . \tag{13.81}$$

If the receive coil is a solenoid (as in the previous section), then the reference field strength B^r at the center of the coil, produced by a unit current I_0 passing through the coil, is given by

$$B^r = \mu_0 I_0 N \, , \tag{13.82}$$

where N is the number of turns per unit length. Substituting (13.82) into (13.81) yields

$$|V| = 2\pi \nu_0 V_s M_0 \sin \alpha \, \mu_0 N \, , \tag{13.83}$$

where we have assumed $|I_0| = 1$, since it is unity reference current.

An expression for SNR is found using (13.78) and (13.83):

$$\text{SNR} = \frac{|V|}{\sqrt{\sigma^2}} = \frac{2\pi \nu_0 V_s M_0 \sin \alpha \, \mu_0 N}{\sqrt{2k\mathcal{T}R/T_A}} \, . \tag{13.84}$$

Substituting for R using (13.79) and M_0 using (12.4), and working through some algebra yields

$$\text{SNR} = \frac{\gamma \hbar^2}{\sqrt{4\pi k^3}} \frac{2\pi \nu_0 P_D \sqrt{\rho}}{r_0^2 \sqrt{L\mathcal{T}^3}} V_s \sin \alpha \, \sqrt{T_A} \, . \tag{13.85}$$

This expression has three basic terms in the product. The first fraction is comprised of physical constants, which cannot be changed or affected. The second term is comprised of terms that are related to the object or the system design and perhaps can be changed or selected. For example, objects with larger proton density P_D, resistivities, or rest in larger static magnetic fields (hence, higher Larmor frequency) will yield larger SNRs. On the other hand, lowering the temperature or making the coil smaller (radius or length) will increase the SNR.

The third term in (13.85) comprises terms that are directly selectable by the user in a given imaging scenario. Here, V_s represents the size of the sample represented by a single reconstructed signal value—i.e., the pixel volume. We can directly select pixel volume through both slice selection, FOV selection, and choosing the size J of the reconstructed image. We have already long recognized that choosing the tip angle $\alpha = \pi/2$ maximizes the received signal. Finally, we can improve SNR by increasing the amount of time T_A we spend gathering Fourier data.

13.4.5 **Artifacts**

The most common form of image distortion is geometric warping. This arises when the gradient strength is not uniform across the entire field of view; for example, if the amplitude of the x-gradient falls off as a function of increasing z, in a coronal image the patient will seem to "pinch in" at the extreme values of z. Because of the demands to image faster, physically shorter, faster switching gradients are now often used; hence, there is often considerable drop-off in the gradients at the edges of a larger field of view. Software is used to correct for this distortion at the time of image reconstruction. However, the slice shape and thickness will change with changing gradient strength, and this effect cannot be corrected in reconstruction.

Ghosts are one of the most common artifacts seen in MR images. They are most commonly caused from changes in the object between Fourier space acquisitions—e.g., motion due to breathing, swallowing, tremor, heart beat or other flow effects. Because of these changes, the acquired data do not correspond to the same object, and reconstruction based on these data produce faint signals in one or more acquisitions that are not present in others—these are the ghosts. Ghosting artifacts are reduced by imaging in a breath-hold, by acquiring data in synchrony with the heartbeat—i.e., gating on the ECG signal—or by acquiring images exceptionally fast. There are disadvantages to these solutions, however. Breath-holding can be uncomfortable or impossible for the patient, gating increases the overall image acquisition time and is sometimes difficult to reliably acquire, and fast imaging always suffers from reduced SNR.

Wraparound artifacts are caused by aliasing in the phase encode direction. It has previously been determined that

$$\text{FOV}_y = \frac{1}{\Delta\nu},$$

where $\Delta\nu$ is the Fourier space separation of the measured Fourier scan lines in the phase encode direction. If the object does not fit within the FOV in the y-direction, then it will wrap-around, in exact analogy to the wraparound of a conventional sampled signal (see Chapter 2). The main difference here is that the sampling is done in one of two Fourier directions (the phase encode direction) and the resulting replication is in the spatial domain rather than the Fourier domain. This does not happen in the readout direction because an antialiasing filter is employed during readout.

To eliminate the wraparound artifact, it is necessary to eliminate the object that is outside the field of view. This can be done by rotating the scan plane to put the shorter dimension of the object in the phase encoding direction, or by using small receiver coils whose sensitivity profile matches, the field of view. Outer volume saturation pulses can be used to eliminate unwanted signals, but they are effective only for a short duration.

One method of reducing scan time is to eliminate phase encoding steps that sample the higher spatial frequencies in Fourier space. Unfortunately, not only does this produce a low resolution image, it also produces a "ringing" artifact.

Gibbs ringing or "truncation artifact" is a classic problem with poorly designed low-pass filters; if the cutoff of the higher frequencies occurs too abruptly, such as with a "top hat" filter, the sinc-shaped point spread function causes large oscillating sidebands at discrete edges in the image. The method to eliminate these artifacts is to apply a filter that has a smooth transition zone, or better yet, to take more data!

In Chapter 12, we saw that chemical shift changes the Larmor frequency of nuclei in a spin system depending on what chemical environment they find themselves in. For proton imaging, we are primarily interested in two species of protons—those found in water and those in fat. At 1.5 T, the resonance frequency of protons in water is approximately 225 Hz higher than that of protons in CH_2 components of fat. Therefore, since position in x is encoded by frequency, fat protons will be positioned slightly different than those in water. There will therefore be a relative position shift in the readout direction between fatty tissues and nonfatty tissues. In extreme cases, this can cause the position of fat to appear in the wrong place relative to the water image, making anatomy look abnormal when it is in fact normal. Finally, magnetic susceptibility differences can cause changes in the local magnetic field felt in the sample. Typically, this leads to signal "dropouts" at air/tissue interfaces.

13.5 Summary and Key Concepts

MRI systems are very popular in medical imaging because they provide high-resolution anatomic information noninvasively. Such systems represent large versions of NMR systems, with significant additions to form images. In this chapter, we presented the following key concepts that you should now understand:

1. An *MRI scanner* consists of five principal components: (1) the main magnet, (2) a set of switchable gradient coils, (3) RF coils, (4) pulse-sequence and receive electronics, and (5) a computer.
2. The most common magnet is a *cylindrical superconducting magnet* with field strength ranging from 0.5–7 T.
3. The *gradient coils* produce the change in local magnetic field necessary to encode spatial location in the MR signal.
4. The *RF coils* or *resonators* receive the MR signals, and may be large (e.g., a body coil) or small (e.g., a surface coil).
5. Manipulation of the gradient coils produces *frequency encoding of location*; manipulation of the RF excitation pulse-sequence produces *varying image contrast*.
6. *MR data* are scans of Fourier space; *MR image reconstruction* is based on the inverse 2-D Fourier transform and represents the distribution of *effective spin density*.
7. *MR image quality* depends on *contrast* (which itself depends on intrinsic tissue parameters and the choice of pulse sequence), *sampling*, and *noise*.

Bibliography

Brown, M. A., and Semelka, R. C. *MRI Basic Principles and Applications.* Hoboken, NJ: Wiley, 2003.

Cho, Z. H., Jones, J. P., and Singh, M. *Foundations of Medical Imaging.* New York: Wiley, 1993.

Haacke, E. M., Brown, R. W., Thompson, M. R., and Venkatesan, R. *Magnetic Resonance Imaging: Physical Principles and Sequence Design.* New York: Wiley, 1999.

Liang, Z. P., and Lauterbur, P. C. *Principles of Magnetic Resonance Imaging: A Signal Processing Perspective.* Piscataway, NJ: IEEE Press, 2000.

NessAiver, M. *All You Really Need to Know About MRI Physics.* Baltimore, MD: Simply Physics, 1997.

Macovski, A. "Noise in MRI." *Magnetic Resonance in Medicine* 36 (1996): 494–97.

Stark, D., and Bradley, W., Eds. *Magnetic Resonance Imaging.* St. Louis, MO: Mosby, 1992.

Problems

Contrast

MR Imaging Instrumentation

13.1 A uniform magnetic field in z direction is applied to a sample. When an x-gradient is applied, what change(s) are made to the field?

13.2 Briefly explain the function of RF coils.

Encoding Spatial Position and MR Imaging Equations

13.3 You intend to image one slice ($\bar{z} = 5$ cm, $\Delta z = 1$ cm) of an off-centered cube having width 10 cm as shown in Figure P13.1. You are given $G_z = 1$ gauss/mm and $\gamma = 4.258$ kHz/gauss.

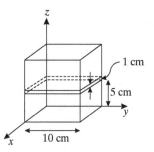

Figure P13.1
Problem 13.3.

(a) Find the bandwidth (in Hz) of the RF waveform needed to perform the slice selection.

(b) Give a mathematical expression for the RF waveform $B_1(t)$ needed to perform slice selection (in the rotating frame).

13.4 (a) What is the slice thickness (defined as full width at half maximum) if a slice-selection gradient amplitude of 1 gauss/cm is used with a gaussian-shaped RF pulse that has a shape given by $A(t) = A_0 \exp\{-t^2/\sigma^2\}$, where $\sigma = 1$ ms.

   (b) What is the slice thickness if we cut the gradient amplitude in half? Suppose we change the shape of the RF pulse so that σ is reduced by a factor of 2.

   (c) What is the new slice thickness (with a 1 gauss/cm gradient)?

   (d) What else is affected by this change?

13.5 An RF signal is given by

$$s(t) = A\Delta v \operatorname{sinc}(\Delta vt)e^{-j2\pi \bar{v}t}.$$

   Show that the tip angle distribution is given by (13.17).

13.6 Explain why a refocusing gradient is usually needed, and why its duration is half the duration of the RF excitation when its waveform is a constant negative gradient of strength $-G_z$.

13.7 Substituting (13.22) into (13.44) and assuming T_2 is constant yields

$$s_0(t, A_y) = \int_{-\infty}^{\infty}\int_{-\infty}^{\infty} Ae^{-t/T_2(x,y)}M(x,y; 0^+)e^{-j2\pi\gamma G_x xt}e^{-j2\pi\gamma A_y y}dx\,dy,$$

$$= Ae^{-t/T_2}\int_{-\infty}^{\infty}\int_{-\infty}^{\infty} M(x,y; 0^+)e^{-j2\pi\gamma G_x xt}e^{-j2\pi\gamma A_y y}dx\,dy.$$

   What is the effect of the term e^{-t/T_2} on the reconstructed image? (Assume that the pulse sequence in Figure 13.15 is used.)

13.8 Suppose a point object is moving along the x-axis with the following trajectory: $x(t) = x_o + vt$.

   (a) Calculate the phase shift that is induced in the transverse magnetization after the application of the gradient waveform shown in Figure P13.2.

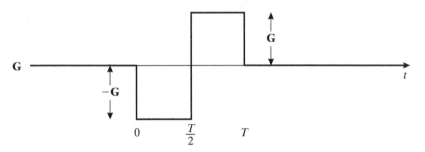

Figure P13.2
Problem 13.8(a).

   (b) Calculate the phase shift induced by the waveform shown Figure P13.3. This is called a *flow compensation* pulse.

   (c) Repeat the calculations for the above two waveforms for a trajectory given by $x(t) = x_o + vt + \frac{1}{2}at^2$.

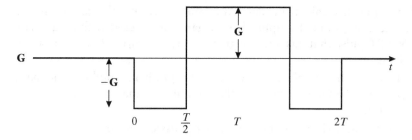

Figure P13.3
Problem 13.8(b).

(d) What gradient waveform could you use as an *acceleration compensation* pulse?

(e) Is it possible to design a gradient waveform that will produce phase shifts that are independent of acceleration and are dependent on the velocity?

13.9 Suppose that a slice has been selected through a $5 \times 5 \times 5$ cm cube. We wish to produce a projection of this slice with the application of a readout gradient. The pulse sequence shown in Figure P13.4 is used to produce the profile.

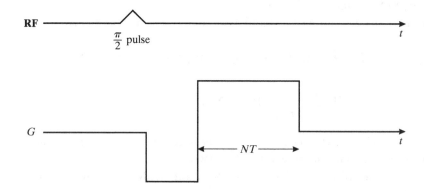

Figure P13.4
Problem 13.9.

If the gradient strength $G = 1$ gauss/cm and $N = 256$ sample points are taken in the total sampling time $NT = 10$ ms (recall T is the time between sampling points in the readout direction).

(a) What is the spatial extent of the profile? [This is referred to the *field of view* (FOV).] What is the bandwidth of the received signal?

(b) How many pixels does the profile of the object span?

(c) Suppose we set $G = 0.5$ gauss/cm, $N = 256$ and $NT = 20$ ms. How does this affect (a) and (b)?

(d) Repeat (a)–(c) for the pulse sequence shown in Figure P13.5.

13.10 In order to produce an actual image of T_1, as opposed to one that is merely T_1-weighted, it is necessary to take at least two separate images and use them to compute a pixel-by-pixel estimate of T_1.

(a) Show that the imaging equation in (13.63) can be written as

$$\frac{f}{\sin \alpha} = e^{-T_R/T_1} \frac{f}{\tan \alpha} + AM_0 e^{-T_E/T_2}(1 - e^{-T_R/T_1}).$$

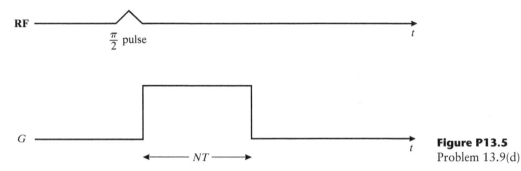

(b) Assume that three images f_1, f_2, and f_3 with tip angles α_1, α_2, and α_3, respectively, are acquired. Show that the points $(f_i/\tan\alpha_i, f_i/\sin\alpha_i)$, $i = 1, 2, 3$ should lie on a line.

(c) Using the fact shown in part (b), find a closed-form expression for an estimate of T_1 given only two images f_1 and f_2 with different tip angles α_1 and α_2.

Sampling the Frequency Space

13.11 A pulse sequence uses the x- and y-gradients shown in Figure P13.6. Draw its Fourier space trajectory and carefully label the axes.

13.12 Consider the pulse sequence shown in Figure 13.20. Suppose we want to sample a radial line from $(-0.25, -0.5)$ mm^{-1} to $(0.25, 0.5)$ mm^{-1} and measure 128 points along the line. Ignore the ramps at the ends of the gradients.

(a) If the duration of the x and y gradients in the interval labeled ② are 0.1 ms, what are the strength of the gradients?

(b) If $T_s = 10$ ms, what are the readout gradients? What is the sampling rate?

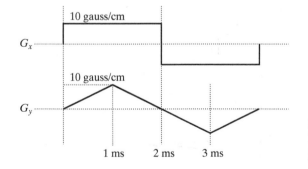

13.13 Draw the pulse sequence that scans 2 lines of k-space after an excitation RF pulse. Start scanning Fourier space at $(-1, 0.5)$ mm^{-1} and end at $(-1, 0.4)$mm^{-1} as shown in Figure P13.7. Assume that the RF pulse is not slice-selective. Label your timing diagram, including the amplitudes of the gradients.

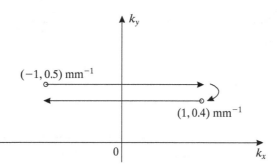

Figure P13.7
Figure for Problem 13.13.

13.14 We have been asked to phase encode two point objects offset in y. Let's first turn to an example in which we have two point samples in the imaging plane that both have the x coordinate $x = x_o$, but different y coordinates. In this case, the signal is the sum of the individual signals from each object; therefore, we cannot determine their individual y coordinates from a single measurement such as that shown in Figure 13.15, and described by (13.42). Show that we can obtain *two* measurements, each measurement having a different value of the phase encoding gradient amplitude G_y, to solve for y the position of the two point objects.

13.15 Explain why the refocusing lobe, the phase encoding gradient, and the gradient echo formation lobe can all overlap in time. (Refer to Figure 13.18.)

MR Image Reconstruction

13.16 2D projection imaging was an early MR imaging technique that is enjoying a comeback because it is very fast. Consider the pulse sequence shown in Figure P13.8.

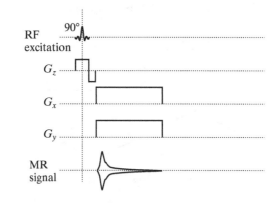

Figure P13.8
Pulse sequence for backprojection imaging.

In the pulse sequence, let

$$G_x = G\cos\theta \qquad \text{and}$$

$$G_y = G\sin\theta.$$

(a) Write down an expression for the signal given θ.

(b) If the cross section being imaged is $f(x,y) = A\delta(x-1,y) + B\delta(x,y+1)$, sketch the recorded signal for $\theta = 0$ and $\theta = 90°$.

(c) How can you obtain an image of the cross section using this pulse sequence?

13.17 We wish to image two "point objects" that are sitting in slightly different static fields due to magnet imperfections. Point object 1 is in a field B_o, and point object 2 is in a field $B_o + \Delta B$. Describe why this is a problem if we are going to use the 2D projection method of MRI.

13.18 Consider the profile of an object in the readout direction. Show that the inverse Fourier transform of the baseband signal $s_0(t)$ given in (13.27) is a projection of the object $M_{xy}(x,y;0^+)$ onto the x-axis convolved with the inverse Fourier transform of the T_2 decay envelope[1] $\mathcal{F}^{-1}\{e^{-t/T_2}\}$. (Assume that T_2 does not depend on x and y.)

MR Image Quality

13.19 We want to acquire an image of a sagittal slice (parallel to the y-z plane).

(a) In what directions should the slice selection gradient, phase encoding gradient, and frequency encoding gradient be?

(b) How can we prevent aliasing in the phase encoding and frequency encoding directions?

13.20 We want to reconstruct a 2-D MR image with a size of 256×256 pixels.

(a) What parameter(s) determine the spatial extent in the phase encoding direction?

(b) What parameter(s) determine the spatial extent in the readout direction?

(c) What parameter(s) determine the spatial resolution in the phase encoding direction?

(d) What parameter(s) determine the spatial resolution in the readout direction?

13.21 A cross section of a sample of protons within a $25.6 \, \text{cm} \times 25.6 \, \text{cm}$ square is being imaged by an MR scanner. We want to reconstruct an image with matrix size 256×256. We use a Nyquist sampling rate in u and v directions.

(a) What is the extent of the object in the diagonal direction? How many samples are there in each (main) diagonal direction?

(b) What is the Fourier space sampling rate for the diagonal directions?

13.22 A solenoid coil with N turns per unit length, a length of L, and a radius of r_0 is used to image an object of radius r_0 with uniform resistivity of ρ. The coil is excited with a current $I(t) = \cos(2\pi v_0 t)$.

(a) What is the average power dissipated in the object? What is the voltage induced for a cylindrical shell of radius r?

---

[1] This is a Lorentzian function.

(b) Given that the differential conductance of a cylindrical shell with length L, radius r, and resistivity ρ is $dG = L\,dr/(2\pi r\rho)$, show that the effective electrical resistance of a solenoid is given by (13.79).

13.23 What repetition time (T_R) should we use to generate the maximum signal difference between the two tissues whose T_1 values are T_1^a and T_1^b? What T_R value will give the maximum signal difference to noise ratio? (*Note:* Assume that you can use a T_E of zero and that all transverse magnetization has decayed before each $\pi/2$ pulse.)

13.24 The resolution of optical imaging methods is equal to the Rayleigh limit, which is one-half of the wavelength of the optical frequency being used. If MRI were performed using radio waves propagating into the body, what would be the predicted resolution for a conventional magnet?

13.25 The location of "fatty" tissues in an MR image is displaced relative to "watery" tissues because of chemical shift. Explain how chemical shift will affect the appearance of images containing both fat and water and develop an approach to "suppress" the fat image using the knowledge that $T_1(H_2O) \approx 4T_1(\text{fat})$.

13.26 It is often useful to consider what happens to image SNR when only a single imaging variable (or minimal group of imaging variables) is changed. Consider making each of the changes in the following parts and determine how the SNR changes as a result.

(a) $G_x \rightarrow 0.5 G_x$.

(b) $N_y \rightarrow 2N_y$.

(c) $f_s \rightarrow 2f_s$ (assuming that $N_x \rightarrow 2N_x$ as well).

(d) $f_s \rightarrow 2f_s$ (assuming that N_x stays constant).

Applications, Extension, and Advanced Topics

13.27 Consider the imaging problem in Problem 13.3. Assume that the effective spin density within the cube is constant and equal to 1.

(a) Referring again to Figure P13.1, sketch the 2-D function $f(x, y)$ that you will image. Also sketch $|F(u, 0)|$. Use the gradient-echo pulse sequence shown in Figure P13.9 to image the slice with gradient $G_x = 0.5$ G/mm and $G_y = 0$.

(b) Find the duration (in seconds) of the x-gradient preceding the readout gradient in order to collect data in the range $-0.4\ \text{cm}^{-1} \le u \le 0.4\ \text{cm}^{-1}$.

(c) How long (in seconds) after the gradient G_x changes sign will the gradient echo occur and why?

(d) After reconstructing the data collected during the acquisition, will we able to get a perfect reconstruction of $g(\ell, 0°)$, where $g(\ell, \theta)$ is the Radon transform of $f(x, y)$? Explain.

13.28 Consider a constant magnetic field $B_0 = 1.5$ tesla is applied on an object. The gyromagnetic ratio of the object is $\gamma = 2\pi \times 4258(\text{rad/s})/\text{gauss}$, 1 tesla =

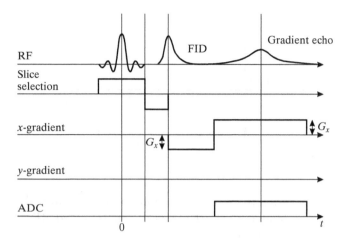

Figure P13.9
Pulse sequence for
Problem 13.27.

10^4 gauss. Suppose a RF pulse is applied. The pulse is given as

$$B_1(t) = A\Delta f \operatorname{sinc}(\Delta f t) e^{j2\pi \nu_0 t} \operatorname{rect}\left(\frac{t}{\tau_p}\right),$$

where $\Delta f = 4.258 \times 10^4$Hz, $A = 2$ gauss/Hz, $\tau_p = 2$ms, and ν_0 is the equivalent Larmor frequency.

(a) Find ν_0. Write down an equation to calculate the tip angle.

(b) Assume τ_p is long enough so that $B_1(t)$ could be approximately considered as

$$B_1(t) = A\Delta f \operatorname{sinc}(\Delta f t) e^{j2\pi \nu_0 t}.$$

What is the spectrum of $B_1(t)$? Sketch it. Suppose a slice selection gradient is applied in the z-direction while $B_1(t)$ is applied, $G_z = 2$ gauss/cm.

(c) Where is this slice centered? What is the slice thickness?

(d) If you want to select another slice *adjacent* to the current slice (move in $+z$ direction) with the same slice thickness, how should you change the RF pulse? Assume G_z stays the same.

(e) Right after the RF pulse in (d), what is the maximum phase difference of the spins inside the current slice? How does one rephase the spins?

(f) Suppose you want to sample the k-space using a sampling trajectory that starts as shown in Figure P13.10. Assume $|G_x| = 2.5$ gauss/cm, and $|G_y| = 1.5$ gauss/cm. Sketch a pulse sequence to achieve this part of the trajectory. Indicate the acquired signal, readout gradient, phase-encoding gradient, and slice selection gradient.

(g) For the sampling indicated in (e), if the FOV$_x$ needs to be set as 50 cm, what is the smallest sampling rate f_s?

13.29 Suppose that we have an object that we wish to image in all three dimensions—that is, we want our final image to be an $N \times N \times N$ array of voxels

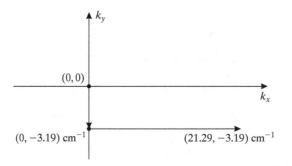

Figure P13.10
Problem 13.28.

with isotropic dimensions. However, our ability to produce RF pulses is limited to short bursts of unmodulated RF; therefore, we cannot use slice selection.

(a) Can you think of a method for imaging the object? (*Hint*: Look at the phase encoding section of the text and extend these ideas.)

(b) What would the equation for the signal be?

(c) How would you reconstruct the image?

13.30 Consider a generalized signal acquisition scenario for spatial encoding. Show that given N point objects with amplitudes A_j (where the A_j are real numbers), we can solve for their y-coordinates y_j from the N equations derived from N separate signal acquisitions having different phase encoding gradient amplitudes G_y^m $(m = 0, 1, 2 \cdots N)$.

Index